MORE ROCKS IN OUR HEADS

Stories of Exploring for Mineral Deposits
in Exotic Lands

MORE ROCKS IN OUR HEADS

Stories of Exploring for Mineral Deposits in Exotic Lands

COLLECTED AND EDITED BY

Andrew Drummond

With a Foreword by Pierpont

Connor Court Publishing Pty Ltd

Published in 2022 by Connor Court Publishing Pty Ltd.

Copyright © Andrew Drummond

Connor Court Publishing Pty Ltd.
PO Box 7257
Redland Bay QLD 4165
sales@connorcourt.com
www.connorcourt.com

ISBN: 9281922449986

Cover Design by Maria Giordano

Printed in Australia.

This book is dedicated to those who died while exploring
for the mineral deposits mankind needs.

Contents

Pierpont/Trevor Sykes **Foreword**   ix

Introduction   xiii

The South West Pacific

1. *Andrew Drummond* Having the Devil of a Time in Godzone
   Country, New Zealand, 1987-1989   1

2. *Sandy Moyle* The Land of the Unexpected, Papua New Guinea,
   1982-2013   15

3. *Roger Langmead* Tales from the South Pacific, The Solomons,
   1980-1986   53

4. *Ian Plimer* Ruby and Rubbery Rules, Vanuatu, Late 1990s   65

5. *Stephen Turner* A Geologist in Paradise, Fiji, 1983-1984   70

6. *Roger Marjoribanks* An Incident in Bougainville, Papua
   New Guinea, 1968   80

South East Asia

7. *Anthony Williamson* Lost in Translation, Various Asian,
   1982-2010   89

8. *Ian Mulholland* A Zinc Oxide Adventure in Laos,
   Laos, 2004-2009   120

9. *Jeff Rayner* You're the Piano Man, Indonesia, 1997   143

China, Russia and the Former USSR

10. *Ian Mulholland* Chinese Burns, Australia and China, 2009   151

11. *Bob Besley* In Search of the Golden Fleece, Georgia, 1994   160

12. *John Garlick* Getaway in Siberia, Russia, early 2000s   167

13. *Michael Fellows* Magadan on Mogadon, Russia, 1994   171

14. *Martin Spence* Romanian Rhapsody, Romania, 2013   183

The Muslim World

15. *Ian Tedder* An Infidel in Saudi Arabia, Saudi Arabia, 1979-1980 189

16. *Jeff Rayner* The Cradle of Gold, Saudi Arabia, 2008   215

17. *Wilson Forte* Revolutionary Rumblings, Iran, 1977-1978   219

18. *John Nethery* Tourist! Tourist!, Turkey, 1989   232

19. *Greg Ambrose* Working for Gaddafi, Libya, 1992-1997   237

20. *John Hammond* A Helicopter Gunship in The Karatau Range,
Kazakhstan, 1997-1999   258

21. *Mike Kitney* "She's Apples in Central Asia, Mate", Tajikistan, 1997 266

22. *Ralph Stagg* "I Will Not Shoot My Fellow Citizens",
Tunisia, 2005-2013   270

Sub-Saharan Africa

23. *Max de Vietri* A Horror Story Of People You Don't Want To Meet,
West Africa, Early 1980s   283

24. *Roger Thomson* Oh No! Manono, Zimbabwe and Zaire, 1981   294

25. *Dan Greig* Finishing School, Zambia, 1969-1970   298

26. *Andrew Drummond* Eating My Words, Namibia, 1997
and 2010-2012   307

27. *Mort Cowan* Real Men Don't Need Visas, West Africa,
Late 1990s   314

28. *Mark Hughes* Humility jn Botswana, Botswana, 2001   317

29. *Cyril Geach* Africa – East and West, Various Africa, 1993-2010   319

The Americas And Beyond

30. *Brian Levet* Ambush In Peru, Peru, 1992   337

31. *Neil Stuart* The Cerro Negro Story, Argentina, 2000-2007   344

32. *Lynda Frewer* What – No Piranhas? Venezuela, 1992   354

33. *Jason Beckton* The Treasure of the Sierra Madre – For Real,
Mexico, 2004-2006 and 2015-2017   371

34. *Ken Russell* "The World Is Your Oyster, Son", Various   388

35. *Phil Fillis* Biggles Goes Exploring, Various   399

Foreword

Field geologists provide the best reading of any scientists I know. Anyone who doubts this only needs to read a few pages of this follow-up to the original of *Rocks In Our Heads* to see my point.

Field geologists are the men – and increasingly women – who go outdoors to look at rocks and decide whether and where a company should explore with the hope of finding a mine. This is difficult enough in Australia, but Andrew Drummond has compiled a second collection of geologist's reports from around the world, which make Australia seem very friendly indeed.

The book is almost entirely written by geologists, all of whom began with ordinary degrees and several of whom have qualified to the point where they have earned a Master's degree or Doctorate. The areas they have plodded around include some of the world's most remote and difficult countries. The difference between geologists and almost every other form of scientist is that scientists, generally, work in cities whereas field geologists have to go out into the country. And the geologists in this book have not only inhabited relatively safe Australia but have gone out into some of the world's more dangerous places.

Working in remote areas means these geologists have to sleep and eat where they can. If they don't speak the local language, those necessities become even more difficult.

Mike Kitney in Tajikstan, had to taste the local vodka, which was made from corn. He said afterwards that paint stripper was probably better tasting. In Ghana, Cyril Geach dodged an offer to eat barbecued rats with their fur still on. Wilson Forte, eating with Baluchis in Iran, said one of his workers suddenly found a goat's eye staring at him from a bowl of soup. Anthony Williamson found himself in a Japanese restaurant eating skewered fish that were still alive.

Republics are bad news because a dictator can change his mind at short notice and take over a mine. Unstable democracies aren't much better because an incoming president can change policy and render a mine unworkable.

Andrew Drummond, who put the collection of geological reminiscences together, tells what happened to South Boulder Mines when they were exploring in the north African state of Eritrea. South Boulder delineated a very large potash deposit, giving their shares a boost in the market. However, the Eritrean Government promptly seized 50 percent equity in the project, which immediately trashed the company's share price. The deposit has still not been brought into production. Government interference on that scale makes it exceedingly difficult to raise capital for mine development.

Even Namibia, which I often thought was the best government in Africa for a mine developer to work with, has a sour record. The country has always welcomed people with the money and ability to bring its uranium mines into production, but it's a different story with undersea phosphate. Namibia is on the west coast of Africa, immediately above South Africa. The Australian company Minemakers and its joint venture partners discovered a huge potash deposit under the Atlantic Ocean in seas just over 100m deep. The phosphate price was good, the banks were keen and all that was needed was government approval for a dredge to suck up the sands, pump them to port, screen them for phosphate and immediately send everything else back into the ocean.

The scheme was opposed by greenies, NIMBYs and others, mainly on the grounds it would impair the quite valuable fishing industry. Quite nonsense, the JV said said. As the phosphate zones already extended into the fishing areas, patently the fish weren't disturbed by them. Minemakers knew where the fishes lived and was happy to situate their dredging some distance away. The company gathered supporting expert advice to help their case, but a hearing in front of government leaders was held anyway.

At the hearing, Andrew circulated a jar of the black sand which was supposed to be poisoning the fishes. The jar was circulated to everyone present. When it returned to Andrew, he unscrewed the lid, licked then stuck his finger into the sand sample, withdrew it with a good coating of the material, then stuck his finger in his mouth and sucked all the sand off and swallowed it.

Andrew writes: "The meeting finished soon after and we all left, with yours truly not frothing at the mouth or having to be carried out; I have not heard of anyone else ever having to eat their intended ore to gain mining approval, but even if others have, it must be a pretty rare event. The things we directors do for our shareholders!"

Nevertheless it wasn't enough to win the government's approval.

Mike Fellows will serve as another example. He was working for a small resource company in the wilds of Kyrgyzstan in 1996 when it had not long separated from the old Soviet Union. Kyrgyzstan is in Central Asia with Kazakhstan to the north, Uzbekistan to the west, Tajikistan to the south and China to the east. The main language is Russian. (*This story will now appear in the next book* (Ed.))

Mike was working for a British company and trying to find gold in Kyrgyzstan. However, his problems were overwhelming. A company called NKGE was his joint venturer, but limited in exploration drilling because they could only sink expensive diamond holes. On one of their projects core recovery averaged only 65 percent, whereas in Australia the rate is closer to 100 percent. In one hole they sank 521 metres and recovered a mere 31 percent. "It was not uncommon for them to set up on a single drill hole for half a year", Mike says.

An Australian company would be sacked on that performance, but NKGE was a joint venturer. Mike considered sacking them but refrained, because he was sure they would find ways to foul him up if he did.

One of the considerable problems which Mike never overcame was that of the Kyrgyzstan government officials. Unused to the capitalist system, they had quickly developed a culture in which nothing was approved without a bribe being paid. This was seen most clearly in field conditions, where the public servants charged $US300 a month to explorers and $US5 a day for food to the workers. What actually happened was that only $US100 a month went to finance the food allotment and the men in the field had to find their own rations. The agreement with the government turned out to be "written with a pitchfork on water".

Another example was the exploration vehicle, a Land Cruiser. It had to be driven in from Iran, which was trouble enough. The Kyrgyzstan government insisted that the vehicle had to have a vehicle department in which it lived. This was simple enough for a vehicle owned by an individual, but posed problems for an exploration company where a vehicle typically roved the countryside and rarely visited any garage. "The bureaucrats must have their piece of paper with a signature and a stamp," Mike wrote.

Currency was yet another problem area. Most businessmen and traders in Kyrgyzstan preferred to deal in $US. Except for the petrol station which would only accept Kazakh tenge. So currency trades had to be done in the local bazaar.

Supplies of good material (essential to mines everywhere) were almost impossible to find. Aluminium was of such poor quality that Mike says a fork would bend cutting into a hard-boiled egg. Insulation of a supposed heavy duty extension lead melted at the first use. When the camp finally received an air compressor, it could only service a single drill and had no pressure gauge nor safety pressure relief valve. The electricity supply was variable and cut out entirely at times. "Shoddy workmanship and products were the norm and nobody seemed to care!" Mark wrote. "I'd reached the conclusion that besides being financially bankrupt, the nation was ethically and spiritually bankrupt as well."

It took 10 months for the team to drill their first hole. In that hole the compressor air hose burst twice in the first 18 metres. That caused scrapes on the legs for the drill operator, who had to scramble over rocks to turn off the compressor. The safety sensors which automatically turned the compressor off weren't working.

Mike spent 18 months working there before the money ran out. He finished tidying up the project, collected a bonus he was due, and left. He doesn't know whether a mine resulted or not.

They are just snippets from four of the 35 chapters in the second edition of *Rocks In Our Heads*. They are stories of intrepid geologists who have gone everywhere from Siberia to Peru and from Laos to Venezuela. It's a great read.

Pip Pip!

Trevor Sykes (aka Pierpont)

Introduction

More Rocks In Our Heads is a follow-up and companion to the original *Rocks In Our Heads* book, which was published in 2020. Despite the difficulties in promotion of the latter due to the widespread Covid restrictions of the last two years, it has been warmly received and has prompted the collection and publication of a second suite of stories in this new book. As previously, most of the stories are by Western-trained veteran earth scientists, predominantly geologists, and who are usually resident in Australia. Collectively, they are professionals who have devoted their working careers to the discovery, proving up, and bringing into production the mineral deposits the world's growing and ever more wealthy population needs.

As the developed countries of the Western world have been subject to scientifically based exploration for a long time, the more easily discovered ore bodies have been found, brought into production where economic, and many have been mined out. This has necessitated a concentration of work in Third World and developing nations. The collection of stories in this book is again about the efforts by our mining industry individuals and companies in those far-flung countries.

Readers will enjoy and perhaps learn more about the difficulties and hazards of trying to find, retain ownership and develop those orebodies in these foreign jurisdictions. Problems can stem from the different types of government, laws, races and religions, politics, underdevelopment of education, geological history, language, wild fauna, insurrection, and so on. Many of these are represented in this book – as told by the often-bruised survivors of those efforts. Endings are often not happy.

In the Introduction to the first book, I explained that, at normal compound rates of growth, the world's population will require the mining of as much product over the next twenty five years as it has used since the dawn of history. On top of that, new demands can arise and will require even more mining of some mineral commodities. A good example is the trend towards electric cars and storage of electricity by batteries, particularly as pushed by European demands, as this will require major new sources of lithium, cobalt, rare earth elements, nickel and more, to be brought into production. One would think that Europe would welcome the development of the discovery of a major new lithium mine within its borders. But, as this work goes to press, the Serbian government has cancelled the exploration and development licences held over the very large Jadar lithium deposit which had been discovered and investigated by Australia's Rio company. It had spent USD450 million or AUD625 million on its pre-development work on the deposit before that government action.

And sadly, there is the present Russian invasion of the Ukraine by Russia, which is resulting in the destruction of the infrastructure of the former, and severe economic sanctions by many countries against the latter. There is likely to be a large impact on world mineral supply for the foreseeable future. Ukraine reportedly ranks fourth in the world for total value of in-ground mineral deposits, and, for some individual commodities, it is understood to rank first in Europe for uranium; second in Europe for tin, and tenth in the world; its iron reserves are the third largest in the world; it is second in Europe for mercury; third in Europe for natural gas; and seventh in the world for coal.

Russia is in the top five in the world for production of gold, platinum, silver, nickel, vanadium, cobalt, sulphur and phosphate. Europe has been relying very heavily on imports of natural gas from there. The mining industry will have to step up even more than it had anticipated, and it will need understanding, support from and lessening of unreasonable hurdles by those in control.

In *More Rocks In Our Heads*, there are 35 stories from 32 contributors. The primary aims are to bring reading enjoyment and, perhaps, an amount of non-technical understanding of what, how and why mineral explorers do what they do in their spearhead niche in the world mineral supply chain; and insights into the personal drivers for many of them. Geologists need to think and operate in that fourth dimension – time. Of all the sciences, geology is unique in that we have to try to work out the how, why and when of processes that may have occurred sporadically over billions of years. These rather freaky processes can result in enrichment of metals and other commodities to the grades and sizes which may make them economic to mine today. And after the formation of these deposits, they may be obscured and changed by deep weathering, folded or faulted to great depths, or covered and concealed by barren younger units, or by ones which have been thrust over them.

The challenges are many, disappointments abound, but sometimes there are much-needed successes. Scientific and technical advances have provided an array of techniques which can be used by the mine hunters. The readers should also gain a better appreciation of the challenges facing individual Westerners and their employers as they set out to apply their determination, technical knowledge, experience and access to development capital in some of the remoter and underdeveloped parts of the world.

Of the 33 stories which describe either a success or a failure for their efforts, only nine are associated with projects that had a favourable outcome for those doing the work and spending the dollars. Guns and violence; and individual, corporate and government misbehaviour are all too common in the stories. Explorers and associated personnel die in about a dozen different ways throughout the book.

We trust you enjoy the stories of some of our adventures as we illustrate that ore deposits don't grow on trees!

The exploration of Lihir Island, refer Sandy Moyle's story, page 15
(below and overpage)

Above: Loading supplies and, below: Peter McNab and local
wearing personal protective equipment, April River, 1983

THE SOUTH WEST PACIFIC

Above: In camp at the end of the day, August 1982, and below:
The celebratory feast

1

Having the Devil of a Time in Godzone Country

New Zealand 1987-1989

Andrew Drummond

Andrew Drummond is 71 years old and graduated with BSc(Honours) as a geologist from Adelaide University in 1972. He has been continuously involved in the mining industry since then in the fields of mineral and energy exploration, feasibility and development, gold mining, and company foundation and management. He was primarily responsible for the ASX listing of Westonia Mines Ltd and Minemakers Ltd, was involved in other floats and has been a director and or manager of numerous companies until his recent retirement.

Minemakers was credited with being the top performing company on any listed stock exchange in the world in the GFC year of 2008. His overseas experiences are in New Zealand, Russia, the Philippines, China, Namibia, and several other African countries.

Australian Consolidated Minerals Ltd (ACM) and its listed subsidiary, ACM Gold Ltd, were doing well by mid-1987 and were on the verge of growing gold production strongly. In Western Australia, their wholly owned Edna May and Golden Crown gold mines, at Westonia and Cue, respectively, were producing successfully. The new discovery at Wirralie in Queensland was in construction phase, as was the 50% joint venture over the old Big Bell mine, also near Cue. In New Zealand, ACM Gold bought a minority position, and management rights, in a joint venture which was to redevelop the famous old Waihi gold mine. Located at the base of the Coromandel Peninsula, Waihi was also under construction.

At that time, I was chief geologist at Edna May and was asked to transfer to New Zealand near the end of the year to initiate a strong gold exploration effort in that country. Reasons for targeting New Zealand for the ACM group were several. The first was its inherent gold prospectivity, especially in the Coromandel and Otago areas. The second was that ACM, through its merger with the Australian arm of the American AMAX group, had accessed the world leading, innovative geochemical exploration technique known as BLEG, which used very low detection limit techniques and which it was anxious to apply on a broad scale in New Zealand. As it was put to me, the third reason was that "it takes just as much management time to run a mine which we quarter own as it does one which is fully owned, so let us try to find or acquire some of the latter".

My family and I transferred a few days after the October '87 share market crash, with my position being general manager of the new local subsidiary, ACM New Zealand. We took up residence in Waihi and I worked from an office within the gold mine administration complex.

With hindsight, full marks to ACM Gold for acquiring its stake in Waihi. However, that was essentially a paper deal on a project under construction, it having been already fully permitted. Hence, ACM Gold management had not suffered the many years of expense and exasperation of getting the project to the end of that permitting stage and perhaps was not fully aware of just how difficult the task had been. That slog involved not just technical issues attendant upon any mine start-up, such as delineation of an ore reserve, mine planning, metallurgical studies, plant design, feasibility and finance, but also the mountainous New Zealand legal and permitting hurdles. With respect to the latter, it can be summarised as trying to re-start a major mine in a region where there was a significant anti-mining element; to working in a jurisdiction where mining was pretty much viewed as a 19th century industry; where government was of a fairly green tinge and ambivalent about whether it wanted a mine anyway; and the general tough permitting system for a development of any kind, mining or not, did not really take into account the necessities of a mining operation. It was all very costly and time-consuming. The original joint venture manager was the US company AMAX. It cost them years and a fortune to attain all the permitting. I once asked one of their executive whether they thought they would ever be able to recoup their investment. He responded that he doubted it, and its ownership share passed to an ACM stable company in due course.

Exploration had to contend with the same hurdles, as I and the company duly found.

When I arrived, New Zealand was still economically backward. Industry was highly protected and consumer durables were very expensive. Getting off a plane in Auckland from Australia, one could generally separate the Kiwis from the Aussies at the baggage carousel: the formers' luggage included obvious items such as electric frying pans or VCRs. Before GST was introduced into Australia, a trip overseas almost automatically included a preceding visit to the duty-free shop in the nearest capital city, or loading up with cameras and sound equipment in Fiji, Hong Kong, or Singapore. The Kiwis were still doing it when I arrived. Cars were enormously expensive and there was a lot of old junk such as Vanguards on the roads; particularly third-rate stuff from Britain which was never good enough to be imported into Australia, and vehicles like a Mini tarted up with a different grille and being flogged as a Wolseley.

Near to Waihi, the kiwifruit industry was booming and making good money for the growers but almost all other sectors of the economy were in a bad way, and this was reflected in house prices. In 1988, I did a deal with NZSX-listed Platinum Group Metals, which owned an excellent five-bedroom house, which they also used as a field office, in a Southlands regional town. They paid less for it than they did for the company Mitsubishi Magna station wagon which, from memory, they had bought for $30,000.

On the more macro scale, with its inherently smaller economy and thinner pool of industrialists and capitalists than had Australia, the October 87 crash hit New Zealand very hard – and particularly its relatively small exploration industry. The number of exploration companies was not large anyway, new investment funds became very difficult to obtain, and they collectively struggled or crashed. This opened many opportunities for a relatively cashed up newcomer such as ACM Gold to deal and to acquire ground and projects which otherwise had to be relinquished by their Kiwi owners.

To summarise to this point, ACM Gold had apparently arrived at the right time. And we were able to offer employment to several excellent geologists, such as Keith Brodie, Lorrence Torckler and Stu Rabone, as well as numerous contractors, all of whom were otherwise suffering in that downturn.

The rest of this story largely focuses upon and illustrates why it proved more difficult to do business than we had anticipated.

Part of the reason was undoubtedly that mining was generally regarded as a fossil industry that had been important in opening up New Zealand shortly after European settlement, but which had run its course and had no future in a modern vision for the country. Besides the rather anomalous Waihi redevelopment, other mining in New Zealand was limited to coal, local iron sands for the steel industry and a few, mainly family or syndicate size, alluvial gold operations in Westland on the South Island.

More critically in my mind, the New Zealand people and its government had convinced themselves that they genuinely lived in 'Godzone' country. To their mind, it was beautiful, which it undoubtedly is; and pretty much pristine, which it certainly isn't, and should be left that way. The northern half of the North Island was buried under volcanic ash after the Taupo eruption in the early AD years but revegetated itself. But mankind has left, and still does leave, a huge imprint on this country. About 1000 years ago, the Maori arrived and killed off the earlier settlers, then it is understood that they burned off a large portion of the eastern part of the South Island in the chase to extinction of the meaty moas.

Europeans arrived and began clearing for sheep and cattle farms. Extensive forests of the magnificent kauri trees of Northland and the Coromandel were pretty much logged out leading to extensive erosion of the soft volcanic ash soils. For instance, Coromandel Bay, into which Captain Cook had sailed, is now just a mud flat. Intensive animal husbandry to this day resulted in chemical and excrement pollution of many streams. By Australian standards, the stream waters look excellent, but tourists introduced Giardia rendering the water rather hazardous to drink. Introduced plant and tree species such as gorse, blackberries and pine trees infest much of the land. Introduced herbivores, such as possums, selectively damage tree species and introduced carnivores such as stoats and ferrets resulted in a drastic drop in the numbers of native birds.

One could reasonably argue that the environmental impact of a large mine, over an area of a square kilometre or two, would have a much smaller impact – a relative drop in the bucket. Bushwalking is encouraged by government and very popular over the length and breadth of the country. Many tracks have been cut and constructed and wood is burned for domestic heating and for campfires. But try to walk alongside a creek to pick up the odd one-kilogram sample of silt from it to assay for gold and watch the proverbial hit the fan. On the exploration side, the main aim of ACM Gold was to acquire relatively large tracts of ground on which we could be the first to apply the innovative BLEG technology to search for the presence of previously unknown mineralisation. The power of that technique is illustrated in the story in *Rocks In Our Heads* by Roger Craddock: the Ovacik gold deposit in Turkey was found in the first dozen or so samples taken in that country!

Where we could deal with a company already owning an exploration licence which was fully permitted for this basic exploration, we could move quite quickly once a team of geologists and field assistants were hired and trained. We managed to conduct this work over quite a few tenements in the Coromandel Peninsula, over extensive holdings around Rotorua and Taupo in the Central Volcanic Zone, and in selected areas in Otago and Southland. While this was going on, my main task was to try to gain approvals for new exploration licence applications over prospective target areas, particularly in the Coromandel. The applications were determined by a Planning Tribunal, presided over by a full Judge, and the process was laborious, costly and emotionally difficult. Absolutely anybody could claim to be a stakeholder in the judgement process simply by lodging an objection. We had to deal with each and every one of them in a court hearing, having secured top-flight, and correspondingly expensive, legal representation. This often led to the situation where objectors, of all views and persuasions, took the opportunity to bear their grievances in a public forum about our applications and way beyond: they were comprised of the usual suspects, and they were abundant.

These included the standard anti-mining brigade, most of whom drive to meetings in their cars made of mined metal and powered by oil industry fuel. There is an old observation that goes that an environmentalist is someone who built their home in the woods last year, and a developer is someone who wants to build their home in those woods this year. They were generally the most rabid, often claiming special and spiritual affinity for the area in which they lived, and conveniently ignoring their own impact upon it while they tried to deny access to others.

Maori people were often another problem. Their opposition was probably rooted in a resentment of past land dispossession, and it is not my place to debate that. But some wanted to flex muscles by opposing 'an Australian company' – never mind that we were a New Zealand-registered entity and directly employing up to 20 Kiwis at the time. The women were particularly intractable. By tradition, Maori women are not allowed to speak in their traditional meeting halls, the marae, but given a chance to speak in a more open forum, it is grabbed with both hands, generally opening with words similar to: "I speak for all Maori…".

At one Tribunal hearing, a couple of guys pretty much performed a haka directed at me from the front of the court. This did not amuse the judge, especially when they threatened that if the judgement did not go their way, they proposed to deal with me in a traditional Maori way!

At another hearing, an elderly Maori lady asked for permission to address the court from the back of the room rather than from the witness stand as she wanted to remain with her people and be supported by them when she gave her evidence. The judge allowed this departure from proceedings, and she opened with words along the lines of: "Your Honour, when I found out that this Australian company applied for a lease over my traditional country, there is no other way to express it – I thought I had been raped". This was then followed by aggressive murmuring and agitation from the tribal people around her. Not a pleasant experience for me.

Often, we had to deal with the plain confused. They either refused to believe, or simply failed to understand, that gaining approval to take some soil or creek sediment samples or to fly an airborne survey was not the same as digging up the ground for a large mine. Alternatively, they would carry on about the impact from drilling when it was quite clear that the permit we were seeking had no request for approval for drilling anyway.

There were also the plain ignorant, such as the apiarist who spent quite some time advising the Tribunal that his bees homed back to their hives using magnetic sensors in their heads and that our aeromagnetic survey would ruin their sense of direction and they would all die. A simple phone call to one of us beforehand would have explained that an airborne survey passively records the Earth's magnetic field over the requested tenement, not influencing it in any way.

The most consistent theme was that our activities would somehow irreparably damage the countryside and inevitably lead to a new mine. If exploration automatically resulted in a mine, our civilisation would be much better off, of course, but most exploration geologists would become redundant.

As our intended principal method of exploration was just to take a sample of about a kilogram of river silts every kilometre or so along a stream, it pretty much defied logic that we could do much damage. This was especially the case in the Coromandel, which had been thoroughly logged, very much eroded, extensively grazed by sheep, and often invaded by gorse or wild pine trees. The Coromandel runs northwards for 100 kms or so from the Waihi gold mine, and is subject to frequent flash flooding and erosion consequent upon heavy rainfall events.

The illogicality maybe exemplified by the following story. Quite some distance inland from the coast and at an elevation of about 1000m, a botanist recognised what was thought to be marine algae growing quite happily. This freak of nature apparently needed protection from our samplers' boots at all cost. Never mind that this little critter was tough enough to adapt to a non-saline environment and then work its way kilometres inland and considerably uphill. The power of the green movement to mobilise opinion, irrespective of how ignorant it is, always amazes me. I was very disappointed to learn that the late and renowned naturalist, David Bellamy, was somehow leaned on to state an opinion that our exploration should not be allowed so as to save these algae. As far as I am aware, he never visited the site, nor undertook any original study of it.

The Fine Young Cannibals were a one- or two-hit wonder band at that time. The lead singer, who was probably not a leading scientist, purchased a property on the Coromandel and readily gave his opinion that exploration should not be allowed. His decidedly non-expert opinion was duly reported in the press, and thank goodness it was in the days before social media! Refer back to my conservationist versus developer comment earlier.

Nonetheless, we usually eventually won in court, often having spent more money in legal fees than was required to undertake our planned first pass exploration programme. There was a learning curve for me, of course. Our first tribunal hearing involved three tenement applications and a huge range of objections to them, some being common, and some being individual. Our appendices to our submissions addressed all of them, but some of that paperwork was inadvertently shuffled to which the judge took a dim view and he deferred the hearing for another day. Fair enough and served me right, but the greenies still managed to have it reported it on the front page of the Auckland daily newspaper along the lines of "judge throws out exploration applications by Australian explorer" – so some egg on my face. However, we won the second time around.

Sadly, when our objectors lost in court and we then headed off to carry out the work to which we were then legally entitled, our opponents fought dirty.

When our guys moved out to the field to undertake a sampling program, there was often a road blockade by a mix of those who were genuinely anti-mining (and I can respect that misguided view), greenies, tree huggers, druggies and dole bludgers. One of our guys quickly found an effective tactic to counter this: in those pre-mobile phone with camera days, we put a loaded instamatic camera in the glove box of each vehicle. When we encountered the human roadblock, we then produced the camera and quite obviously began photographing all those in our way. A considerable proportion were wary of being snapped for whatever reason and usually dived behind trees or ferns. We were then able to drive through and get on with the job.

Marijuana crops were a hazard, and we tried to get the message out that we would not report on any we found in the bush. While we may have not been perfect corporate citizens in this regard, I couldn't expose our workers to the chance of injury or worse. The danger was real: to the south of Waihi two of our guys were met by a pair of heavies carrying shotguns. When we explained what we were trying to do, we were told we could freely sample the left fork of the creek, but not the right. We got the message.

I did once make a tactical error in this regard. We hired a helicopter to undertake an aeromagnetic survey and it only undertook the contract on the proviso that we advertised in local papers that we were undertaking a mineral survey and not searching for marijuana plantations. The header on the ad was a purposely rather eye-catching "CANNABIS GROWERS", and part of the following text read: "the machine is contracted for the purpose of mineral exploration and is not operating in the area for any other reason whatsoever. Please do not obstruct the operation in any way as this may put people's lives at risk." I should have requested that the contractor run the ad, the cost of which could be added to our bill. We ended up as a front-page story in the Auckland Herald, and that probably stirred up more opposition than we might have otherwise had.

More seriously, murder could have resulted. In those pre-GPS days, the helicopter mounted a downwards looking video camera which video recorded the flight path. At the end of each day, or at the end of the survey, that video was compared to known mapped features such as roads and creeks so that the path of the survey was then able to be plotted. One evening the contractor reported that he thought somebody with a rifle had emerged from a dwelling that was overflown and had fired at the helicopter. No damage was evident on the helicopter and, unfortunately, the shooter was found to be laterally beyond the range of the recording camera. I was advised subsequently that a bullet hole was in fact found in the helicopter at the end of the survey. I was rather upset that the camera failed to capture that man in the act: I would have loved to have seen him face attempted murder charges.

Subsequent drilling programs needed additional approvals by Tribunals under a different form of tenure. The greenies hated them as they feared that we might just get lucky and find mineralisation at interesting levels – even though statistically we would not – and then they would be up for a huge fight to protect their already degraded environment from those horrible miners. At about the time I arrived, another explorer, Heritage Mining, learned the hard way that they should not leave the drilling rig unattended at night. Their very expensively acquired samples were destroyed, thus invalidating the program.

We didn't carry out much drilling in my time and the first program went without incident. The next one was further north into the New Zealand Ozarks and police escorted the rig to site. We realised we needed to fence off our operations and have them patrolled 24/7 by security guards. The fencing was justified ostensibly to keep bystanders safe from potentially dangerous machinery. The program was completed, and the additional expense of the security added around 50% to the total costs for the program.

Despite the day to day battles in New Zealand, there were some funny times too. At morning teatime, the mining staff and explorers gathered in the common room and shared a few stories and laughs. Growing up in South Australia and then moving to Western Australia, I really did not know anything about rugby, nor care about it. This caused quite some consternation amongst the Kiwis who essentially could not understand why anybody did not eat, live and breathe rugby. Whenever the Wallabies lost to New Zealand (which they did more often than not), there was incomprehension that I was not upset. My response was usually along the lines of: "So what? Big deal. Your gorillas are bigger gorillas than our gorillas and I don't care."

Towards the end of my stay, we decided we would apply for exploration rights over the town of Waihi and this entailed sending an individual letter to each landowner. I travelled to Wellington to personally sign each of the approximately 800 letters. Next morning over instant coffee in the meeting room, somebody asked me how my day went in Wellington. I raised a few eyebrows and then provoked incredulity when I told them with a straight face that I had a face-to-face meeting with the Prime Minister, David Lange, and I had put him right in the picture and told him the score.

I had to enlarge upon that to explain. In the evening, I was having a beer in the Wellington Ansett Golden Wing lounge (and this dates the story doesn't it!) waiting for my flight back to Auckland. I was sitting in a chair nearest to a door of a meeting room, while there was a rugby test between Australia and New Zealand live on the TV. I wasn't paying much attention, but did happen to know that Australia was in front and by how much. The door opened and David Lange stuck his head out, made eye contact with me as the nearest person and asked for a rugby update. My reply was: "Your mob are losing, and the score is 14 – 6" – and there endeth the story of my one-on-one educational advice to the NZ Numero Uno!

Although my initial focus was on the Coromandel Peninsula, it eventually extended southwards along the target geological system towards the Central Volcanic Zone around Rotorua and Taupo. These areas are geologically alive and, as a geologist only experienced before my assignment in New Zealand to rocks which were generally one to three billion years old, I found it all fascinating. I had some learning to do. One evening at home, the pendant lights began to oscillate from side to side and I casually remarked to the children that we must be experiencing an earthquake. One raced to a doorway and stood in it and the other dived under the table. I afterwards asked how they had learned to respond like that and so quickly, and they said they had earthquake drills regularly at school. I never imagined them being taught such a thing.

ACM Gold began to reap some exploration successes. A couple of days of battling tenacious blackberry bushes in the Taupo and Rotorua areas by Stan Harrison and myself resulted in being able to demonstrate that our BLEG geochemical method would work in that area. It also worked near known gold mineralisation in the alpine region of the South Island. Our guys undertook surveys in the Takitimu Mountains in the far south of the South Island where, as far as we were aware, this was the first time it was used to search for platinum mineralisation, and we successfully found anomalous values to follow-up.

I enjoyed reviewing several gold prospects in Westland, again in the South Island. Quite a bit of small-scale alluvial mining was going on and I felt rather at home there. There was the same 'up and at 'em, can do' attitude there that I had experienced in the gold renaissance in Western Australia preceding my transfer to New Zealand. There were plenty of opportunities, large and small, which kept us all busy.

I spent a couple of nights in a small timber country pub at a place called Mahinapoua which was even featured in some TV commercials at the time because it was so antiquated and quaint. The publican advised me that if I were to borrow his shovel and take a track down to the beach a few kilometres away, I could recover gold from the black sand beach by gently sloshing seawater sideways back and forth across a shovel of it. I was quite surprised when this proved to be the case, with several fine gold grains becoming quite obvious as I panned the lighter stuff away. He advised that this had been known for a long while but had defied all attempts at commercial recovery, even during the Depression. At today's gold price, perhaps someone is having another look at it.

Some areas in Otago proved difficult for us to carry out our geochemical sampling. The beds of the creeks (all called streams in New Zealand) were a mix of mud and slimy ooze and home to myriads of fat worms. All that could be done, we thought, was to get a shovel full of it and the worms would die in due course and could later be sieved out. We always used calico bags, which allowed the water in the samples to drain out, but we found that, as the worms died and decomposed, they generated some sort of acid which attacked the calico and rotted the bags which then they fell apart. We then had to re-sample, this time using plastic ones: we left the laboratory in Perth to sort out the dead worm problem.

I looked at several larger targets with a view to possibly dealing in but, for various reasons, we didn't. Not trying to acquire the McRae's operation, which became a very large mining operation over subsequent decades, was possibly a corporate error, but by then the ACM group was starting to suffer growth indigestion.

A lofty attainment of the joint venture which managed the Waihi gold mine was its official opening. Directors and senior management of the various joint venture partners attended, along with invited dignitaries such as the Minister for Resources. ACM Gold's Chairman, Sir Gordon Freeth, who had been a cabinet minister in the Australian government from 1959 for a decade and then an ambassador, was given the honour of making the dedication speech. Heady times for ACM Gold: its directors and management were to fly the next day to Queensland to officially open the new Wirralie gold mine; probably a unique event in

Australian gold mining history, with two official openings in different countries in the same working week. The diligent company secretary, Daryl Gore, wrote the speeches for both openings and passed them on to Sir Gordon. At the Waihi opening, the latter got down to business and shortly into his speech people were starting to look at each other with varying degrees of perplexity and confusion. I soon realised that he was reading the Wirralie speech, so it was a complete muck up concerning identities of dignitaries, thanking of the various parties responsible for getting the project over the line, and so on. I never heard whether Wirralie got that same speech, or the Waihi one!

Sometime later, I took the opportunity to drive the joint managing director of ACM Gold, Ken Fletcher, from Waihi to Auckland airport to enable some rare one-on-one time with him. Amongst other things, he told me that, at that mine opening, the company had booked all the rooms in the town's two motels to accommodate the invited guests (not the Minister I hasten to add) for the night. He said that all the beer and the booze miniatures and every little bottle of shampoo and soap, et cetera, had been cleaned out of every room. His advice to me was: "Watch out for those Kiwis, they're all thieving bastards!".

Happily, I did not find that was the case and genuinely enjoyed working with them in their non-flat, non-desert, non-warm country.

I left New Zealand at the end of 1989, which was also a time when the gold industry was slowing down somewhat, and the ACM group was suffering that indigestion from probably expanding too far and too fast. Despite the hurdles, I considered we had progressed well. One of the projects I had dealt into was over a small old gold mining centre named Wharekirauponga or, understandably, WKP for short. It was not far in a straight line from Waihi but was located in some sort of conservation area where exploration access was difficult. I see from the recent mining press that subsequent drilling by the current owners of the Waihi operation have now delineated a respectable gold deposit in the area. It seems they are currently working out how they may be able to mine it. While exploration and mining matters do not always proceed at a rapid pace in Australia, they generally do, subject to market forces. In New Zealand, 30 years from acquisition to delineation of a resource is a good indicator of just how difficult it is to progress evaluation towards mining and, naturally, I believe it is to the detriment of that country. ACM chose to work and invest in New Zealand and, of course, we had to abide by their laws – frustrating as they seemed to us. My personal belief is that they work to disadvantage New Zealand but, let's face it, I wasn't even a voter anyway.

2

Exploring in the Land of the Unexpected
Papua New Guinea 1982-2013

Sandy Moyle

Alexander ("Sandy") Moyle was raised in country South Australia, before attending Adelaide University to graduate with a BSc (Honours) in Economic Geology in 1980. He has completed the Macquarie Graduate School of Management – Advanced Management Program and is a graduate of the AICD.

Sandy has 40 years exploration experience throughout Australia, SW Pacific and SE Asia in various mineral commodities, primarily gold and base metals, but also iron ore, nickel, uranium, PGE, tin, tungsten, rare earth elements and diamond exploration. Direct discovery team involvement in two economic gold deposits in PNG and one in Indonesia has been amongst the highlights of his career and he has enjoyed the diversity of working for a variety of junior, mid-tier and major exploration companies.

Introduction

Tour operators often refer to Papua New Guinea (PNG) as the "land of the unexpected" and over the years of exploration in that country I found it to be true in almost every aspect. As background, statistics in 1982 indicated that PNG had a population of 3.5 million people. The 2022 population is estimated to be nine million people, a massive increase. Traditionally, the population comprised more than 600 different tribal groups and over 800 languages were spoken. While coastal areas of Papua New Guinea are reported to have been exposed to passing traders since the 1500s, it was gold prospector-led expeditions into the highlands in the 1930s that discovered more than a million people living in fertile mountain valleys, their way of life apparently unchanged since the Stone Age. Now PNG is split into 22 Provinces for government. During my intermittent in-country exploration assignments covering most regions within this stunning country, it certainly delivered the "unexpected" in every way.

My initial exposure to PNG was with Kennecott Explorations Australia Ltd, Sydney, from mid-1982 to 1985 as an Exploration Geologist, then as Country Manager – PNG in 1988 and Regional Exploration Manager Southwest Pacific from 1989 to 1992. My second was from 1998 to 1999 as General Manager – Exploration for Aurora Gold Pty Ltd, Perth, and the final was as Principal Geologist and then CEO of Goldminex Resources Pty Ltd, Melbourne and Sydney, from 2009 to 2013.

The following summary sets the stage as to how I ended up spending numerous years exploring in PNG.

Being raised on a sheep and cattle property in South Australia, my innate instinct was a career associated with the land and incorporating scope for travel and adventure. Various descriptions of the activities of exploration geologists indicated that it could fulfil my ambitions.

My exploration career commenced with Kennecott Explorations Australia Ltd as a part-time student geologist while undertaking studies in Adelaide University during 1979. This work involved various intermittent office and field assignments on the Gawler Craton, SA, and tungsten exploration in the New England Belt, New South Wales. Following completion of my BSc (Honours) year in 1980 I commenced full-time work based out of Kennecott's head office in Sydney in 1981. As a result of the major increase in the gold price at that time, my initial exploration programmes were targeting alluvial and epithermal gold near Mt Coolan, Qld and tin in the Tenterfield Region, NSW.

During 1981 intense trade union activity and a decline in the copper price resulted in the Bingham Canyon Porphyry Copper Mine, Kennecott's main income source, becoming uneconomic and it was forced to close. A major restructure of Kennecott's activities ensued. Worldwide exploration budget cuts caused Mike Turbott, General Manager – Southwest Pacific Region to reduce Kennecott's Australian exploration activities substantially. With no gold or copper projects at the Resource stage, the Western Australian office was closed. Various staff were encouraged to donate two weeks' annual leave and take a 10% pay cut to salary, which I did. Many found this unacceptable and left the company. Others were dismissed or declined to relocate to Sydney. Difficult times, but a reality check of just how tough it can be to survive in the mineral exploration business.

Kennecott's re-entry into Papua New Guinea

With a very limited budget, Mike reduced his team to three geologists, being Gavin Thomas, Exploration Manager; Peter Morrissey, Senior Geologist; and myself, Geologist. Mike and Gavin both had experience in PNG and saw the exploration potential for major gold and copper deposits there. They felt that this country had the best chance of delivering Kennecott a world-class deposit for its limited exploration budget. PNG was, however, a politically sensitive jurisdiction for Kennecott due to their withdrawal from the Ok Tedi project in 1975 following unsatisfactory development terms being offered by the PNG Government at that time.

Without sufficient funding for Kennecott to set up a full PNG exploration operation in its own right within that country, as several other major exploration companies had done, Mike and Gavin teamed up with Geoff Louden, a geologist, stockbroker and company maker with PNG experience, and Peter Macnab, a PNG Geological Survey Geologist with an extensive knowledge of the local gold mineralisation and exploration scene. Geoff's company, Niugini Mining Ltd, was independently seeking funds to explore its own PNG exploration targets, and already had the capability of exploring in PNG. The Kennecott–Niugini Mining Joint Venture (KNMJV) was formed in 1981, to enable Kennecott to explore gold targets in that country in a cost-effective manner. Under the terms of the Joint Venture (JV), Kennecott Exploration Australia funded 100% of exploration costs for an 88% interest in agreed targets, and Niugini Mining retained a 12% carried interest to be repaid, with interest, from 80% of its share of profits from an economic discovery. Peter Macnab retained a financial interest through his consulting arrangement with Niugini Mining for his contribution of PNG knowledge and field capabilities.

The other three key initial contributors to the Kennecott–Niugini Mining Joint Venture were Ken Rehder, a very experienced PNG prospector associated with Niugini Mining, and geologists Peter Morrissey and me from Kennecott. Our focus in 1982 was on the field-work aspects of the JV exploration programmes.

In PNG we travelled by boat, plane, helicopter and foot to remote and isolated villages in one of the world's most challenging and inaccessible regions. The terrain encountered included raging rivers amongst rugged mountainous jungle in the Highlands, vast floodplains in Western Province, and scattered islands and rough seas in New Ireland Province. Rarely did we have the opportunity to use a motor vehicle when undertaking greenfields exploration during the 1980s.

One of the highlights in any exploration geologist's career is an economic mineral deposit discovery. For me that 'discovery dream' was fulfilled twice in PNG during the early 1980s when I was part of the discovery team of two gold deposits. The first was Simberi which has subsequently been developed and mined by Allied Gold Mining Ltd, and currently by St Barbara Mines Ltd. The second was the massive 40 million ounce Ladolam gold deposit, Lihir Island, which was initially developed and mined by Lihir Gold Limited (partly owned and initially managed by Rio Tinto following its acquisition of Kennecott), and is currently being mined by Newcrest Mining.

Simberi 1982

The PNG adventure commenced when first flying to Rabaul from Sydney on Wednesday, 11 August 1982. Rabaul is situated on the northeast tip of the Island of New Britain within a collapsed volcanic caldera which forms a very protected harbour. In 1982 the Rabaul Harbour area was very seismically active, had minor recent ash eruption activity from a daughter volcano, and contained several areas of hot springs. Volcanologists identified the biggest threat being the swelling Matupit Island, situated within the main caldera and which hosted Rabaul's sealed commercial airport at that time.

The seismic activity and imminent threat of an eruption were very evident just after we had checked into the Kaivuna Hotel. I was placing my bags on the bedroom floor when the room started to shake. The cars on the road outside were bouncing across the moving ripples in the bitumen road. Then there were screams from the swimming pool as the water was thrown from side to side and slopped out over sun-baking guests relaxing by the pool. This was a real eye opener for me being a South Australian country boy with no previous exposure to the Pacific Rim of Fire. The Kaivuna was subsequently placed in the second to highest danger rating zone (pink) but kept operating. The bar once advertised to "have a gin in the pink", most appropriate at that time.

Rabaul was the provincial capital and most important settlement in the province. It was destroyed in 1994 by falling ash from a volcanic eruption in its harbour. During the eruption, ash was sent thousands of metres into the air, and the subsequent rain of ash and small stones caused 80% of the buildings in Rabaul to collapse. Authorities declared a state of emergency and, after the eruption, the capital was moved to Kokopo, about 20 kilometres away.

The Rabaul township and harbour are steeped in war history. The northern half of PNG was a German colony up until the First World War when the British took command of the region. During the Second World War the Australian / English/ American military base there was bombed and captured by the Japanese in 1942. They dug tunnels into the soft volcanic tephra of the wall of the main caldera and built bunkers. These housed troops, munitions and submarines which were on rails. On occasions during our transits in Rabaul, we were able to inspect these tunnels and scuba dive on the Japanese war fleet wrecks and planes that were sunk in the harbour. The Americans were unable to remove the Japanese from Rabaul until the end of the war.

Experienced Papua New Guinea residents, Peter Macnab and Ken, had the task of introducing Peter Morrisey and me to conducting effective exploration and handling local nationals.

The JV considered the Bougainville – Tabar Island trend, situated in the New Ireland Province, to be the highest priority for initial exploration efforts based on gold deposit target size potential and ease of development. In 1981/1982 there was a moratorium on issuing mineral Prospecting Authorities ('PAs', which were licences to conduct exploration) in PNG, so the JV decided that initially it would focus on a prospect covered by an existing one. The Tabar Island Group was chosen as prospectors had reported minor gold occurrences, but no substantial exploration had occurred on these islands. Nord Resources held the PA over the Island group. A joint venture was formed with Nord, and KN-MJV were the operators.

Food and field supplies were purchased from Steamships, Burns Philp and various Chinese trade stores in Rabaul. Several cartons of South Pacific Lager were also procured. A high-winged twin engine Islander aircraft was chartered to carry the four of us and gear to Mapua on Tatau Island, the middle island of the three largest islands of the Tabar Group. Mapua was a Catholic Church Mission outpost with a school and simple medical centre a *haus sik*. The only expat at the Mission was a rather rotund priest, Father Bernie. One notable feature of the corrugated iron Mission buildings were the bullet holes in the timber floors from strafing by the Japanese air force prior to capturing the New Ireland and New Britain provinces during World War II.

Berthold, a cantankerous PNG German, met us with his old 14.5' aluminium boat powered by an antiquated 40HP Evinrude outboard. Much of the gear and the four of us were loaded into the small boat which was stacked to the gunnels. Our destination was Pigiput Bay on the east side of Simberi Island and situated 35km to the north of Mapua.

Simberi Island is approximately 10km in diameter, with moderate topography up to 250m above sea level and steeply incised radial drainages covered by dense tropical vegetation with a high forest canopy. The island weather is nearly always hot, humid and wet, being 400km south of the Equator. Pigiput Bay is one of several beach access locations on the island and is surrounded by coral reef and an elevated palaeoreef. An abandoned coconut plantation belonging to a New Ireland Chinese family is situated near the bay so was chosen as the exploration base. It consisted of a couple of very basic, elevated rusty corrugated iron and timber buildings with a rainwater tank. We lugged the food and field supplies for the six-week exploration trip up the hill to the hut. Berthold took his boat back to Mapua to ferry the second load of supplies over.

Bedding was set up on the timber floor. I was introduced to the concept of a 'bed sleeve' which is basically a 2m long cylinder of canvas which was pushed out into a camp stretcher by inserting two long bamboo poles and holding them apart with bamboo struts at the head and foot of the sleeve. Bedding was a pillow and two sheets. A blanket was rarely needed at sea level this close to the Equator as temperatures were generally 22-24°C overnight and 32-34°C during the day. Humidity was intense and regular tropical downpours heavy. Then there was the obligatory suspended mosquito net in the endeavour to prevent getting malaria, which is rife amongst the PNG tropics. One memory of that first night were the rats scampering across the floor, chewing anything including my hair and the mosquito net.

The next morning Peter Morrissey and Ken walked south with some locals to commence sampling and mapping a drainage. Peter Macnab and I headed north to Munun village to recruit some locals to assist with exploration activities. As we were walking through the abandoned coconut plantation, a coconut fell from above and missed my head by 5cm. Since that close shave I have never again walked under a coconut tree without looking up. At the village, we explained our planned activities and hired six local men who were keen to give us a hand with surveying and sampling. Access to most of the areas of interest was by traversing up creeks. While all expats normally wear boots in the bush, Peter always had bare feet, like the locals, to minimise the chance of slipping over on slimy rock surfaces and to reduce the chance of fungus from wearing wet boots all day.

Betel nut (*buai*) chewing was rife on islands such as Simberi and Lihir, and regularly chewed by the locals in PNG. The nuts, which come from the Areca palm, are chewed for their stimulating properties, and are reported as being one of the most popular mind-altering substances in the world. On Simberi, men and women regularly chew buai to stay awake, particularly when working. Buai is usually chewed as a mixture with slaked lime and betel leaf, forming a red paste which they blow out after chewing, spraying a mist of orange spray, often over walls of buildings in towns. It results in high rates of oral cancer and ultimately early death.

Kennecott had a particularly good systematic sampling and mapping system that was followed. For detailed project work, every creek tributary and the main creek was sampled upstream from the confluence to eliminate the chance of floodwater contamination. Samples collected included fixed volume panned concentrate and -80# stream sediment;

mineralised rock float was described and sampled for assay. With every rock sample a second representative sample was collected for future reference should the original sample be gold anomalous. All creeks and tributaries were surveyed and mapped at 1:2,500 scale using 30m rope with knots every 10m. Semi-continuous rock chip samples were collected at every altered and mineralised outcrop.

Initially the rope method appeared odd, but it soon became obvious that normal tapes would rapidly become damaged in the rough environment and couldn't be replaced locally. The downside of the rope was that the local workers from the villages would cut a couple of metres off to use as belts for their *lap lap* (cloth material wrapped around their waist like a skirt). The national field assistant and I would have to keep checking the rope length to ensure mapping and sampling accuracy!

Following our introduction to conducting exploration in PNG in August 1982, Peter Macnab and Ken left by boat to conduct reconnaissance sampling around the Lihir, Feni and Tanga island groups further southeast. Peter Morrissey and I were on Simberi by ourselves to get on with the mapping and sampling. Macnab and Rehder returned with food after a few weeks and were joined by Mike and Gavin to review our progress and provide direction. We spent much of the seven weeks on the island mapping by ourselves. Prior to returning to Sydney for a break, we were rewarded with a trip to, and a mine tour of, the Bougainville Copper Mine, located on the island of the same name. This was a very impressive porphyry copper-gold mining operation conducted by CRA prior to it being forced to close in 1989 due to civil unrest.

The initial drainage sampling and mapping on Simberi revealed alluvial gold in the creeks and widespread pyritic volcanic breccias. The creeks were surveyed and mapped in detail. Ridge and spur soil samples were collected wherever drainage samples returned visible gold or gold anomalous geochemistry. Some areas with surface gold anomalies were trenched and sampled. Diamond drilling of delineated gold anomalies followed.

Once the initial gold mineralisation discoveries had been made on Simberi Island, the northernmost island in the Tabar Islands Group, similar greenfields mapping and sampling activities commenced on the west side of Tatau and Tabar Islands in 1983, the middle and southern main islands of the group. A large ocean-going single-hull dugout canoe, or *mon*, was used for transport on occasions as it had the ability to navigate through small passages in the thick mangroves. The western sides of these islands were low-lying, and swamps generally were found behind mangroves. These stagnant swamps needed to be traversed to access drainages for sampling. The water, mud and weed in some were up to chest height in depth and always infested with mosquitoes. When the going got tough during the crossings, I always looked at the palm trees and was grateful that there were no Japanese soldiers shooting at us, as they did to PNG and allied soldiers during the war. The Tabar islands were captured by the Japanese during the war and elders who were around as children described the shootings, raping and other atrocities by the Japanese soldiers in the New Ireland Province.

Our ongoing exploration drainage and soil sampling in 1983 and 1984 located gold mineralisation over a substantial area on the western side of Tatau. Later these targets were trench sampled and drill tested.

The standard Kennecott field roster was six weeks on, six days off. We worked seven days a week in the field without any field loading, so the Sunday was for 'love'. Fortunately, mineral exploration is such a diverse environment that it was rare to be doing the same task in the same location a fortnight later. Basically, it was arranging programme logistics and field crew, then map and sample for the six weeks, despatch samples and return to the head office in Sydney. Take the six days off and spend a couple of weeks in the Sydney office drawing maps, interpreting assays and observations then writing reports prior to returning to the field; also developing a strategy for advancing or dropping exploration targets and planning for the next field trip.

Greenfields and some brownfields exploration performed by the JV was normally conducted using fly camps. My preference was to set camps up adjacent to a creek for ease of obtaining fresh water and washing. Simberi is a relatively small island, having moderate topography, and with all the population residing in coastal villages.

The geologist and field assistant generally slept on our canvas bed sleeves, covered with a mosquito net suspended under a small tarp.

My ideal location at a camp site was upstream, away from the rest of the field crew to ensure I had access to the cleaner water and avoided noise. An open fire was used for cooking. Utensils and cooking gear were basic, but adequate. A Codan HF radio, battery, and solar panel for re-charging were carried to report into a base or the PNG office every evening. A kerosene pressure lamp for light, spare mantles and a drum of kerosene were also essential. Food was also basic, usually consisting of tinned beef or fish, tinned vegetables, and rice. It was stored and carried in a locked patrol box as theft occurred by village labourers. According to Ken, the other issue with locals having access to unlimited food is that they will eat all of it in one sitting, will be too uncomfortable to work the next day, and too hungry the following day because there is no food left. When possible, I took stubbies of SP Lager Beer in amongst gear in my patrol box.

Lunch for labourers and me consisted of several *boi biskits* (similar to a massive, tough Sao biscuit) and a small can of *tin pis* (tinned fish) and it was usually consumed on a rock bar or bank in the creek if there was sun, or under trees if it was raining.

The general routine when returning from the field each evening was regimented. It consisted of checking-in using the Codan radio; attending to any labour medical issues; issuing food, sugar and tobacco to each labourer for their evening meal; sorting out samples; followed by a relaxing wash in the creek and drinking a single stubbie of beer, which I'd carefully hidden in the water to keep it cool, and then cooking dinner.

When moving camps everything, except the drum of kerosene, would be packed into metal patrol boxes, which had large elongate handles on either end to enable carrying by two men using a single bamboo or sapling pole to suspend it. Tarps and bedding were packed in another, while field equipment was usually bound together for transportation.

Minor earthquakes in the PNG islands situated along the Pacific Plate were not uncommon. One evening when working out of a fly camp on Simberi, I heard a rustle of trees which turned into a loud roar as if a train was about to appear. The ground shook for about 10 seconds and the noise disappeared off into the distance. Another fly camping experience in the "land of the unexpected".

Lihir gold deposit discovery

Lihir Island is located 80km southeast of Simberi Island. It has an area of 192km² and a rugged, deeply incised, youthful volcanic topography rising to 700m.

The moratorium on the grant of new PAs was lifted by the PNG Government in 1983 and PA485, covering Lihir Island, was granted to KNMJV on 19 June 1983. We were keen to commence exploration on Lihir to determine the extent of the gold anomalous rock chip samples collected by Peter Macnab and Ken in 1982, while I was exploring on Simberi.

On 20 May 1983, I was sent to Lihir to commence the initial exploration programme. My first task was to survey in a base-line reference route around the collapsed caldera feature in the Luise Volcano. All initial exploration activities would be tied into this survey. The base-line reference route survey was done at 1:2,500 scale using the same 30m long rope and compass technique and was plotted on gridded mylar film field sheets, as practised on Simberi Island. The loop comprised traversing 7km of meandering creeks and steep hills, while applying slope corrections. Amazingly I was only 20m out when all the field sheets were plotted on the master plan. If only the GPS units of today had been developed at that time, life would have been so much easier.

Ken set up a basic exploration camp near the ocean prior to my arrival. Constructing camp buildings in this environment was simple as there was abundant bamboo and vine for binding. Once the framework was built, plastic was used to line the walls, flywire filled the windows and tarps were used for roofing.

Although minor food supplies had been brought in by helicopter or speed boat, fuel, bulk food, a generator, and other supplies to set up the camp were delivered on a small coastal ship and ferried ashore in tenders filled to the gunnels. It's a wonder none sank. Whilst it was a very wet, hot and humid climate for physical work at Lihir, there was always the knowledge of a swim in the ocean when we returned to the camp each evening. When topped off with a cold beer supplied by the company and a good meal, the Lihir project was ticking a lot of boxes at that stage.

Gavin, and contract geologists John McManus and Dave Shatwell, arrived the day after me and the exploration began in earnest. That afternoon Gavin inspected the hot springs and while traversing them, the surface crust gave way and he went up to his knees in boiling mud. His burns were extremely painful and the risk of infection in the tropics is high, so he took the helicopter into Rabaul to seek medical attention.

Detailed drainage sampling and ridge-line soil sampling exploration, like that completed on Simberi, was conducted within the Luise Caldera during the following couple of weeks. The initial exploration revealed extensive areas of clay alteration making meaningful geological mapping difficult. Trenching was introduced in these areas for mapping and sampling purposes. Drainage samples within the caldera revealed several tributaries shedding visible gold. Exciting times! We had covered a lot of the caldera in the first three weeks so, after despatching the samples to the assay laboratory in Lae, compiling the exploration data onto a master plan for drafting at Kennecott's office in Sydney, and putting a caretaker in the camp, I went to Madang to complete another exploration assignment while awaiting Lihir assay results.

After a short field break it was time to interpret the recently received Lihir assay results and write a report on our findings in the Sydney office. This initial exploration programme had produced sufficient encouraging results, geological understanding, and size potential to justify a drill programme.

Diamond drilling began in September 1983 and returned excellent shallow gold intercepts. The drilling continued while I was undertaking greenfields exploration at April River, Manus Island and Tatau Island in the Tabar Islands group. In June 1994 I returned to Lihir to manage the field activities for five weeks. Thirteen diamond drill holes had been completed at that stage; however, it was challenging drilling with the thick argillic and advanced-argillic clay alteration continually bogging rod strings. PNG Drillers was granted the initial drilling contract on Lihir. The hands-on owner, Henry Vox, a tough driller of German heritage and with a heap of PNG experience, could recover most drill strings when other drillers had given up. The thick zones of alteration clays and regular heavy rains were causing constant difficulties with earthworks and drill rig movement due to deep mud. A D-5 swamp track dozer was brought in to improve productivity. In addition to managing the project while Peter Morrissey was on break, much of my time was spent planning holes and logging core.

The use of epithermal gold consultants, including Greg Corbett and Ray Merchant, had assisted in developing a geological model and our understanding of the deposit was becoming more refined.

As Lihir was developing into a potentially economic deposit I was sent over to Kennecott's USA operations in Tucson and Reno, and their headquarters in Salt Lake City, to be briefed on geological modelling techniques, geostatistics, and computer-based resource estimation techniques A tour of the Bingham Canyon copper-gold Mine in Utah and a weekend of snow skiing in Salt Lake were also amongst the highlights of the trip.

When back in Sydney, Lihir reporting, planning, budgets, geological interpretation and financial modelling were occupying most of my time. I was given an IBM portable computer the size of a sewing machine and a screen less than A5 in size to do the work with.

In 1985 most of my activity was Lihir-focussed and the benefits of being directly involved with large discoveries continued to come. Gavin, Peter Morrissey and I visited the Rotorua geothermal zones, volcanoes, and epithermal gold deposits in the north island of New Zealand with geothermal consultants to better understand the mineralisation setting at Lihir.

In early March 1985, I returned to Lihir and spent nine weeks managing the exploration. There were 40 expatriates, 70 drill offsiders and 165 casual labourers working on site. PNG Drillers had nine diamond rigs drilling, and Wallis Drilling had been successfully using their air-core system drill rigs on site to get representative core samples through the thick clay zones where the diamond drilling was failing to achieve adequate sample recovery for assay.

Following a break and a brief time in the Sydney office, I returned to Lihir for a month commencing early June with the task of relogging all the diamond drill holes to finalise a geological model to be used for the inaugural resource estimation model. This was a rewarding assignment as I was able to lay out all 54 of the diamond holes drilled, section by section and log from the bottom up instead of the top down. Relationships between rock types and structures were a lot easier to understand by starting with the least altered rocks rather than the intensely altered clays and weathered core at the top of the holes, which often also had poor recoveries. The original logging of most holes was good, and it was variations of lithology and alteration descriptions and interpretations used by different geologists that needed to be altered. The end result was a cohesive sectional geological model, most of which reflected the original individual logging, but now with consistent terminology.

A bit of geology for those interested. The rocks more frequently intersected in drill holes are volcanics, intrusives or breccias containing clasts of these two types. Most volcanics are latites, andesites and trachytes. The most abundant intrusives are fine to medium grained monzonitic types ranging in composition from pyroxene microdiorite to biotite syenite. Two distinct phases of alteration were recognised. The first phase of porphyry-style potassic alteration was followed by violent caldera formation, producing extensive brecciation and veining. The second alteration phase partly overprinted the first phase and was an epithermal one, evidenced by argillic, advanced argillic and phyllic assemblages. Boiling and mixing of circulating fluids in the open breccias is thought to have resulted in deposition of the gold.

The sectional geological model and assays for the first 54 diamond drill holes and the assays for the 165 aircore holes were sent to Pincock, Allen and Holt, resource and mining consultants in Tucson Arizona. I followed a week later in early August 1985 and spent several weeks there working on the data with them. This is where I got a good appreciation for Mexican food and where western movies are made, visiting several towns often mentioned, like Tombstone and Bisbee.

While Pincock, Allen and Holt were crunching the numbers, instead of flying me back to Australia for a break, Kennecott flew my wife to Tucson. From there, we flew to Salt Lake City, drove up to Yellowstone National Park and then from Seattle south to Los Angeles. Steve Craig, Kennecott's Exploration Manager in Nevada at that time, took us on a whirlwind, three-day trip of gold mines in Nevada, including the Carlin Trend.

Following that brief but action-packed break it was back to Tucson to continue working on the Mineral Resource and Ore Reserve block models. We received information from several additional holes to incorporate in the model. By early October 1985 we had an initial Kriged Mineral Resource estimate for the oxide and sulphide mineralisation at the Coastal and Lienetz Zones of 712,000 oz Au. This was in less than two years from starting exploration. I returned to Australia and continued resource and reserve estimation work in Sydney with others and ongoing logging reviews on Lihir which resulted in a large increase in contained gold being reported in December 1985 by the KNMJV.

In 1986 and 1987 my involvement in Lihir was minimal as I was managing Kennecott's re-entry into gold exploration in Western Australia. When returning to PNG in 1988 as Regional Exploration Manager – PNG, my key focus at Lihir was exploration of targets on the island within remnant calderas, peripheral to the massive Ladolam deposit, which was progressing with our excellent exploration team. Most of the work conducted on Ladolam was directed by Kennecott's engineering group.

Life on Lihir was not all hard work

While managing the exploration on Lihir in June 1985, the landowners from the Lataul village on the southeast coast of Lihir Island invited me down to a *Sing Sing* (pidgin for celebration dance). It took about 20 minutes to get there using the company tractor. There were about 30 pigs tied up by vine rope to bamboo poles for carrying, and they were lying adjacent to fires.

Numerous groups of teenagers, then adults, from the nearby villages then appeared and performed impressive displays of their clan dances wearing full traditional dress of grass skirts, various head bands and shell or vegetation necklaces. Ochre paints and arm bands were used by some for decoration. Most groups had leaf wands or carved and painted story sticks. The dances were to support the spirits of their dead people.

Killing and cooking of the pigs followed. Suffocation was by twitching up the vine rope already over their snouts to prevent squealing during the prior ceremonies. They were thrown onto the fires, the hair burnt off and the flesh partially cooked. The pigs were then gutted and cut up, portions wrapped in leaves and buried with root vegetables in coals. Known as a *mumu*, it is a traditional ground oven method of cooking large quantities of food for celebrations in PNG. It involves digging a pit, filling it with hot coals or rocks, adding food, then burying the whole lot for hours so it can cook, like the traditional Maori *hangi*.

Being seated with the village chiefs, we were the first to be served portions of lukewarm ultra-rare pig on cracked and chipped porcelain plates. I was trying to look appreciative, but all the time was concerned about contracting leprosy, tuberculosis or dysentery, all of which Lihir Island had been well known for in the past. Overall, it was a great privilege to be invited to the ceremony, and it reflected the good relationship that we had developed with the Lihir people at that time.

Another one of the less demanding tasks at Lihir was diving on the underwater monitors which were part of baseline environmental studies Kennecott had initiated within Luise Harbour, which lies on the margin of the collapsed Luise volcanic caldera, and immediately adjacent to the gold deposit.

One Sunday distraught villagers from the nearby Malie Island raced into the camp asking us to search for their outboard motor which had fallen off their mon canoe. Malie is situated 10km north-northeast of our camp. Geologist John Miles, who had a similar passion to me for scuba diving, and I took a canoe out to the island to look for it. Volcanic islands are typically fringed by coral reef with very steep sides dropping into the abyss below. Although realising that the search was likely to be futile, we covered the area they wanted searched. The water was crystal clear, and we ended up at about 50m depth which was far deeper than we intended. At one stage John realised that we were being closely followed by an enormous groper with its mouth open, spanning one metre wide. Its length appeared to be at least 2.5m so we made a rapid exit without incident, but sadly had to report that we hadn't located the motor. The outboard-powered mon ocean-going dugout canoes used around the islands and Sepik regions of PNG were constructed from a single tree and range between 10m and 16m in length and can carry more than a dozen people or a couple of 200 litre drums of fuel.

Endemic malaria and foot rot

Operating in the tropics was fraught with medical risks. Kennecott was proactive with providing staff with the best tropical medical advice and necessary vaccinations prior to commencing work in PNG, and anti-malaria tablet regimes when in the country. Unfortunately, several of the tropical medical risks which have been around for centuries are yet to be totally mitigated. Although diligent with malarial prophylaxis medication, I contracted malaria in 1982 and subsequently had 16 bouts of it with my last occurrence being in 1996. The malaria parasite resides in the liver and is very difficult to purge from the body. It is transmitted by the female Anopheles mosquito and is a major cause of death in the tropics. At one stage Kennecott had 16 geologists in PNG and only one hadn't contracted malaria. Of the four types of malaria reported, I had Vivax, which was the common malaria reported during World War II. Several of the geologists contracted Falciparum, a strain which results in cerebral malaria, with obviously more severe implications.

Another common medical issue was severe fungal tinea in the feet, which generally is caused by constantly having wet feet when walking up creeks or due to tropical down-pours. Infection of wounds and swelling without known cause were also a common occurrence in the tropics and hard to remediate however, fortunately, penicillin sorted out most.

Unusual airport departure

Air safety was not the only risk when travelling in PNG during the early 1980s. On 29 July 1983, I was scheduled to commence a programme of reconnaissance sampling and mapping in the headwaters of the April River, one of the major rivers draining north out of the Highlands into the Sepik Valley. Getting there involved a commercial flight on an Air Niugini Fokker F28 from Port Moresby departing at 6.00am to Wewak on the north coast. This flight was due to land there with just enough time to connect with a Talair Islander aircraft which ran a bi-weekly flight to Ambunti, a small village on the banks of the mighty Sepik River. From there, arrangements were that a chartered four-seater Hughes 500 helicopter would be meeting me to take me to a remote village located on the April River, where the rugged slopes of the Highlands meet the flat Sepik plains.

All was going as planned, arrived in Port Moresby the night before, checked into the Airways Hotel close to the airport, had a good meal and night's sleep. Arose early to check out of the hotel at 4.30am. The PNG factor complications began. Behind the front desk was a guy who fitted the description of the maintenance man or security guard, not a front desk attendant. Communications were not wonderful as he could speak no English and had limited pidgin. He took my credit card, and looking very important, attempted to process the payment. After 10 minutes he realised that it was all too complicated, so he disappeared out the back to get help. After another 15 minutes he returned and had another attempt. At this stage, both time and my fuse were very short. I was just about to grab my card and leave without paying when another person came in and processed the transaction.

Fortunately, a taxi was waiting, so the short trip to the airport was simple. However, when going to the Air Niugini counter, I was told the flight was closed. At this stage it was 20 minutes prior to the scheduled flight departure time and the plane had not started its taxiing to the runway. Despite pleading with the young local man behind the counter to speak to the manager, he said I couldn't. Rather desperate at this stage and thinking of the chaos that would follow if I didn't get on this flight and of the helicopter charter costs; the only thing I could do was find the manager, so I jumped the counter and ran into the offices behind, with the counter attendant chasing me.

At this stage the Fokker F28 had started its engines. I found the manager, explained my predicament, and typical of the can-do attitude of many of the expats in PNG at that time, he said he would get me on that flight. Meanwhile the aircraft had commenced to taxi toward the runway. He radioed the pilot requesting that the plane be stopped, which it did halfway to the runway. He also asked the counter attendant to open the door to the concourse, but the attendant couldn't find the key, so I jumped out the baggage chute with my bag and patrol box full of field gear. On the other side, a baggage handler and I grabbed the gear, threw it on a trolley and ran to the plane as its door and baggage hold were opening. Thankfully by 6am, the planned departure time, I was on board with my gear, and bound for Wewak.

The Wewak airstrip, like several others including Vanimo, Aitape,

Angoram and Bogia on the north coast of PNG, still had visible craters on either side of the runway from Japanese bombings during the war. Fortunately, boarding the Talair flight was less eventful. After waiting for a couple of hours at Ambunti, Phil Cotter, a highly experienced ex-Vietnam veteran pilot, arrived and landed the Hughes 500 helicopter near the Ambunti Lodge guest house on the banks of the Sepik River. Phil trained originally as a helicopter aircraft maintenance engineer, and after several years became a pilot. He had thousands of hours of rotary-wing flying experience, many in PNG. I flew hundreds of hours in helicopters with him as pilot during the 1980s and always felt safe when he was at the controls. We loaded my gear and other supplies into the helicopter and flew 70km southwest to the exploration base camp that had been established next to the April River village. Numerous massive crocodiles scurried from their nests on weed islands within swamps as we flew low over the Sepik River flood plains. The sparse plains were broken up by heavily vegetated, sharply protruding ranges with near-vertical 300m cliffs on their southern sides.

April River adventure

The village adjacent to the April River is relatively remote and isolated, and is located where the river flows northward out of the ranges onto the Sepik plains. Apparently, the site was chosen by the missionaries because of the flat land allowed the clearing of an airstrip. The village consisted of about 20 houses at the time, each elevated approximately 3m off the ground to prevent inundation at times when flood waters of the river flowed over its banks. They were solidly built with numerous relatively thin tree trunks supporting the floor and roof. Walls were constructed with vertical sapling trunks interwoven with bamboo and bush vegetation. The roofs were typically thick, sloping, bedded, coarse grass and fronds to repel the regular heavy rain downpours.

There were also a few very high tree houses with supporting stilts and retractable ladders. The floors can be approximately 20m high above ground level. The houses utilised several branches to fasten it at the top of a single living tree with a large trunk. The houses varied in size but generally appeared to be of similar size to those closer to the ground. While I don't recall seeing families living in the tree houses at the time, the houses were in good order. It doesn't pay to ponder over what they did for ablutions at night from those heights. Apparently, this style of tree house was a lot more common in the decades prior to the 1980s and were designed to prevent killings and theft by marauding neighbouring tribes.

The locals were relatively friendly, but communication was difficult as they spoke their local language and knew little pidgin. Fortunately, Peter Macnab had passed through this area previously and was well aware of the language challenges, so he provided experienced national field assistants to help manage the local labour.

With an elevation of 70m above sea level and being less than 5° S from the Equator, the heat and humidity are oppressive. Malaria and 'grille', a fungal skin infection which looks like ring worm, are prevalent at this low altitude. While grille occurs in other low areas of PNG, it was unusually prolific in this part of the Sepik. Many children had distended bellies, apparently from a combination of a high starch, low protein, diet, and had enlarged spleens from numerous doses of malaria.

The people from the lower reaches of the April River were relatively primitive. Seeing a piglet suckling on a *meri's* (native female) bosom was not uncommon. Other juvenile pigs were tethered with a bamboo lead and accompanied the meris. Generally, the female dress was a grass skirt, a woven arm band, one or two large platy shells suspended on a woven necklace, and pierced ears often with large holes supporting a hollow slice of a bone. A woven *bilum* (net bag) suspended from their heads was used to carry anything from food to belongings.

Men often wore *tanke* (native grass) covering their buttocks. Quite a few of the men would tuck their penis into a cowry shell or a hollow gourd, alternatively they would wear tanke or other native vegetation on a woven belt to cover their genitals. All men had bones or pig tusks through the noses and the majority had spikes, which appeared to be from rhinoceros beetles, sticking through the tops of their noses. Ear piercing was also common amongst men. Western clothing items were worn by relatively few in the area. Peter Macnab arrived by helicopter to my camp in the Leonard Schultze River area in 1983 armed with a massive, recently invented, video camera. One of the local men was nearby in standard traditional attire as mentioned above, including the cowry shell. I managed to photograph the two of them standing next to each other exemplifying the contrast between cultures and technology.

Bows and arrows, shields spears and *kundus* (small elongate hollow drums) were magnificently constructed and used for hunting, tribal fighting and ceremonies. Bows were all made from flexible black palm and used thin bamboo or braided plant fibre as strings. Most bows were decorated with braided fibre wraps, evenly spaced along the bow's limbs. The arrows were real works of art consisting of perfectly straight cane stems with elaborate patterns burnt into them, carving on the head end with perfectly woven plant fibre webbing fastening the heads onto the arrow shaft. A dozen different head types were observed, including the single fire-hardened wooden blade; various types of pointed, barbed hardwood heads; several styles of multi-pronged wooden heads for birds and fish; and bladed and sharpened bone heads. No man-made fibres were used in implements constructed at the villages, unlike those produced at that time on parts of the Highlands' spine, which used synthetic string binding.

Shields in the April River and neighbouring Leonard Schultze River catchment region were usually large, moderately convex types with handles on the back. The dimensions varied but 2m high by 40cm wide were common with the extra width, giving far better protection against arrows and spears compared to the general highland shields, although I haven't seen them used in combat.

The JV exploration base camp at April River was relatively large with six beds with separate kitchen/living, shower and long-drop toilet buildings. All were constructed with tarp roofs and plastic, timber and bush material walls being able to withstand the regular heavy downpours and occasional strong winds. The separation of living and sleeping quarters proved beneficial at times when all the geologists returned from fly camps in the hills and were enjoying many stories of fly camp incidents and a beer or six. Several geologists were working as contractors for the KNMJV at that time. This included Jim Allen, Graeme Minnifie, Greg Drake, Dave Shatwell, John Wier and Jim Lockhardt. The programme was scoped out by the KNMJV committee of Gavin, Mike, Geoff and Peter Macnab as usual.

We were exploring the headwaters of the April River drainage for porphyry copper-gold and epithermal gold deposits using rope and compass surveying and the standard systematic Kennecott drainage and rock float/insitu chip techniques.

In bigger fly camps at major tributary junctions, field crews consisted of two geologists, each with an experienced national field assistant managing six national local men from the nearest villages to assist with field work. Fly camps in tighter areas usually only had one geologist accompanied by a national field assistant and labourers, due to more frequent moving.

Luxury was not usually a word that was synonymous with fly camps. Most of my greenfields and some brownfields exploration campaigns in PNG were undertaken using them.

When working near the maximum efficient walking distance from each fly camp location, the field assistant and a couple of labourers would clear another helipad and camp site at twice that distance up stream. Sites on a spur ridge suitable for helicopter access, and near running water, were preferred but not always possible. Once the new site was cleared, the samples were sorted for despatch and left on the pad for collection by the helicopter pilot. If fine weather, the helicopter would bring additional food and supplies, assist with the camp move and collect the samples from the original pad prior to returning to the April River base camp. If the pads were clouded in, however, as often was the case, the camp move would have to be on foot and the heavy patrol boxes carried. This was always tricky due to steep drainages, often with waterfalls and damp slippery rainforest adjacent to the creeks.

It was very rare to see other people in the upper reaches of the April River. Most of the locals living at the April River village could not speak the same language as those in living in the upper reaches. Some could translate if a family member had intermarried in the area. On one occasion, I noticed a very fit man amongst the workers. It turned out that he lived with his family a few kilometres from where we were camped and came down to see what the helicopter activity was about. The only thing he had which wasn't produced from animals, vegetation or stone was an old medieval-shaped double-bladed axe, a type that I previously had never seen in PNG, so its origin was puzzling. It is unusual to see men not carrying a bush knife everywhere they go but I guess the axe served the same purpose. I asked the locals to encourage him to stay and assist with the work and, more importantly, show us any existing tracks that may exist in the bush. He did so and then, after a couple of days, we came across his lone house and garden surrounded by thick jungle in every direction. As we approached, a woman and children disappeared into the jungle, and I didn't see them again. It appeared that they were scared as I was probably the first white man they had seen in this area and possibly ever. The man showed me the inside of his low-roofed, bush material house. There were several beds surrounding a smouldering fire in the centre of the room, rudimentary clay pots on the dirt floor but nothing European at all, not even a bush knife.

The Sepik highlander continued to work with us for a week and his knowledge of the bush tracks was extremely useful. The real dilemma I had was how do I pay this man, without dramatically altering his traditional lifestyle, nor being detrimental to his wellbeing. He wouldn't know what money was and there were no shops or trade stores nearby to spend it in anyway. It was the first time ever that, whichever way I compensated him, it would have a major influence, and probably have a beneficial impact to his daily lifestyle, but it may be detrimental to the family's cultural wellbeing. In the end he was given a couple of bush knives, pots, plates, bandages, a roll of cloth and some food. I still think about the ultimate impact that it would have had on his family's subsistence lifestyle, as prior generations had lived and survived like this for the previous thousands of years.

Our exploration activities at April River were supported by a Hughes 500D helicopter, again piloted by Phil. There was one event, though, which exemplified the calculated tenacity that he had when managing most difficult situations. I had been up in the head waters of April River conducting drainage sampling and was not only running out of food, but due to go on field break. Phil had been trying to get up the river to my camp for three days but had to turn back each time due to low blanket cloud. Then one morning he saw a break in the cloud, took off from the base camp and made it halfway up to my camp, but had to land on a previously cut pad at an abandoned camp site due to bad weather. After waiting for several hours, another break in the clouds appeared so he continued up toward my camp. He got to within a couple of kilometres of the camp but came up against another cloud bank. We were talking to each other on the HF radio, so Phil asked me to continually update him on the compass bearing that the helicopter sound was coming from. He progressed slowly up the valley flying from treetop to treetop through the thick cloud. It is critical not to lose sight of the tree canopy in this situation otherwise disorientation and crashing is relatively common. Then, after 15 minutes of flying, he appeared through the cloud and was able to land safely at my camp.

Generally, Phil flew without a door on when working at altitude to ensure that his visibility was not impeded. I was extremely appreciative of his efforts and managed to take a fantastic photograph of the helicopter arriving just above the treetops through the thick cloud. My estimation of visibility in the clouds at that time would have been around 40m.

Another recollection from April River fly camps was listening to the final America's Cup yachting race on shortwave AM radio early on the morning of 27 September 1983. At 5.20am, Alan Bond's *Australia II* won the America's Cup. The elation was too much, so I turned on the Codan HF radio standard exploration communication frequency, which was used for checking-in by all exploration companies during the 1980s, and announced the win. The channel went from silence to constant cheering and chatter. There was much rejoicing from expatriate Australian explorers in all regions of PNG that morning.

Manus Island expedition

Manus Island lies about 350km to the northeast of mainland PNG and is 110km long and 30km wide. Geologically it comprises Miocene aged volcanics, intrusives and associated sediments fringed by coral reef. Several occurrences of sub-economic copper and gold are found there.

One of the JV's exploration targets was located on this island. Our initial reconnaissance investigation was by helicopter in August 1983 and was followed by a scouting exploration traverse across the centre of it in October 1983.

Coming straight out of a rough highland fly camp environment at the April River Project, the offer of a reconnaissance trip to Manus definitely sounded attractive. Peter Macnab and I were flown to Madang in a Squirrel helicopter by Phil, via Ambunti. In Madang, an Islander aircraft was chartered to carry us to Manus as aviation laws prevented passenger transport by helicopter for long distances over water. Our flight in the plane took two hours and, shortly after landing in Lorengau, the capital of Manus Province, Phil appeared in the Squirrel. Following a relaxing swim in the afternoon we had an enjoyable night at the Kohai Lodge in Lorengau, quite a contrast to the hardships and fly camps of April River.

We were in the air at the crack of dawn the following morning to investigate a couple of potentially mineralised areas. It involved flying up meandering creeks attempting to find sites large enough to land the helicopter so we could collect drainage samples. Being thrown around in the back of a helicopter by continuous steep banking while seeking a landing site isn't something my stomach could handle. Fortunately, my 35mm camera was in a large plastic bag so once my motion sickness vomiting started, I was able to contain it. The violent turning continued until a landing site was found shortly after. By that time, I had been sick three times and the bag was filling rapidly. I have never been so pleased to exit a helicopter when we finally landed, but I am sure they would have been just as keen to get me out of it! So tight was the landing site, that Phil had to finally rest the helicopter partially under a large tree to avoid rock outcrop. It was the first and last time I have seen a pilot manage to do that. After completing three days of reconnaissance drainage sampling over much of the island, we flew back to Madang by plane and then on with Phil and the helicopter to Pacific Helicopter's hangers in Goroka.

The second exploration reconnaissance trip to Manus Island was more challenging. In October 1983, following the completion of a field programme at April River, Ken Rehder and I flew directly to Madang by helicopter and transferred to an Air Niugini flight to Lorengau. While I purchased food, bedding, patrol boxes and other supplies, Ken organised the logistics for the trip. Fortunately, he had several useful contacts on Manus, so was able to recruit some carriers and charter a fishing boat to take us to Tulu No. 1 village, on the north coast of the island, for the following morning. After departing early with supplies and labour on board, the boat had mechanical problems and turned back to Lorengau. Eventually it was decided that hiring another boat was the only option. After arriving at the Tulu No. 1 village at 4pm and hiring five local village labourers, much of our gear was loaded onto a motorised outrigger canoe to transport it up through the deeper parts of the Harlu River. That night we slept on the floor of an abandoned hut in the village.

The following day we commenced walking along the riverbank while the large canoe carried the gear until the river became too shallow. Most gear was then transferred to two smaller canoes and the rest carried. Kennecott's standard reconnaissance drainage sampling was carried out at tributary confluences and panned concentrates revealed gold colours at a couple of sites. During the first day the locals guided up the walking track which crossed the river several times in waist-deep water. Heavy rain caused flooding during several consecutive days preventing us from sampling, so we would stop walking early, choose a site and erect a camp for the night. The further we progressed up the tributaries of the river, the less the national labour knew about the area being traversed. This is not unusual for coastal dwellers and, the greater the distance from the coast, the more scared they became. We had to walk upstream in the creeks all day because the locals could not find walking tracks. The rocks were interesting so the 12 locals, Ken and I had heavy loads of samples collected. We resorted to navigating by compass as we progressed, as the locals were totally lost and scared. On the fifth day we crossed over the top of the central range of the island and commenced heading down tributaries of the Watani River, toward the south coast. On the sixth day we reached the coast but were confronted by mangrove swamps and could not progress through. One of the boys went off to find assistance as he had previously stayed at a nearby village. He returned with a small canoe, and we paddled into the Timoenai village to hire a larger motorised canoe to carry us, our gear and samples to Pelipowai, a larger village on the south coast.

There we paid our labourers for their assistance and hired the only motorboat that could take us the 65km back to Lorengau. Unfortunately, it was only a 13-footer, but it had a reasonable motor on it. The passage was slow going due to our heavy load and windy conditions. After five hours travelling, the boat ran out of fuel at the bridge between the island of Los Negros and mainland Manus. Fortunately, we were able to offload our gear, pay the boat skipper for his services, and hire a PMV (public motor vehicle) van to complete the journey back to Lorengau at dusk. The following morning, we dried our clothes and equipment, and packed and despatched our samples to Pilbara Laboratories in Lae for assay then caught an Air Niuguini flight back to Port Moresby. The nine-day Manus exploration expedition had certainly fitted "the land of the unexpected" quote.

Aviation safety

Air transport was always an issue in PNG because of the generally rugged topography and the short steep landing strips, due to limited flat country in the Highlands. The rapidly changing weather and dense jungle vegetation compounded the problem and multiplied the risk of flying in both fixed-wing aircraft and helicopters. Minor and major accidents occur as a result of mechanical failure, weather, pilot error or a combination of these factors. Not a year goes by in PNG without some form of serious aircraft accident, usually resulting in fatalities. During the 1980s and early 1990s, three such accidents involved helicopters chartered by the JV to enable exploration to be conducted in the Highlands. Due to the sensitivities of these events, they will only be touched on briefly.

The first helicopter accident occurred in 1986 during reconnaissance drainage sampling. This programme involved three geologists and a field assistant, each of whom would be dropped at a drainage confluence for around an hour to collect drainage samples and record the geology from the rivers and tributaries. In the meantime, the helicopter would transfer other geologists to new sites prior to doing the same again. On this particular day the helicopter pilot was picking up several of the crew and was about to return them to their base camp for the night. While flying back at speed along a river in the late afternoon, it is reported that the helicopter's skids hit an unmarked depth recording cable strung across the river causing it to crash. Sadly, two geologists and the pilot were fatally injured and the other passenger, a PNG national, was severely injured.

The second accident occurred in August 1990 near Mt Wamtakin, which is located 60km east of the town of Tabubil and west of the Strick-land Gorge in the rugged, high altitude, Western Highlands. In this case the accident was due to a mechanical fuel line failure which caused the Hughes 500D helicopter to lose power and crash at low speed into the forest canopy, becoming suspended upside down above the ground. Fortunately, the emergency locator beacon was activated, the geologist and pilot were not seriously injured and survived the freezing night. They were rescued by the Porgera Mine rescue crew during the morning of the following day. This was considered a potentially catastrophic accident and taken very seriously. A subsequent investigation found that the helicopter pilot was also not making his compulsory search and rescue calls to the Department of Civil Aviation and his licence was subsequently revoked.

The third accident occurred in March 1991 in a Hughes 500D helicopter in which I had been travelling the previous three days. This helicopter departed Mt Hagen and on board were a Kennecott geologist, a Kennecott community relations officer, a PNG Department of Mines and Energy mining warden, and the pilot. The intention was to conduct landowner hearings to obtain permission to undertake exploration in the Tari and Lake Kopiago areas in the Highlands. The helicopter crashed into the side of Mt Giluwe, approximately 50km southwest of Mt Hagen. It was travelling at high speed when it came through the jungle canopy and hit the ground killing all on board. The search for it took three days to locate the wreck and recover the bodies. I was in Mt Hagen at the time of the accident and directly involved with the search and recovery of the bodies, and subsequent events. Several of the Kennecott geologists and the field logistics supervisor were also directly involved with the recovery. Again, the Porgera Mine rescue crew offered their support and played a critical role in recovering the bodies from the crash site.

Survival training

Kennecott took aviation safety very seriously, as did many of the PNG explorers. Annual aviation audits were conducted on helicopter companies and, in the late 1980s, all staff had the right to return to base or request to land if they did not feel comfortable with the pilot or machine.

Kennecott also issued personal survival kits to all exploration personnel and had the policy that they must be carried when travelling in helicopters or light planes. These kits consisted of a medium size 'bumbag' containing basic gear including a light canvas fly tarp, flares, food, basic first aid supplies, insect repellent, signal mirror, small torch, waterproof matches, fuel tablets, compass, whistle and survival handbook. Too often staff were blasé about taking them and this rule was not always adhered to, so I requested that Peter Adamson, an ex-SAS soldier, through his company Adventure West, conduct a seven-day survival training course for 13 of Kennecott's geologists and me. While being aware of the scope of the training and that it would start in Mt Hagen in the Highlands, I didn't know how it would be conducted.

Over time we had built up a team of very capable geologists who were carefully chosen for their capabilities in handling the unexpected and supporting each other in a tough environment like PNG. However, prior to attending the survival exercise, most of the team were scathing about it, inferring that the exercise was not needed.

The 16 of us checked into Haus Poroman, a guest house on the outskirts of Mt Hagen on a Sunday afternoon in July 1991 with our survival kits. Peter and his off-sider, Greg McNee, began the course with safety and first aid that night, stressing the importance of always carrying your survival pack with you when travelling on aircraft, no matter how short the anticipated flight may be. The following morning, we covered using EPIRBs, flares and various emergency signals that can be used to attract search aircraft attention. We were told that we would be inspecting examples of these signal types from the air and returning to Mt Hagen to continue the course in the afternoon. Various aircraft were inspected at the airport to familiarise us with the locations of first aid kits and safety equipment, and how to use them. The action started that afternoon when we boarded two helicopters to fly over the signals. Instead of the anticipated destination, we were all dropped in rugged jungle 25km northeast of Mt Hagen and left there as if we had just crashed. Only three of us had our survival kits, but we did have a couple of 'bush knives' (machetes). Working frantically in the rain, we managed to build a rough lean-to shelter prior to nightfall. The three of us who had carried our survival kits were able to sleep in the large body-bag sized plastic bags they contained, and so stayed relatively dry. We used our three small tent flies on the ground for the others to sleep on. After getting drinking water from a creek, which is never a problem in PNG, and lighting a fire we attempted to get some sleep. It rained heavily at times during the night, it was cold, the place was infested with ants and mosquitoes, and the shelter leaked, so very little sleep was had.

The following day the instructors returned with the survival kits which most of the team had left behind. Signal fire construction, obtaining bush food, how to boil water in a plastic bag, and equipment improvisation were among the activities prior to each of us spending the night alone in separate locations with our survival kits.

The following morning there were some interesting stories about the night including two of the team having wild pigs crashing through their survival set ups. Hunger had set in and it was only going to get worse with only beef OXO cubes and a couple of muesli bars in our kits, all of which were eaten in the first couple of days. Various physical first aid and bush carrying activities were completed. The team was divided into four groups. Two of the groups were flown by helicopter, dropped in a creek at an unknown location and each group hiked in different directions to locations in uncleared jungle. The helicopter returned and did the same for the second two groups in a different drainage. We were told that we had crashed, were on our own with our kits, and a search would be mounted in a couple of days. We set up the camp and built signal fires, et cetera.

The following day our two groups were taken back into Mt Hagen, planned and mounted a fixed wing plane grid search for the other two groups. Hunger had really set in by then as it was our third full day without food. We weren't allowed to buy, beg or steal food when back in Mt Hagen, but were given a bag of food to drop to the other groups if we found them during our search. Unfortunately, even though the seven of us were desperately looking, we didn't spot the other two groups due to the thick jungle canopy and dismal signal fires, despite flying over them. We were then given their location and did a food drop. Even with our best intentions, we missed the target location and were later told that, despite searching, they failed to find the food!

We were dropped back at our pick-up location and despondently walked back to our respective camps with the intention of building better signal fires than the other two teams did, so at least they could drop us food. While we were all still well, lethargy had definitely set in due to the lack of food and every aspect of physical activity was getting tougher. The following day we built additional signal fires and had them ready to light. Our flare guns were close on hand. That afternoon we heard a plane in the distance which we assumed was our search plane so frantically lit the fires and, as it got closer, started firing off our flares, sounding like 12-gauge shotguns at a duck shoot. Fortuitously, we were spotted but unfortunately, their food drop landed in the Kaugel River and disappeared down-stream. We packed up our camp and were helicoptered back into Mt Hagen on the Friday afternoon for a debriefing session. The shower and food were appreciated like never before. It had been five days since breakfast on the Monday morning. Further training and refining Standard Operating Procedures continued the next day. All of the geologists thanked me for the opportunity to do the survival training except one, who had been recently recruited. The instructors said that while the rest of the team members performed well, this same bloke was not suited for this type of exploration: he was subsequently retrenched.

Enga Province 1988

Another event that is firmly embedded in my mind occurred soon after taking up the role of PNG Exploration Manager in 1988. Kennecott was offered a farm-in proposal on a gold exploration project in Enga Province, located in the Highland region. The massive Porgera Gold Mine lies within that province and had had ongoing trouble from the aggressive people fighting, murdering, stealing and destroying property in the local town site, as well as breaking into the mine. The Engans are known for their ruthless behaviour and for tribal fighting.

At that time Kennecott had recently been acquired by British Petroleum (BP) which had a policy of only allowing travel in twin engine helicopters. The size of these helicopters is a real impediment to effective exploration. The smallest available twin turbine engine helicopter available at that time was a Bell 212 from Hevilift Ltd, which was based in Mt

Hagen. The Bell 212 was used extensively in the Vietnam war and has a massive blade span compared to the eight-bladed Hughes 500s. The latter were the preferred exploration helicopters during the 1980s due to their small size and their favourable turbine power to weight ratio, allowing them to be land on small pads and creek landing sites. Regardless, it was a Bell 212 for this Enga task.

We flew north from Mt Hagen and landed at a large helicopter pad located quite a distance from the project site. Waiting for us were several locals from a nearby village to guide us to the area of previous exploration activity. The maps and sampling results given to us by the farm-out company made the assessment relatively easy. While the gold results were encouraging, there were no easily identifiable alteration or mineralisation signatures of potential gold deposits.

When alighting the helicopter, we had set a time with the pilot for us to meet the helicopter back at the pad, as it had fly to a more open site prior to shutting down just in case there were issues restarting the turbines. Although the drop-off pad was large, there would not have been enough space for another helicopter to bring in secondary batteries if necessary.

As we were leaving the prospect to return to the helicopter, one of the local tribesmen said that he would not let us depart and we had to bring in supplies and people. We were still a fair way from the pad, and we could hear the helicopter land while all this was going on. Helicopter turbines spinning always means exploration dollars being rapidly spent. Recalling all the pidgin negotiation lines was a challenge after spending two years on Western Australian exploration. It suddenly became more critical when I decided enough was enough and commenced heading back toward the helicopter. The tribesmen lifted their bows and arrows and had us firmly in their sights. Their hands were quivering and, knowing the history of the Engans, I knew the situation was very serious. I immediately backed off and endeavoured to calmly negotiate, explaining that I was the exploration manager and no-one else would come to conduct exploration until I brought the people and supplies. After about 15 minutes of discussion, I had convinced them to let us go so that we could bring supplies and geologists. After rapidly high-tailing it to the helicopter, I swore that I was never returning to that prospect. In fact, it was the last time that I have considered exploration within the Enga Province.

New Hanover canoe trip and wrap-up

A Kennecott team was exploring for gold at the Kuliuta Prospect on New Hanover Island, in New Ireland Province. Access was by helicopter or boat. In March 1989 I was in PNG to review exploration activities on Simberi, Tatau and Lihir, which are in the same province, so I set aside a day to reviewing the Kuliuta Prospect first and then to stay overnight.

Only being 60km west from Kavieng, chartering a good-sized boat each way to the coastal village of Metevoe appeared to be a more appropriate option, rather than justifying the cost of mobilising a helicopter from Rabaul or Lihir. In Kaveing I hired a banana boat which was in reasonable condition, had life jackets, a two-way radio and plenty of fuel. We had a good crossing and reached Metevoe in 2.5 hours without incident. After reviewing the team's work at the Kuliuta copper-gold mineralisation I stayed overnight, completed the review early and waited for the same boat at the prearranged time of 9am at Metevoe. No boat arrived. Very heavy rain may have been the reason. Attempts to find an alternative boat at short notice were unsuccessful. Due to the lack of transport and time, it was decided that staying another night at Kuliuta camp was the prudent option. A message was sent to the nearby Mission hoping to arrange charter of their boat for a morning run to Kavieng.

The following morning the Mission boat was nowhere to be seen so the only available vessel to cover the 60km to Kavieng was a *mon* powered by a 25hp outboard. While I was concerned about travelling in the mon over that distance, the ramifications of extending the one-day delay further were more worrying at the time. The mon was large but very basic. No seats, so I procured an empty 20 litre plastic drum to sit on and to use as a float should we turn over. The fuel situation appeared adequate, and I was assured that the national skipper and off-sider were experienced in making this trip. It was pouring with rain so I used a couple of garbags as raincoats. The seas appeared reasonable when viewed from land so off I went in the hands of two unknown locals in their dugout canoe.

The initial 25km along the south coast of New Hanover were moderately rough but we were relatively protected by land. I was totally saturated due to the torrential rain and spray from waves. When we left the New Hanover coastline, we had 35km of open water, with a few islands, to cover to get to Kavieng. The seas were a lot rougher than anticipated, however there hadn't been any major issues so far so I put my life in the skipper's hands. The higher the waves, the more the mon was rolling increasing the volume of water splashing into the boat, and the faster the off-sider was bailing to keep us afloat. Visibility was very poor in the extreme rain, so I grabbed my compass to check the direction of travel. Thankfully the skipper was still on the correct course.

We were about 15km from land and the engine stopped. The very agitated skipper was yelling instructions in a local language so I had no idea of the issue, but suspected fuel. Losing power in a large boat in a big sea is bad enough, but this is far worse, and it was now obvious that I had pushed the safety boundaries way too far. In between bailing water, they managed to transfer fuel from a spare drum into the fuel tank. There was a lot of water, both rain and ocean, getting splashed into the tank as well. After 15 minutes they managed to start the outboard and continued, arriving in Kavieng 3.5 hours after departing. Another one of those unexpected spontaneous adventures that shouldn't be repeated without due safety precautions.

There were many other extraordinary events that occurred while exploring in PNG that are worth mentioning, not only in the 1980s but also right through to my last visit in 2013, however, this is not the forum to list them all. Although challenging, like many foreign jurisdictions, PNG is certainly well endowed with undiscovered mineral deposit potential warranting further exploration effort.

Many of the technical and non-technical PNG national and expatriate personnel who work in the mineral exploration industry in PNG develop exceptional tenacity and camaraderie to overcome the hurdles in the "land of the unexpected" which makes much of the rest of the world appear mundane.

3

Tales from the South Pacific
The Solomons 1980-1986

Roger Langmead

Roger Langmead began studying Geology at the SA Institute of Technology (now UniSA), Adelaide, at the height of the nickel boom in 1971 and completed studies in 1974 during the post-boom crash. Despite this, he journeyed to the WA Goldfields and began working at Anaconda Australia's developing Redross nickel mine near Widgiemooltha for three years until late 1977. Following an 18-month sojourn travelling Asia and Europe in 1978-79, Roger joined Newmont Pty Ltd as a Project Geologist at the Baal Gammon tin-copper-silver prospect near Herberton in far north Queensland for 18 months, before switching to gold exploration in Queensland, PNG and the Solomon Islands throughout the 1980s. The Tolukuma epithermal gold-silver deposit in the Owen Stanley Ranges north of Port Moresby was discovered by the Newmont team during this period.

Newmont Australia merged with BHP Gold in 1990 to form Newcrest Mining Ltd and Roger played an increasing role in project generation studies in Australia, Ireland, Greece and Eastern Europe as global exploration opportunities arose with the fall of the 'Iron Curtain'. In 1994 Roger joined the Newcrest/Aneka Tambang team exploring Halmahera Island in eastern Indonesia for two years and assisted in the discovery and delineation of the bonanza Gosowong gold-silver discovery. A family move to Perth in 1996 facilitated an exploration focus on the WA cratons and subsequently a broad 'early project' identification role suitable for Newcrest investment through the Asia-Pacific region, which included a return to PNG after 20 years and Newcrest's involvement with the Wafi-Golpu project. Roger continues to live in Perth with his family, retiring in 2014 after a 34-year association with Newmont / Newcrest, pursuing a love of family history, gardening and sampling Margaret River region wines.

I was first introduced to the tropics in early 1980 when my wife Diane and I moved to Atherton in far north Queensland where I joined

Newmont Mining Corporation as a young Project Geologist working on the Baal Gammon tin – copper – silver prospect at nearby Herberton. Being raised on a wheat-sheep farm near the historic 'copper-triangle' towns of Kadina-Wallaroo-Moonta in the mid-north of South Australia, where the average annual rainfall is 350mm per year, it was a bit of a shock to arrive in the middle of the wet season when 2000mm of rain fell in January alone! In South Australia, rain was mostly a gentle precipitation and a welcome lullaby on the corrugated iron roof, whilst in the tropics it was a loud incessant drumming demanding to be let in, often with windows and doors shaking in unison with the thunderclaps.

By the end of 1980, our main exploration project at Baal Gammon was in economic trouble. The world tin price was collapsing,and the mineralisation proved to be metallurgically complex, requiring a significant capital expenditure in order to extract the majority of the contained metals.

We left the relatively cool, pleasant climate of Atherton for the more sweltering climate of the Cairns coastal strip, a more truly tropical city that had its own subtle delights to discover. For the next couple of years I as exploring for gold in tropical Queensland and thoroughly enjoyed the experience.

During 1982, Newmont became keen to maximise their exposure to gold opportunities in the younger volcanic terranes of the South Pacific islands. David Jones, our Chief Geologist in the Melbourne head-office had been researching such possibilities in Fiji and the Solomon Islands. Papua New Guinea was still subject to a moratorium on new prospecting licences, so was given a lower priority at that stage. Of particular interest were the re-considered theories concerning the formation of bonanza grade, epithermal gold-silver deposits associated with near surface volcanic processes and hot springs.

The British Geological Survey (BGS) had been supervising and assisting the Solomon Islands government in completing a geochemical sampling survey of the New Georgia group of islands further north and west of Guadalcanal. David organised a small group of Newmont geologists to visit Honiara, the capital, in September 1982, comprising Dr Jacques Claveau, a French global explorationist based in Canada with Newmont, David and myself, for whom this was to be my very first overseas geological trip. My anticipation and excitement were profound to be visiting the fabled Solomon Islands, so named by Spanish explorer Alvaro de Mendana in 1568 after reportedly finding gold on the beaches of Guadalcanal. I had read a number of books including James A Michener's "Tales of the South Pacific" and numerous World War II military accounts. My mind was full of visions of tropical island paradises, aka Bali Hai, inhabited by grass-skirted women.

We met at Brisbane airport and I was introduced to Jacques, who looked and sounded like the debonair Frenchman that he was, whilst David and I knew each other from university days in South Australia where he had been my Economic Geology lecturer. After several hours of flight time, our Air Pacific plane banked offshore of Guadalcanal for the approach to the famous Henderson Field airstrip, the scene of bitter fighting between the US and Japanese military forces for six months in 1942. These days it is the Honiara International Airport. From my window I could see lumpy hills with scattered houses and gardens in the foreground, rising steeply to high jungle-clad mountains further inland. Taxiing along the runway I half expected to see wrecked fighter planes and gun emplacements lining the airfield, but alas there was no evidence of its wretched history. The airport terminal appeared to be not much more than a large tin shed; the immigration official quickly stamped our passports and, after gathering our luggage, we were off on the five km journey along the coast westward to downtown Honiara in a rattly old taxi. We soon crossed the Lungga River over what looked like a large 'Bailey' bridge and a prominent sign on the roadside proclaimed, "Welcome to the Happy Isles". Smartly dressed children in school uniforms waved to us as they walked home, and older women were mostly dressed in brightly coloured sari-wraps and often carrying babies in cloth slings wrapped across their shoulders – no evidence of grass skirts here! Downtown Honiara was a bit of a disappointment of grimy stores with barred windows and doors, and pot-holed roads in very poor condition.

The Mendana Hotel was a grey cement brick and dark-stained timber building right on the waterfront. The check-in lobby fronted a large open bar and dining area with scattered tables beneath slowly revolving ceiling fans. The walls of this area and the attached souvenir shop were adorned with beautiful ebony carvings of traditional figures and totems, all inlaid with mother-of-pearl and other shells. Next morning, we presented ourselves at the Solomon Islands Geological Survey, a somewhat dilapidated, colonial era building, and were introduced to Director Frank Coulston, Dr Peter Dunkley who was leading the regional geochemical sampling program, and other local Geological Survey staff. We were able to review some of the preliminary results of stream sediment sampling from Vangunu, New Georgia and Vella La Vella islands in the Western Province. There were some obvious clusters of anomalous copper, lead and zinc; with gold in some panned concentrate samples at a number of localities from all three islands. Discussion turned to the known gold occurrences on Guadalcanal at Gold Ridge, only 25 kilometres from Honiara, and at Sutakiki further inland and high in the central mountain range. We learnt that the Government was planning to put the Gold Ridge prospect up for international tender at the end of the year and copies of the tender application forms were obtained for Newmont to consider.

Given its proximity to Honiara, we arranged for a site inspection the following day, as we were assured it was only a short walk from the end of the road and could be completed in a single day with an early start along well marked trails used by the local artisanal miners. So bright and early next morning, we hired a couple of local villagers and, with Geological Survey geologists, were guided into the foothills along a narrow, single file walking path which quickly became draped in the shadow of the surrounding rainforest and with steaming humidity. Our local geologist guides had insisted we take extra drinking water bottles in our backpacks and, in no time at all, our clothes were drenched in sweat, even though the track was not steep. In places, the path traversed along 'knife-back' ridges, less than a metre wide at the top and plunging away steeply on both sides, while at another locality further on the path we came to a huge sheer rock-face where we had to carefully step along a narrow ledge. A last short, sharp, climb to a ridge-top, then we descended into a small drainage where some local people were digging holes in the stream gravels and panning for gold. They showed us a gram or two of small grains they had in a tobacco tin that they had collected over two days, and a smouldering fire under a rough bush canopy nearby indicated their lodgings! One of my first lessons in tropical exploration was that it was very difficult to find in-situ rock outcrops away from the stream bed exposures, and so it was at Gold Ridge. Clay altered volcanic sediments appeared to host small, widely spaced narrow quartz-carbonate veins over an extensive area so it appeared to have reasonable potential for an open pit mine if the grade were sufficiently high. We sampled a number of outcrops that appeared typical for later analysis and petrological examination, ate a snack lunch of Wopa biscuits and tinned fish with our feet soaking in the cool stream waters. By the time we had returned to our vehicles in the late afternoon, our drinking water supply was practically exhausted and that was a lesson not forgotten for future exploits in the tropics.

Upon our return to Australia and reporting our evaluation of the geological potential and investment climate to senior management, Newmont tendered an offer for the Gold Ridge project, and made four Prospecting Licence applications covering the geochemical anomalies on Vangunu Island ("Kele River" project), New Georgia ("Humbe River"), the complete island of Vella La Vella, and also covering the gold occurrence at Sutakiki in the rugged interior of Guadalcanal. To our surprise and great disappointment, we were not successful in our bid for the Gold Ridge tender, losing out to Amoco Minerals, but the four Prospecting Licence applications were granted soon after.

In early 1983, Newmont began to seriously consider gold exploration in Papua New Guinea following the end of the exploration embargo on new licence applications late in 1982. My exploration manager in Brisbane, Richard ('Dick') Keevers, had attended an exploration seminar in Port Moresby the previous November in order to prepare the way for our activities. Everything was organised for Dick and me to fly from Cairns to Port Moresby for prearranged meetings with the Mines Department, the Geological Survey and to arrange logistical and storage facilities. I awaited Dick's arrival in Cairns on the morning of our departure, somewhat apprehensive, as I had never been to PNG, but comforted with the fact that Dick was familiar with Port Moresby from his earlier visit. A phone call from his flustered secretary in Brisbane let me know that he had left his passport behind in Brisbane and would not be able to travel further. Oh shit! So, with a true geologist's step into the unknown, I was given all necessary documents at the airport when Dick arrived and his credit cards for rental car and hotel payments. My wife later told me what a forlorn figure I appeared to be, walking across the tarmac to the plane in my shorts, sandshoes and plastic shopping bag full of documents.

The noisy, cramped old terminal building at Port Moresby was a bit of a culture shock and, after waiting a while to let the bulk of passengers be processed, I approached the Avis booth with the biggest smile and greeting I could muster, before trying to explain the circumstances whereby the names on the credit card and driving licence did not match. I have since learnt what wonderful people Melanesians can be when one is in a pickle, and this lovely Avis lady let me have a car and a map of Port Moresby without a hassle. Would not happen today! Same explanation at the Travelodge hotel overlooking Ela Beach and the same result. Over the next few days, Prospecting Authority applications were finalised, logistic facilitators arranged, and Tony Williamson at the Geological Survey, whom I had previously known at the Baal Gammon project, allowed me to store a large aluminium 'patrol box' under his house containing our rudimentary exploration supplies. I returned to Cairns, mission accomplished and with a boost to my standing in the company.

A short return trip was made to the Solomon Islands in mid-1983 with David and Jacques once more, along with Dick Keevers, primarily to review early results from mapping and sampling at the Humbe River project on New Georgia. From Honiara we caught a SolAir Twin Otter flight to Seghe airstrip on the southern coastline of New Georgia adjacent to the world-famous Marovo Lagoon. Seghe airstrip was begun by the Japanese during the World War II battle for Guadalcanal and they attempted to hide their construction activities with vegetation and other camouflage strung on wires across the construction site. However, their activities were observed by the Australian coast-watcher network and the Americans attacked and eventually captured New Georgia at terrible cost to both sides. Seghe was then completed by a US construction battalion and became a forward fighter base in the ongoing battles. We were met at the airstrip by project geologists Steve Turner and Colin Morrow. Steve had been working in the Eastern Goldfields of Western

Australia after a two-year stint at the Telfer mine in the Great Sandy Desert. As soon as Newmont announced that exploration was to proceed in the Solomon Islands, Steve let his desire to be involved well known to all and sundry. Colin was unattached and happy to live in the Solomons after transferring from NSW. Our luggage was loaded into two fibreglass 'banana boats' to carry us the 15 kilometres across the Marovo Lagoon to Uepi Island resort where a logistics base had been established. Before we set off, the boats idled across the shallow reef at the end of the airstrip until we could clearly see the outline of a twin tailed American P38 fighter lying on the bottom in about six metres of water. The speedy and robust V12 turbocharged P38s helped swing the air-war in favour of the Allies in the South West Pacific during 1943. The aftermath of the battles of WWll seemed to be all around us in the Solomon Islands. We motored at speed across the lagoon with the lofty volcanic peaks of Vangunu to the south, and the coastline of New Georgia to the west. Our local boatman pointed out flocks of seabirds wheeling overhead in front of us and the surface of the lagoon was churning beneath them with schools of feeding fish. We slowed on approach while he unfurled a thick fishing line with ganged hooks and coloured ribbon on the end. In no time at all there was a good-sized bonito thrashing about on the floor. I guessed that was dinner easily sorted!

Uepi Island is part of the outer reef system protecting the extensive Marovo Lagoon; it is banana-shaped, approximately 1.5km long by 300m wide, and lies on the western edge of the narrow and deep Charapoana channel that enters into the lagoon. We enjoyed a cold beer in the afternoon whilst sitting on the small wharf overlooking the channel, watching the small brightly coloured reef fish darting about. A giant trevally was also cruising back and forth near the surface along the deep water drop off, when suddenly it surged into the shallows and snatched one of the small fish in front of us. Its speed and power of movement was quite impressive. After a meal of fresh fish and a few more beers we retired to our thatched cabins for an early morning start. I was awoken during the night by an alien throbbing noise that seemed totally out of place in this island paradise. I ventured outside to see a very large fishing trawler entering the lagoon, to net baitfish for the fleet of Japanese tuna boats working further out to sea I was later told.

Next morning, we recrossed the lagoon to a sheltered canoe landing site on New Georgia that was closest to the Humbe River project where we were met by some of the local workforce and began the trek into base camp. I was soon learning a few facts about working in all tropical rainforests. They tended to smell the same, a sort of musty, decaying organic matter kind of odour, not at all unpleasant. There were trip hazards present with every step and one had to eye every footfall carefully, and not rush to grab any nearby tree or shrub if in danger of falling as it might be covered with nasty thorns. Don't expect your footwear, or anything else for that matter, to stay dry for any length of time.

The Humbe River camp consisted of two separate tarpaulins stretched over timber frames, with rudimentary lighting provided by a couple of pressured kerosene lamps. We reviewed the preliminary geological and alteration maps and sampling results to date, before traversing portions of the project area during the day. Lunch was the ubiquitous Wopa biscuits and tinned meat, with rice and tinned fish for dinner, livened up with some Tabasco sauce. Before retiring for the night, we were warned to make sure our boots were inverted on wooden stakes well off the ground and to carefully check them in the morning for the presence of crawlies, especially the much-feared large centipedes whose bite would lead to 24 hours of excruciating pain, during which you might wish you were dead! Pleasant dreams everyone! After another two such days at camp, we overnighted at Uepi to freshen up, then returned to Honiara and home.

Fast forward two years to early 1986.

I had been mostly occupied during this interval with Newmont's exploration programs in Papua New Guinea where we had discovered encouraging signs of gold-silver mineralisation at Tolukuma in the Owen Stanley mountains between Port Moresby and Lae. Both the Humbe River and Kele River projects in the Solomon Islands had been progressed to the drill testing stage, but assay results were unfortunately uneconomic at prevailing metal prices, and exploration had been temporarily suspended. More significantly, major budget cuts had been imposed on the Pacific and Queensland exploration programs, and the Cairns office was to be closed and staff transferred to Brisbane. My new Exploration Manager, David Royle, observed that no exploration activities had been undertaken at either SPL 001 (Sutakiki) or SPL 004 (Vella

La Vella) since they had been lodged, due to landowner or logistical issues that had arisen. The budget cutbacks required that a decision be made on these properties, so a last-ditch effort was planned for a reconnaissance of both properties.

David and I, along with David Jones and Chief Geologist Arnold Offenberg, flew to Honiara and once again stayed at the venerable Mendana Hotel. Whilst there we met Nick Swingler, a Sydney-based geologist working for a small company assessing a porphyry style copper-gold prospect on the steep, rugged south coast of Guadalcanal. Our meeting with the Sutakiki landowners, led by Mr 'Whisky' Kuba, was somewhat strained due to different landowners' interests, but Whisky assured us he would have it all sorted within a week or so after our return from Vella La Vella. As it would take a full day to hike into site, he suggested we stay at Valebaibai village during our visit. We then turned our attention to the Vella La Vella project. Pacific Helicopters had a JetRanger based at Honiara to service the many exploration projects at that time, and Nick Swingler was using it to set up a fly-camp across the mountains that day with the help of some local field assistants. We chartered it for several days afterwards to get to Liapari Island, located off the southern tip of Vella La Vella where we learnt of a young Australian couple running a copra/cocoa plantation and boat building business. We arranged to stay with them and hopefully learn more of the local politics and persons of influence to contact.

Liapari is a small, tear-drop shaped island of 85 hectares, separated from Vella only by the 150-metre wide Rovomburi passage and neighbouring coral reefs, which form a beautiful sheltered double lagoonal anchorage. A deep entrance passage to the inner lagoon had been blasted through the coral reef by the US Marines in World War II when it was used as a PT boat base and probably frequented by JFK, the future US President who was sunk in his PT boat and rescued nearby. We were made welcome by the owners and given information on who and where the important landowners resided whilst we ate dinner. The next morning, we landed the helicopter in one of the nearby villages, which caused a great commotion and a crowd quickly gathered. We explained to the headman that we wished to take small samples of sand and rocks from the rivers to test for signs of copper, gold mineralisation and we answered many questions about how we do this, and what else was involved. After trying to explain everything in some detail and with many locals still looking confused, someone from our group suggested, "why don't we drop into a couple of creeks with the chiefs and show them". The village elders were chuffed at the helicopter ride, and I was chuffed to see senior Newcrest executives getting down and dirty whilst sampling in the rivers!

That evening it was agreed that I would stay on at Liapari and attempt to sample more of Vella La Vella's streams, while the Newcrest executives would return to Uepi Island to inspect the Humbe and Kele River projects' drill core in storage there. The Liapari owners arranged for the hire of a small aluminium runabout with local experienced boatman Peter, and use of one of their plantation huts as a base, while I sampled the southern portion of Vella. I had learnt in discussions that the northern part of Vella La Vella was predominantly Seventh Day Adventists, the southern part mostly Catholic, and that they did not intermingle a great deal. A message was sent to the SDA pastor at Irringgila village in the north kindly requesting accommodation for Peter and me over a couple of days.

Peter and I quickly settled into a daily routine of heading up the coast early in the morning in our small boat, with my backpack and sampling supplies in the front, myself sitting on the centre aluminium cross-seat and Peter in the rear controlling the outboard and scanning the waters ahead. At each village we would "tok sav" with the headmen for permission to sample, then collect samples with the help of a couple of local lads, return them to the village to be invariably offered refreshing coconuts and paw-paws. The drainages along the west coast were commonly estuarine and I did wonder about the presence of crocodiles as we gently motored up the narrowing waterways until we could go no further, whereupon I and the local help would traverse upstream with my backpack and sampling shovel and sieves until there was a good flow of clean water and a sandy creek-bed that I could sieve and collect my nominal 5 kg of fine sediment. A quick look around at all the creek pebbles and boulders for signs of clay alteration or quartz veining which were sampled into numbered plastic bags, then back to Peter and our boat ready to move on to the next drainage.

In one estuary, a large dead tree had fallen into the water and, thinking that it a likely lair for predatory fish, we ran out our fishing line and lure and sure enough a large mangrove-jack was soon flapping on the floor of the boat. Each night back at Liapari, after a cleansing shower and cold beer from the trade-store, I would sort and label the day's samples and describe the rock samples in detail in my waterproof field notebook, often as the evening rain pelted down. Gazing out from my hut amongst the coconut trees, I could see the towering jungle clad slopes of Kolombangara Island close by to the south and tried to imagine the colossal events of history in this remote area. During the war, Kolombangara was very strongly fortified by the Japanese after they lost the battle for New Georgia, so the American and NZ forces bypassed it and captured the relatively weakly defended Vella La Vella instead. Within two months, the US 58th Naval Construction Battalion had built a 3600 feet by 150 feet operational airstrip at Borokama, just to the north-east of Liapari along the flat south-east coastal fringe of Vella. This serene vista was the subject of terrible conflict on the land, sea and in the air for many months in late 1943.

Our daily commute became less efficient the further we had to travel to each sample site, so I warned Peter that next day we would leave Liapari for at least several days and stay overnight in villages to the north. We organised extra rice and tinned fish and departed next morning with a change of clothes and mosquito nets added to our luggage. We were able to sample some short drainages on the way to the SDA village of Irringgila in the north-west, where the local SDA pastor and family kindly invited us into their home for meals and a place to sleep for two nights. There was a complication however the next day as it was Saturday, the SDA sabbath, so further sampling was out of the question. I felt obliged to accept an invitation to the morning service at the local church out of deference to local custom. I dressed in my cleanest tattered shorts/shirt and seated myself on the aisle near the rear of the church, gazing over the half walls to the village outside. The church soon filled with smartly dressed families and the Pastor began to present his sermon, partly in English, partly in Pidgin, but my mind was wandering off. I caught the phrase, "and strangers will walk amongst us before the second coming of the Lord", upon which everyone turned around to stare at me, then nodded knowingly to their neighbour. The things I do for Newmont, I thought to myself!

After sampling some small drainages the next day, we pushed on to Paraso village where I was most interested in gaining access to potentially the most geologically interesting area on Vella La Vella. A large NE-SW trending rifted valley cut across the northeast side of the island and, within it, were areas of recent geothermal activity including hot springs, steaming ground and sulphurous fumes, all encouraging signs for potential gold mineralisation. The meeting with the local landowners was however a bit awkward, the ownership situation being very complicated with relatively small parcels of land bounded by watercourses and creek junctions. The meeting broke up without permission to sample being granted, and I sat alone somewhat dispirited and pondering how to overcome this setback whilst eating my rice and fish dinner in a small hut allocated to me on the edge of the village.

Shortly after dark a young man approached the door, introduced himself and asked to talk. By the reflected glow of a kerosene lamp, he proceeded to tell me about an amazing gold rush that had occurred at Paraso some months earlier. By his account, a former employee of Bougainville Copper had returned to his village at Paraso and reportedly found gold in the geothermal area. This had caused a mad rush of people to the site to dig and pan for the precious metal; gardens were abandoned, children left school to help their parents and cooking pots were turned into rudimentary gold-pans, everyone excited at their new found wealth. My new acquaintance produced a jar of glittering substance from his woven carry-bag and asked my opinion of his gold hoard. It was full of pyrite – 'fool's gold' – and completely valueless. Now I began to understand the reluctance for outsiders to be allowed into the geothermal area. Although crestfallen, I thought the young man was probably smart enough to suspect this was all too good to be true and I took the opportunity to quiz him further about the key landowners. It transpired that he was the son of one of the elders and was prepared to show me the geothermal area next morning as long as we stayed on his side of the river. I went to sleep that night in a much better frame of mind.

After breakfast, we hiked a kilometre or so down a well-used path to a broad, clear flowing, shallow river that was easily traversable on foot. Upstream the river banks gradually turned to shades of white, yellow and brown, a sure sign that the hot acidic waters had changed the hard rocks into soft clay that contained much visible pyrite. Here was the source of the great gold rush! We collected stream sediment samples where the river divided into two equal sized tributaries, and also from some of the altered rock outcrops nearby. I was tempted to venture onto the dividing ridge between the two tributaries where hot steaming ground could be observed, but a previous incident in PNG where I scalded my ankle in a similar situation led to caution over valour. Besides, it was outside my guide's land claim and I did not wish to provoke any further complications. That night at Paraso village, I was closely questioned by the headmen about my exact movements, and whether pyrite had any value at all.

We bid our farewell next morning and motored directly back to Liapari, bouncing along at pace inside the fringing outer reefs. I had pre-booked a light plane charter back to Honiara with MAF Airways, colloquially known as the "Missionary Air Force", in two days hence, and spent this time sorting, drying, relabelling and describing the various rock samples collected. The Liapari owners had just completed the construction of a sizeable steel-hulled work-boat and we all witnessed the launch down the concrete slipway into the lagoon with celebratory drinks. The flight back to Honiara in a small Cessna was quite spectacular, flying over the narrow stretch of water known as "The Slot" in the war, with Santa Isabel Island to the east, and New Georgia – Vangunu to the west. It was afternoon, and as we flew through one thunderstorm along the way I found it disconcerting to watch the pilot's vertical climb/descent gauge wildly gyrate as we passed through the central updraught and peripheral down-draughts of air; quite violent in a small aircraft and I was thankful we were at a safe altitude. At Honiara the only hire car available at the airport was a tiny Hyundai Pony, looking well past its prime. Driving into town I could see the roadway through a hole in the driver's floor beneath the pedals, and one of the windscreen wipers was not working. Back at the Mendana Hotel once again, after a nice refreshing shower and change out of some smelly clothes, I felt like a new person. I sent a message to Whisky Kuba from the Sutakiki land-owners group and arranged to meet with him next morning for the long-anticipated trek up to the Sutakiki goldfield.

Whisky arrived somewhat agitated, explaining that there had been complications about the trip, and that I would not be able to visit. I was shocked and thoroughly surprised, having thought that everything was to be sorted and organised by Whisky. Despite my protestations, cajoling and everything else I could think of, Whisky would not budge, and we parted without agreement. Terribly disappointed and still puzzled, I drove to the freight forward depot at the airport and completed the paperwork for despatch of the Vella La Vella samples to the laboratory in Brisbane. My boss David Royle was still reviewing drill core at Uepi Island and was due to fly back to Honiara on a scheduled SolAir flight the next morning. I would have disappointing news for him! After dinner that night at the Mendana, I noticed staff beginning to stack and carry away the tables and chairs to a rear storeroom, which was not usual practice. Placed under the door of my first floor seaward-facing room was a note to guests that strong winds and heavy rain was expected in coming days and to secure doors, windows and any loose items. I then noticed people at the yacht club next door retrieving boats or securing them with extra mooring lines. The sky was overcast and the humidity very high, although it remained quite calm.

In the morning the rain was falling steadily from a leaden sky and the wind gradually increasing. I doubted that the scheduled flight from Seghe with David would arrive, but I felt I had to go to the airport to meet him just in case. The drive to the airport was uneventful, although there were few people about. The airport was almost deserted, but I eventually found a SolAir employee who confirmed that all flights were cancelled for the day. By this time the rain had become very heavy, so I jumped into the little Hyundai to head back to the hotel. The rain then became torrential, and I had trouble seeing through the windscreen, and water was sloshing around my ankles from the hole in the floor. Even more concerting, the road surface disappeared beneath a sea of water, so I concentrated on following the tail lights of a large 4WD vehicle proceeding slowly in front of me, hoping that they could see the road better than I. I really hoped that the electrics on the hire car would not fail me at this point. Thankfully I made it safely back to the Mendana where I hunkered down in my room looking out across the bay into 'Iron Bottom Sound'. The wind strengthened throughout the day to gale force and the heavy rain did not let up. Boats began to break from their moorings outside the yacht club and wash up along the shore, along with all sorts of debris. The storm continued until after midnight, when the wind and rain that had been lashing the hotel doors and windows abated somewhat.

The view over the bay from my room next morning was incredible, the shoreline packed with marooned dinghies, yachts and a large amount of vegetation debris. Further out, the ocean appeared to contain a large number of whole trees half floating in the water. I ventured to the airport once again and looked incredulously at the timber logs piled metres high along the coastline, along with two large coastal trading ships heaved over in the shallows. Local people were already tackling the logs with axes and machetes, presumably securing future firewood supplies. Most disconcerting though were the numerous scars visible on the high inland mountains due to landslips on the steep sodden slopes. The airport was largely deserted, but I soon heard the approach of the Pacific Helicopters JetRanger, so I waited to talk with the pilot. Out of the machine jumped a bedraggled Nick Swingler and his field assistant, relating harrowing tales of surviving the night in the storm, holding onto their flimsy tarpaulin cover for shelter. The pilot shook his head when I asked what he could see from the air, "bad, very bad situation", he said. The multitude of landslides had blocked and dammed many of the rivers, before the trapped waters burst forth and carried all before them downstream. That explained the multitude of logs in the ocean and shoreline, and it appeared that most of the Guadalcanal coastal plains were under water, bridges swept away and many houses destroyed. Henderson Field was not deemed safe for commercial planes and all flights were cancelled until safety inspections could be completed. Back at the hotel there were radio reports of widespread damage on Guadalcanal and Malaita Island to the east in particular. The storm was of cyclonic intensity and was named "Namu". It was apparent that local emergency services were overwhelmed, so Australian and New Zealand relief services were being mobilised. The following morning at the airport I watched as two Iroquois helicopters were unloaded from a RAAF C130 "Hercules" aircraft and quickly made airworthy, along with emergency generators, foodstuffs and personnel. By mid-afternoon there was a flurry of helicopter flights carrying supplies to and from the airport.

Rumours began to circulate about a substantial loss of life, especially in some of the more remote villages and in flood waters along the densely populated coastal plain. Rescue coordinators decided early on to evacuate foreign nationals back to NZ and Australia due to the breakdown of electricity and disruption to food supply and other services. Each day the relief efforts were ramping up with continuous helicopter flights carrying sling loads of cargo ferrying back and forth from Henderson Field. All foreigners were asked to register for flights, with priority given to families with children. David was able to fly into Honiara from Seghe after a couple of days, the airstrip considered safe at that stage. We were booked onto a RAAF C130 evacuation flight to Sydney the next day, and we were warned that the flight could be quite cold so were advised to wear warm clothing. No one travels to the Solomon Islands with warm clothing! I searched many of the Chinese-owned trade-stores without success, until I finally purchased an undersized Slazenger tracksuit from a sports store. We eventually boarded our C130 aircraft for Sydney, were given safety earplugs, and seated in meshed sling seats along the aircraft sides. The RAAF crew entertained us with their typical dry Australian humour and everyone was in good spirits to be heading home, but also thinking about the devastation and the people remaining in the Solomon Islands. At Sydney I was able to quickly connect with a flight to Brisbane, arriving home coincidently on my wife's birthday, and with a few stories to relate about an extraordinary field trip.

Post Script

Cyclone Namu was one of the biggest natural disasters ever recorded to hit the Solomon Islands. At least 150 people lost their lives, including 38 from the village of Valebaibai in the Sutakiki area, where there were only five survivors. A large mudslide had carried away the village. One of those casualties could have been me.

Some of the samples collected from my circumnavigation of Vella La Vella were anomalous for gold and silver, necessitating a more detailed follow-up survey by others later in 1985, before Newmont relinquished the prospecting licence. Years later, another company drill-tested portions of the Paraso graben but without economic success.

4

Ruby and Rubbery Rules
Vanuatu Late 1990s

Ian Plimer

Professor Ian Plimer is Australia's best-known geologist. He is Emeritus Professor of Earth Sciences at the University of Melbourne, and has served as Professor and Head of Geology at the University of Newcastle, as Professor of Mining Geology at the University of Adelaide and German Research Foundation Professor at the Ludwig Maximilians University at Munich. He has published more than 120 scientific papers on geology and was one of the trinity of editors for the five-volume Encyclopedia of Geology.

He has been granted numerous Australian and international medals and awards for his scientific and educational works. He began his geological career in Broken Hill, and has extensive exploration and mining experience. After leaving academia, he has served on the boards of several listed and unlisted mining and exploration companies.

He has been an advisor to governments and is a regular public broadcaster. He has written and spoken extensively and fearlessly on climate change matters, with emphasis on the need for scientific rigour and veracity, common sense, and the perspective and relevance to be gained from the long-term inputs from the geological timescale. His recently published book Green Murder *is a best-seller.*

It was nearly 25 years ago. A cold call from Australian Customs in Brisbane had the mind racing about which rocks and minerals I had brought into Australia that might have been troublesome or even illegal. Was it the huge number of arsenic-rich samples that I had brought in from Zarshuran in NW Iran? Was it due to carrying half the island of Milos in my luggage? Was it gold-rich samples from all over the world? Was it the radioactive minerals from the Soviet uranium mines in Czechoslovakia? Why would Customs ring me out of the blue? Australian Customs clearly knew that I was a man of minerals, all of which I had declared. I had a former student working in Customs at Tullamarine Airport who often checked the rocks and minerals I carried as we casually chatted about them. At that stage in my life, I was the Professor of Geology at the University of Melbourne with a special interest in mineral deposits.

The Customs chap explained that he was leading a joint operation with NZ and Vanuatu Customs about a 38 kg ruby that appeared at Port Vila, and which Vanuatu Customs had confiscated from its honorary consul to Thailand. Customs wanted to know where the stone came from, how the consul could have sourced such a stone, what it's worth and what they could do with it? "Of course I would help", I said. However, the Customs chap then told me the bad news.

There was a political bun fight taking place in Vanuatu between various branches of the bureaucracy and government. The Police wanted the stone, but it was held securely by Customs. The jungle drums were saying that the stone was worth $240 million and would be the answer to a maiden's prayer for insolvent Vanuatu. The consul had fishing interests in Vanuatu, wanted to build a fish-processing factory and expand his fishing interests and, coincidentally, wanted to give the stone to Vanuatu as a token of goodwill. That same consul was known to be exceptionally generous and had recently gifted the Police a fleet of motorcycles. What a kind man! He was normally met by his Police friends at the international terminal and didn't bother with the time-consuming annoyance of Immigration and Customs.

NZ and Australian Customs had been tipped off about a large ruby of uncertain provenance coming into Vanuatu from South East Asia. In a special operation, Vanuatu Customs got to the consul before the police did, and discovered the ruby in his luggage. Police and Vanuatu Customs had been at war for years and I was warned that, if I went to Vanuatu, it could be a tad dangerous because the Police wanted the stone, wanted a high valuation of it, and supported the consul: whereas Customs held the stone, held a different view on the character of the consul to that of the Police, and wanted the consul to face charges.

The rumours of a hugely valuable ruby created excitement in the community, crowds of people wanted to touch the stone in order to give them great prosperity and Customs didn't want to be pestered by armies of people. Religion in Vanuatu is a blend of Christianity and voodoo superstition and the stone was manna from Heaven. I was also told that the Police were on the side of the Government – whereas Customs were on the side of the Opposition. Customs in Brisbane stated that they would understand completely if I did not go to Vanuatu to evaluate the ruby.

I was curious. Magmatic rubies are small and high-quality gems, whereas metamorphic rubies are larger and less likely to be gem quality (for example, those found in the Hartz Ranges of the Northern Territory). It was a couple of days before Christmas, I was winding down, and agreed to fly to Port Vila. Upon arrival, there was a swarm of Police waiting for me. Vanuatu Customs had booked my ticket, so how did the Police know which flight I was on? Vanuatu Customs were everywhere; they got to me first, I was whisked off like greased lightning, and it was the quickest passage through Customs and Immigration I have ever experienced after an international flight. A fleet of Customs cars took me to a motel that had been booked out by it and they borrowed my passport, laptop and mobile phone for the duration of my visit. I was not allowed to contact anyone. The Australian chap who owned the motel comforted me with the greeting: "So you're the mug!" Too late, I was there. Customs officers were assigned to stay with me day and night, collect food and fermented fluids and drive me here and there. I was not allowed to go anywhere alone. I have shared rooms with snorers elsewhere, but the obese night guard sharing my room took the cake.

The next day I went down to Customs House. There was a crowd outside who wanted to touch the ruby with bank notes in order to bring them riches. As the great professor, I was respectfully ushered into a vault that held the 38 kg ruby. It took less than a second to realise the origin, mineralogy and value of a stone that was supposedly worth $240 million. It was clearly not a valuable, high clarity, gem Burmese, Thai, Tanzanian, Sri Lankan, or even Anakie, ruby. I thought "What the Hell am I going to do now?" I had a far smaller similar specimen in my personal mineral collection and knew that this was a ruby-zoisite-chrome diopside-phlogopite rock from Letaba in Transvaal.

I pretended to be intelligent, inspected the stone with my hand lens, poked it with a magnetic pencil and scratched here and there with my diamond-tipped pencil while I was thinking how to get out of the pickle. I smelled the specimen, licked it, tapped it and rotated it in the light. I thoughtfully muttered "Hmmmm", wrote notes in a field book and slowly looked in detail at all six sides of the Besser brick-sized block of ruby-bearing rock. On one side of the stone, there had been a futile attempt to polish the ruby. I spent half a day being the expert and then asked Customs to take me back to the motel, give me back my laptop and give me clear air to write my report.

Most rubies are heat-treated to increase colour intensity, homogenise zonation and increase clarity. By heating to near melting point at 1200-1600°C, inclusions of rutile and pyroxenes are dissolved back into solid solution, fluid inclusions are vented, and fractures are closed. Some rubies that do not respond to heat treatment are doped with beryllium or lead glass. Gamma ray irradiation does not change the colour of ruby as it does with other gems. No matter how a Letaba ruby is treated, it remains opaque, has a very prominent cleavage, hence will only take a cabochon cut, and is purplish red. The 38 kg rock was a mineral specimen suitable for a bookend or a museum glass case display.

It took a couple of hours to write my report wherein I stated that the Letaba ruby cannot be enhanced to gem quality, that it may fetch tens of thousands at the mineral trading centre of Idar-Oberstein in Germany, and it should be sold to clear the decks. While I was writing the report, my Customs minders told me that the Prime Minister had invited me to a garden party at his home; that no Customs people were invited, and that I was strongly advised to attend rather than insult him. This party was just for me. Despite having met a number of Australian Prime Ministers, not one of them has ever invited me to a garden party just for me! Shame.

At the garden party there were Police everywhere; I was warmly welcomed by the Prime Minister and his cabinet, treated to wonderful pork, seafood and tropical fruits, and offered beer, spirits and kava. I claimed to be a teetotaller. Readers who know me are well aware of my giant evening thirst, know that the cash flow I generate for breweries keeps them solvent, and will understand the crippling sacrifice I made to remain sharp in that snake pit. Various cabinet ministers sidled up to me and asked me what I thought: I was told that Vanuatu would be in serious financial trouble if the stone was not worth $240 million and it was hinted that, if I got things wrong, I could have a long stay in Vanuatu. I was also told what a good bloke I was, was there anything I needed, that I was always welcome and that I would be looked after whenever I visited this nation. I pointed out time and time again that I was still working on my report which Customs would receive in due course. I gave nothing away. It was two days before Christmas and the Prime Minister pressured me to stay in Vanuatu until after New Year and promised to have me well "entertained". I reminded myself that this was a joint Australian-NZ-Vanuatu operation, which probably offered me some sort of protection, but there were no supporting documents. When the Customs pickup car came, I could not escape quickly enough.

Back at the motel, I insisted that Customs get me on the morning direct flight back to Melbourne and that I would give them my report on a floppy disc upon departure. At Port Vila airport, I had Team Customs swan me through a back gate away from the eyes of Police, do my Immigration and Customs paperwork for me and escort me onto the plane. Vanuatu gentlemen are large for their size and heavy for their weight and, at the airport, I was presented with a Vanuatu Customs shirt as a token of their appreciation. Some 25 years later it now fits me! By the time I landed in Melbourne, my report would have been read, the conclusions understood and, whether they liked my report or not, I would be back in Australia.

Customs clearly did not take my advice and, a few years later after a change of government, I was invited to travel to Vanuatu again to reassess the stone and write a new report. Mineralogy does not change with a change in government. I declined the invitation.

I did not invoice Australian, NZ or Vanuatu Customs because I was pleased to be back home after what could have been a dangerous operation that consumed a couple of days of my life. For the following few years, every time I went through Australian and NZ Immigration and Customs, officers would see something on the screen, lift their heads quickly, look at me as if I was one of theirs and wave me through.

I have never been back to Vanuatu.

5

A Geologist in Paradise

Fiji 1983-1984

Stephen Turner

Stephen Turner is currently Newmont's Chief Geologist for the Australasian region, based in Perth, Western Australia. He graduated with a BSc (Honours) degree from the University of Western Australia (1979), and subsequently obtained an MSc degree (Distinction) through the Geothermal Institute at the University of Auckland, New Zealand (1986) with studies on the Mt Kasi high-sulfidation epithermal system in Fiji. He gained his doctorate at the Colorado School of Mines (1997), based on a field-based study of the giant Yanacocha high-sulfidation epithermal gold deposits in Peru. He has worked with Newmont as an exploration geologist for over 40 years and conducted exploration and reviews in over 40 countries.

There was a real buzz of excitement amongst Newmont's Australian geological team. Rock chip samples from a series of float boulders in Done Creek on the Mt Kasi Project in Vanua Levu, Fiji, returned several assays of +1000 g/t gold, including a high of 11,000 g/t or 1.1% gold! Surprisingly, this was from a sample with no visible gold. Indeed, no gold could be panned from the creek in which it was found. The samples were re-assayed with similarly high values.

As a consequence of this spectacular assay, I was re-assigned from a porphyry exploration program in the sweaty tropics of the Solomon Islands to the relatively sub-tropical climes of Fiji. As the latter was a popular holiday destination, especially for Australians, being able to work and conduct exploration in Fiji generated a degree of suspicion amongst family and friends, and even work colleagues back in Perth, that it was a boondoggle. It is still difficult to convince exploration leadership, perhaps unfamiliar with the very real prospectivity of Fiji, on the merits of exploration there.

From a geological point of view, Fiji is located at a unique conjunction of oceanic plates which have subducted to create volcanic arcs and jostled and rotated and pulled them apart, contributing to a special style of alkaline volcanism which is highly favourable for the development of gold-rich ore deposits. The leading example is the old, world-class Vatukoula gold mine which has produced gold since 1935. But there are a significant number of other gold and copper, base metal, manganese and bauxite deposits throughout the islands, with a variety of different styles of mineralisation. This is all great for an exploration geologist.

Mt Kasi hosted yet another style of gold mineralisation termed high-sulfidation epithermal; it is associated with a distinctive suite of alteration minerals and, as it turned out, some very high-grade gold, together with silver and baryte.

Savusavu, a little town on a bay of the same name on the southern side of Vanua Levu, was the staging post for our trips to Mt Kasi. It was a 40-minute flight from Suva, flying over the tiny chiefly island of Bau and many other small islands and scenic coral atolls. If the plane wasn't flying, the alternative was an overnight ferry ride operated by Indians and showing non-stop Bollywood movies. Once was enough. From Savusavu, it was either a 40-minute powerboat ride across the bay to the dock which serviced the camp, or a one-and-a-half-hour drive around the scenic northern margin of the bay. When we drove past the little villages on the way, the kids would come out laughing and yelling. We joked with the Fijian crew that they were yelling "daddy, daddy"!

Savusavu is a little paradise with the local Hot Springs hotel overlooking a picture-perfect bay. Bubbling hot springs emerged from the rocks near the edge of the water and often held hessian bags of vegetables, and maybe fish, to be steam cooked. Along the foreshore was the Planters Club, the local watering hole for generations of mixed-race copra planters, some of whom had spent too long in the tropics and were a little 'troppo'. We can't forget the cook in the local little Chinese restaurant earnestly asking, "taste good, taste good?", as we ate his monosodium glutamate-laden meals.

Geoff Taylor, our jovial senior geologist and long-term Fiji resident, settled in Savusavu with a local girl and has produced a dynasty of European-Fijian children and grandchildren. Geoff is a natural storyteller, which is also the meaning of the Fijian name of his ketch '*Talanoa*', and his story is told in his own words in *Not My Fault: Memories of a Field Geologist in Fiji (Star Printery Ltd)*.

To counter the impression that we were being dispatched to such a tropical paradise we made it a point of emphasizing the 'realities' of exploring in Fiji, like that Mt Kasi was in reputedly the wettest part of Fiji, on the southern, rainy side of the second island of Vanua Levu. Or that cyclones that were spawned in the warm tropical seas around the Solomon Islands would follow their arcuate courses to wreak occasional havoc in Fiji. And that, although malaria was not known, there were plenty of other tropical ailments and hazards that could strike you down.

For example, any little scratch obtained while strolling over the raised coral beaches could very quickly infect and become a dreaded tropical ulcer. It was very discomforting to see a small mosquito bite turn into a gaping hole in your arm, which needed an immediate dose of antibiotics. A fellow geologist, on getting repeated infections, commenced treatment with antibiotic tablets, which graduated to visits to the local clinic for injections in his rear end when the tablets stopped working; and eventually he had to leave Fiji when the injections also failed him. Goodbye tropical paradise.

For myself it was waterborne parasites which proved to be stubborn residents of my gut, eventually diagnosed by a doctor in Auckland after almost two years in Fiji. At that stage, being only a few years into the offshore exploration programs, neither I nor the company had recognized the value of regular medical checkups by doctors familiar with tropical diseases.

Mt Kasi was located on a hill about 400 meters above sea level, so the nights were cool, and even cold after rain, and mosquito-free which limited exposure to that other curse of the tropics, dengue fever.

Being on the wet side of the island, the rainforest was quite thick and the constant rainfall contributed to a strong dissected landscape. One of our prospects was given the fanciful name '6000N Plateau', which referred to a sharp ridge which plunged away into the jungle on either side. When we sent a geophysical crew out to survey the prospect, they came back cursing and complaining about what they must have thought would be an easy trek and wanting to know where the plateau was. Most of our drill tracks quickly turned to thick mud which restricted access by our 4WD vehicles, so we used a tractor with a platform on the back to transport equipment, trays of drill core and people. Several times, while walking around, I struggled to retrieve my gumboots which were stubbornly stuck in deep mud.

I experienced my first cyclone in Fiji while spending time in the Suva office and staying at our guest house in the suburb of Lami on the south side of Suva. I remember the howling and shrieking wind, and the sharp cracks as surrounding trees gave up their branches. I stood by the windows at the front of the house watching as the glass flexed under the pressure of the wind and thinking, "I shouldn't be standing here". The house survived quite well, and it turned out Suva didn't get a direct hit, but there was some damage around the capital and on the docks, with a lot of trees down and the power out for days. On another occasion I returned to the camp at Mt Kasi after the passage of another cyclone to find that the toilet had blown down, which was inconvenient.

Of course, none of this information appears in the glossy tourist brochures. Despite our efforts to earn the additional overseas allowance as compensation for tough living conditions it was hard to escape the fact that working in Fiji was really quite amazing. James A. Michener in 'Return to Paradise' commenced his description of Fiji as follows: "Imagine a group of islands blessed by heaven, rich in all good things needed to build a good life, plus gold mines and a good climate. Picture a native population carefree, delightful and happy". The weather is pleasant most of the time, the jungle is mostly quite benign, the geology is fascinating, the food is generally good and fresh, and the time spent on coconut-tree and coral fringed beaches learning to windsurf were memorable. Island time is a real thing. When I first arrived in Fiji and walked down the streets in Suva I would be constantly running into the back of the slower-moving Fijians. After a year in Fiji and returning to Australia, people would run into the back of me on the street. Life at a slower pace is definitely a good thing.

The best part of Fiji was working with Fijians. They are strongly built, friendly, easy-going people who are fun to be with and are simply some of the best people on earth to have as friends. Except perhaps for the system of *kerekere*, an accepted part of Fijian culture whereby someone with a need can just *kerekere* that item or favour from anybody in the village community without any expectation of repayment. It is a great system of communal welfare and sharing but as a 'rich' foreigner who is accepted as part of the local community they can literally *kerekere* the shirt right off your back, as happened once when I returned from Australia with a T-shirt that they liked.

The smiling, happy-go-lucky, native Fijian shares the country with Indo-Fijians, originally brought in by the British as indentured labourers for the sugar cane fields. The Indians are strongly family-focused and hard-working, often as tenant farmers in the sugarcane fields, or as businessmen and shopkeepers throughout Fiji; roles that the average Fijian finds challenging due to the system of *kerekere*.

The problem comes with governance. The native Fijians still want to govern their land and most of the land in Fiji is still owned by traditional community or tribal groups termed *mataqali*. Even the sugar cane fields operated by Indian families are located on leased land of the *mataqali*. This conflict over governance has led to several military coups, with the military controlled by native Fijians. I happened to be back in Fiji, in Savusavu, for exploration review meetings during the 'second coup'. Apart from an evening curfew there was little evidence of the coup and certainly no threat to Westerners. I wanted a T-shirt with the slogan, "I survived Coup II". However, the Indo-Fijians would not have felt as safe. The graffiti on the walls during the coup included wording like "we will pluck you like chickens', which is a fearful thought. The average Fijian man is still a warrior at heart, and they can be fierce and loyal soldiers, something that the United Nations recognized, and they were retained for peacekeeping missions in global hotspots, despite later criticism. Several of our field assistants had served in the military.

If I were ever to need a bodyguard I would want a Fijian, not only because of their strength and loyalty, but also because they are fun to be around. The warrior spirit can often be seen in rugby games, especially the Fijian Sevens, when played against their neighbours from Tonga and Samoa. The games could rapidly descend into chaos with posts ripped out and beaten over woolly heads. At our Mt Kasi camp a flat area was bulldozed to provide space for late afternoon games of 'touch' rugby. It was a lively, fast-moving game and there was never any doubt when you were 'touched' by a Fijian and had to give up the ball.

During our exploration program at Mt Kasi an U-16 rugby team from Victoria was invited out to the local village nearest Mt Kasi, Dawara on the Yanawai River. They were treated to a spear dance by the local boys, a traditional *meke* dance by the village woman and a Fijian-style feast, before being humiliated on the rugby field! There was also an all-night dance, where the guitar players would simply pass the guitar to the next guy after taking their turn. I turned up at the docks at 8.30 am the next morning and the band was still playing, albeit a little more slowly and less tuneful.

The warrior spirit was also evident in their movie preferences, the more violent the better. Strangely enough, when we took a Star Wars movie to watch in camp they really didn't like it. The science fiction story and weird characters were too much for their world view and it made them fearful.

On a traverse along a stream in central Viti Levu we collected stream sediment, float and outcrop samples at regular intervals. Between sampling points the Fijian field assistants, despite their size, would move surefootedly and quickly over slippery rocks to the next location, leaving us expat geologists to flounder along behind. During one break we decided to slow down the lead assistant, Saimone Tuibau, the son of a chieftain, by placing a large stream boulder in his backpack when he wasn't watching. He must have known his pack was significantly heavier but he gave no indication, tossed it onto his shoulders and took it all the way back to the campsite, not slowing down at all. My respect for Saimone and the Fijian team increased hugely.

My favourite memory of this particular sampling program is the evening meals. We had brought our usual cook from the Mt Kasi camp, Bill Namino, who had worked as a ship's cook. He was nicknamed Mr Quantity because of the large meals he would prepare for us, being accustomed to trying to sate the appetite of our Fijian crew and their preference for large quantities of starchy food, such as the root staples, dalo (taro) and cassava (tavioka). But on this occasion the Fijian crew had caught a large number of river prawns and he cooked them together with potato chips. Absolutely delicious and the one time I could have eaten more.

At one time we visited several prospects by vehicle, travelling along the south coast of Viti Levu, which required crossing several one-lane bridges. We were about halfway across one such bridge when an Indian driver, coming the other way, decided that, since his vehicle was larger, he could force us to back up. A Fijian driller travelling with us got out of our vehicle and started to make his way towards the truck, which resulted in the truck driver grating the gears in his hurry to reverse up, and probably saved himself from being pulled from the cabin and thrown off the bridge! On another occasion, travelling towards Labasa, we crested a hill to be met by two cane-laden trucks coming at us side by side, travelling just a few kilometres an hour, one trying to overtake the other. We had to drive onto the side of the road but, since they were travelling so slowly, our driver was able to pull a length of sugar cane from the back of the truck, reverse up and proceed to threaten the nearest driver.

The native Fijian connection to the land is a sacred heritage and an immensely important part of their culture. To visit and explore any village-owned land in Fiji required a formal request to the village elders, a gift of *yaqona*, and a *sevusevu* welcoming ceremony with kava. *Yaqona* is the roots of a plant that is crushed with water (in times past chewed by village maidens) in a large, carved wooden bowl termed a *tanoa*. The resulting kava is a murky liquid that looks a bit like dirty dishwater, which has a slight peppery but not unpleasant taste. The 'kicker' is that the kava has a mild narcotic effect such that after a couple of *bilos*, half coconut shells, quaffed in one go, your tongue and lips start to go numb. After several hours sitting cross-legged and drinking kava, it becomes increasingly difficult to 'wake up' your legs to totter off to relieve yourself. Kava drinking is great for a good night's sleep with no ill effects in the morning, and is a much better and safer alternative to alcohol.

The ceremony can be a very formal affair when it was part of the protocol to visit a new area, but it was also a nightly social activity in the camp accompanied by a card game called Fijian Trumps. The server is meant to scoop the liquid from the top of the *tanoa* where the kava is more dilute, but our local Fijians would scoop the stronger kava from the bottom of the bowl for unsuspecting visitors. When we wised up to this deception, and it was noticed, we were permitted to send the *bilo* back for a refill. We soon learned the clapping routine, one-handed holding of the *bilo*, and drinking the entire contents all at once. We spent many evenings in camp in a grass-roofed meeting hut called a *bure*, drinking kava and socializing.

Our Fijian team had a great sense of humour. The drill rigs were manned by a very competent and boisterous crew of Fijians. Drilling being a dirty business, with grease and oil and drilling muds, required a lot of rags for cleaning up. The drillers had obtained a large bag of secondhand clothing to be used for this purpose. I arrived at the rig one morning to find the entire drill crew decked out in old suits, jackets and even dresses! Just the time that I didn't bring a camera.

Fijian names are melodic and roll easily off the tongue so that, even to this day, I can remember many of their names. Such as Niko Naibi-takele, our draftsperson, who grew up in a village adjacent to the golf course amongst the mangrove swamps near Suva. He learned to play golf by sneaking onto the course as a kid. We played a few games with him and, as a giant of a man with a huge swing, he would send the golf ball off into the mangroves never to be seen again. He would probably have done really well playing on a more open course.

Kaiyava Davui was previously a sergeant in the army and from him we coined the term a 'Kaiyava shortcut'. "I know a shortcut", he would tell us as we made our way back from a field traverse. We would go crashing down a steep incline, through thick undergrowth, shinny over a waterfall, scramble up a rocky face and then through more thick bush before we emerged, puffing and sweaty, onto a perfectly fine contour trail, which was the long way back. He also tried to take one of our vehicles on an excursion off the trail, tipping it over in the process and fortunately not injuring himself or anyone else.

Iliape Ralago was unusual in being more studious and serious than his compatriots. He had a strong faith, was very smart and eventually spent time in Australia to complete studies and become a geologist. He also played a mean game of table tennis, another evening activity at the Mt Kasi camp.

Of course, to maintain illusion that we were hard at work in a hostile environment, we shouldn't mention that we ended up sharing office space with Pacific Crown Helicopters at the Lami helipad. It was almost too convenient and probably meant we had a few more helicopter trips than strictly necessary, including being 'dropped' at the airport one time because the helicopter happened to be heading that way. Or that we hired a float plane right out of the Tradewinds Resort hotel, located on the Bay of Isles, to fly south past Beqa, Island of the Firewalkers, to the southern island of Kadavu – famous for the Great Astrolabe Reef, breeding grounds for large fish and sharks. The local villagers of Soledavu came out to the float plane in their canoes to pick us up and return to the village for the welcoming *sevusevu*. We spent several days sampling altered rocks along the coast, coming back late afternoon to bathe in waterfall-fed pools near the village. Later, after I had left Fiji, the company constructed a drill platform on the fringing coral ledge to be able to drill back into the silicified cliffs.

I won't dwell on the local highlights in Suva during our office down times; Lucky Eddie's disco, the Red Lion and Lifeboat restaurants, the dance shows at the President Hotel or the Tradewinds Resort Hotel, where the front section was slowly sinking into the mud. It was just a short drive south from Suva to the Pacific Harbour complex where we could watch the famous firewalkers in action and relax in a resort setting as occasional tourists. Such distractions might keep us from our exploration mission.

So, what transpired with the Mt Kasi project and its promise of bonanza gold? The zone with the spectacular grades was drilled from every angle but, like many an exploration project, the high gold grades were restricted to a very small shoot controlled by two intersecting silicified structural trends we termed silica ribs. It had no size potential. We drilled under the old open cut, originally operated by the British between 1932 and 1946, but the better grades were limited by a footwall fault. Other prospects in the district were mapped and sampled including old tunnels at Waidamudamu (red water). By the end of my second year working in Fiji the chances of a significant project were diminishing. But the mineralisation at Mt Kasi was fascinating, so I collected a batch of samples plus extensive notes from logging drill core and applied for a Master's degree study at the Geothermal Institute in the University of Auckland, New Zealand. Remarkably, the father of my supervisor, Professor Pat Browne, had been a mining engineer at the Mt Kasi mine during its heyday, so I was able to copy some old photos of the mining days for him.

Many years after Newmont relinquished the project, Pacific Island Gold picked up the tenement and, in 1997, turned Mt Kasi into a small-scale mine again, treating the oxide material, and picking up the high-grade boulders from Done Creek to boost the mill grades.

It was time to leave Fiji. The memory that remains very strong and guaranteed to bring all that was Fiji back to me in a rush is the Fijian Farewell Song, the *Isa Lei*. *Isa Lei* has no direct translation in English but is best described as a sad sigh or lament. Like when the tractor backed over the camp cat at Mt Kasi, the *mataqali* chief's wife cried *"isa lei"*. The words of the song are very moving but, when sung with the deep timbre of Fijian men's voices harmoniously joined with the high, lilting voices of the Fijian women, it becomes transcendent, a heartfelt farewell that is not forgotten.

> *Isa, isa you are my only treasure*
> *Must you leave me so lonely and forsaken*
> *As the roses will miss the sun at dawning*
> *Every moment my heart for you is yearning*
>
> *Isa lei, the purple shadows fall*
> *Sad the morrow will dawn upon my sorrow*
> *Oh! Forget not, when you're far away*
> *Precious moments, beside dear Suva Bay.*

(English translation of the *Fijian Farewell Song*)

6

An Incident in Bougainville

Papua New Guinea 1968

Roger Marjoribanks

After graduating in 1966 from the University of Glasgow, Roger Mar-joribanks migrated to Australia and worked on base metal exploration programs throughout Australia and Papua Niugini for Conzinc Rio Tinto of Australia, before gaining a PhD (1972) in structural geology from the Australian National University. Following this, as a Research Fellow at the University of Adelaide, he studied the structure of the giant Broken Hill silver/lead/zinc deposit. Returning (thankfully) to Industry after this spell in academia, Roger worked on mines and exploration programs for gold, copper, nickel, zinc, iron ore and coal in Australia and South East Asia as an employee of the Anaconda Copper Company and of Renison Goldfields Corporation, becoming Chief Geologist for the latter in 1989. In 1994, Roger set up his own independent consultancy offering services to industry on exploration and mining projects in Australia, South East Asia, North America, Africa, and Europe. Roger's book, Geological Methods in Mineral Exploration and Mining *(Springer, 1997, 2010. Mandarin edi-tion, 2016) has sold well, over more than 20 years.*

Now retired, Roger fills in his idle hours making and selling core orien-tation frames and writing (mostly technical) posts for his blog at www. rogermarjoribanks.info

At 5am on Saturday, 6 July 1968, thirty-six men quietly assembled at the Sohano government wharf, near the Provincial capital of Ki-eta on the southeast coast of Bougainville Island. Most were uniformed and helmeted members of the Royal Papua Niugini Constabulary, all armed with wooden pickaxe handles. An Inspector from Melbourne and a Sergeant from Mt Hagen, each carrying a pump-action shotgun and side arms, led the uniformed force. In overall charge was a young Australian civilian Patrol Officer, known in Papua Niugini as a "Kiap".

Tied to the wharf behind the group, with its strange profile becoming clearer as the tropical dawn crept in, was the *MV Craestar*, owned by Conzinc Rio Tinto of Australia Exploration (CRAE). She was a 35m long exploration ship, converted from a Japanese fishing vessel through the addition of an assay lab in the hold and a helicopter landing pad at the rear. From the ship, three company geologists – armed with geology hammers – joined the group. They were Warren Atkinson (the Senior Geologist), Geoff Scott and me. Once assembled, we all boarded open-topped trucks or light four-wheel-drive vehicles and set off along the rutted coast road. We faced two hours driving on ever smaller tracks followed by a two-hour trek on jungle paths before we would arrive at our destination – a village deep in the interior, about 10 km to the southeast of the porphyry copper-gold prospect of Panguna, then at advanced feasibility stage.

As you have probably guessed, we were expecting trouble.

That trouble had been brewing for a couple of years. The Panguna deposit was found by CRA in 1964 near the crest of the Crown Prince Range which forms the central spine of Bougainville Island. By 1968, through a massive drilling campaign, CRA had defined a giant open-pittable porphyry copper-gold deposit. The company needed to acquire land for mine, roads, port and townships. The Australian colonial administration oversaw the land purchases, restricting prices so as not to upset existing stakeholders in the coastal copra industry. Subsistence farmers in and around Panguna thus saw land worth billions of dollars in contained metal value being taken from them and compensation offered based on its value in growing coconuts, the whole issue being exacerbated by long-held political resentment in Bougainville Island to rule from Port Moresby on the PNG mainland.

Underemployed and unsophisticated young men muttered about Bougainville independence and play-trained in the forest with mock wooden rifles. I was told that only a few months before I arrived, a Kiap had been ambushed in a remote jungle area and hacked to death with machetes. I have since tried unsuccessfully to verify the tale I was told of the murder of this chap. The story may have been apocryphal – the kind often recounted by old hands to scare naïve newcomers. I am more suspicious and cynical now than the young geologist of 1968 then was. But the point is, I believed the story then.

Before the 1968 field season, the Australian colonial administration held meetings with village leaders to explain the CRAE stream-sediment sampling program that was planned for that year throughout the island. Most islanders welcomed the incoming exploration team from the *Craestar* – they wanted a mine to be found in their district that would bring infrastructure and jobs. Throughout Bougainville, this friendliness was our usual experience. Many times, our helicopter could barely take off from the ground due to the weight of baskets of local produce (pineapple, citrus, banana, melon, sweet potato and sweet corn) tied to our skids, which we were unable to refuse without giving offence. But in an area of 10-15 km radius around Panguna, many locals resolved that they would repel any incursions by CRA explorers. In the event, we geologists, using a helicopter to leapfrog each other up and down the drainage to collect stream sediment samples, managed to pass through the country so quickly that we were usually gone by the time most locals were able to organise, or even be aware of our presence on the ground.

That is, until 5 July. On that day, Warren, Geoff, and I landed in some overgrown native gardens near a large village located about 10 km SW of Panguna. Anticipating potential trouble, Warren and Geoff headed off to wet sieve a fine silt sample from the bank of a small river which flowed past the village. I undertook the safer task – as we thought – of sampling a small tributary stream that came down from the jungle. This involved following a forest track that ran along the bank of the stream to collect a sample upstream from its confluence. The track was faint and overgrown, constantly looping away from the stream to bypass fallen trees or thickets of thorn. I had to travel at least a kilometre before I found a suitable sample site. As I was kneeling in the stream bed, angry shouts suddenly broke out behind me. Three men stood on the bank, brandishing machetes. Their rapid pidgin English was beyond my limited command of the language, but their message was very clear, and Anglo-Saxon swear words are a universal language.

"*No kissim sample long dis pela ples!*
Itambu!
Fuck off long Australia!"

It was clear they were not happy. As they said "Itambu", they were pointing to an inconspicuous stick which had been stuck in the ground – a marker of some sort. The word "tambu" (as I found out later) means a sacred, forbidden place. A taboo. That partly explains their extreme anger.

I made an exaggerated pantomime of dumping my sample back into the stream, waved my hands in what I hoped were apologetic and deprecatory gestures, and hurried back the way I had come. The men followed close behind, continuing to harangue and swear at me and flicking small branches onto the back of my head and shoulders to hurry me along. I thought of the Kiap, hacked to death just a few months before in circumstances very much like my own. If attacked, my geology pick would be a poor defensive weapon. As unobtrusively as I could, I slipped my hand into my shoulder bag and hooked a finger into the ring-pull ignition of a rocket-propelled parachute flare. We always carried a couple of these flares as part of our emergency equipment to signal to the helicopter from the ground: fired vertically, the device ignited and released a rocket that, with impressive noise, smoke, and showers of sparks, shot 100-150m into the air before releasing a bright magnesium flare which would slowly descend on a parachute. I hoped that, fired horizontally, it would create enough confusion and noise for me to escape. I was not optimistic: fired too early, it might escalate a threatened attack into a genuine deadly one; but waiting for a deadly attack would probably be too late. In any case, there was no guarantee that my one-shot horizontal missile would have worked as I hoped. A smoke flare would have been better – we usually carried them too, but I had none with me that day. Thankfully, although the aggression continued at the same low level, I did not have to make that choice. After twenty minutes of pure terror, I made it safely back to the clearing by the main river.

I had been away for over an hour. At the helicopter, Warren, Geoff, and the pilot had been waiting with increasing anxiety for my return. They themselves had not got far, having almost immediately been chased back to the landing site by a large crowd of village men who were now surrounding the helicopter. My three aggressors joined this crowd and added their voices to the cacophony of angry shouts.

We quickly flew off, glad to get away so easily and, in my case, to be still alive.

Once safely landed on the *Craestar* tied to the wharf at Sohano, Warren hurried off to the regional capital Kieta to inform the District Commissioner of what had happened. It was determined that we had to go back to the village – this time with police protection. Government authority had to be maintained: CRAE had to be seen to take a sample at this site. The Commissioner had over 50 riot police housed in nearby barracks at his disposal. They had been specially flown in at the beginning of the season from the PNG mainland and New Britain for just such an eventuality.

So, next day, it came to pass that a long file of men climbed steadily into the mountains along a narrow foot track. The Kiap led, followed by Warren, Geoff, and me, then the Sergeant at the head of his 30 men, many of whom were soon limping due to a recent issue of new boots. The Inspector from Melbourne brought up the rear. En route, as we passed each village, crowds of locals appeared: men, women and children, the numbers swelling as the march wore on until, at one stage, at least 200 people were advancing through the jungle and into the hills. Raucous obscenities were shouted, but no one tried to impede our passage – most were obviously just along for the fun, some may even have been cheering us on. After about two hours of steady climbing, the head of the group, consisting of the Kiap, we three geologists, the Sergeant and four of his men (the rest of the police party then nowhere to be seen) reached the overgrown gardens beside the river from which we had been evicted the previous day. By this time most of the spectators who had followed us from the coast had melted away, but around two dozen locals, now ominously all men, were waiting for us. Most of them carried machetes.

"Okay guys, take your sample now, then we can all go home", said the Kiap.

Warren and Geoff waded into the stream with their sieves and scooped up some silt from its bed. There was a shout of anger from the crowd which pressed forward – the Sergeant and his four men tried to hold them back, swinging their pickaxe handles like broad swords, but several men evaded the weak cordon and started to wrestle with Warren in the stream for control of his sieve. The police were badly outnumbered and bloody mayhem threatened: but just at that moment another half dozen police arrived at the top of the ridge overlooking the little valley and, seeing the developing battle below, charged down the hill yelling and swinging their clubs in the manner no doubt instilled in them at the Port Moresby Police Academy. This unexpected flank attack won the day and the angry crowd pulled back. All except one – a skinny, white-haired old man – who continued to grapple with Warren for possession of his sieve. The Sergeant promptly arrested and handcuffed him (the old man, that is). By the time the remainder of the police contingent – the Inspector from Melbourne still bringing up the rear – had trickled in by ones and twos over the next half hour, everything was quiet, and most of the ex-belligerents and spectators had gone. The return journey to the coast with our handcuffed prisoner was uneventful.

The little old man, I was told, was found guilty of civil affray and sentenced by the magistrate to a week in the Kieta jail.

There were more riots and violent protests in Bougainville in 1968 and 1969 before the Australian administration managed to secure a deal with the locals that enabled CRA to proceed with mine construction. The Panguna mine, through CRA subsidiary Bougainville Copper Limited, operated successfully through the 1970s and 1980s and became a major contributor to the coffers of both the company and the government of Papua Niugini. But disquiet over returns to Bougainville from the super profits of the mine, and continuing political unrest over Bougainville independence, did not go away. In 1989, the outbreak of the Bougainville Civil War precipitously forced Bougainville Copper to walk away from the mine. Rio Tinto has had no access to the mine site since, which today lies abandoned to the tropical rains.

The now largely forgotten, ten-year-long, civil war that started in Bougainville in 1989 and led to the deaths of several thousand people, was much more violent than anything experienced by me more than 50 years ago. But even today, I still sometimes wake in fright after a nightmare of being confronted and violently attacked in a dark jungle clearing by three half-naked, machete-wielding locals.

Refer to Roger Marjoribanks story,
"An Incident in Bougainville", page 80

A peaceful confrontation near the Panguna Copper prospect between Bougainville villagers and geologists from the MV Craestar that took place a few days before the incident described in this story. The spokesman in the white T-shirt spoke good English and is making forceful points about land rights and expropriation. In this instance, we were eventually given grudging permission to collect our samples.

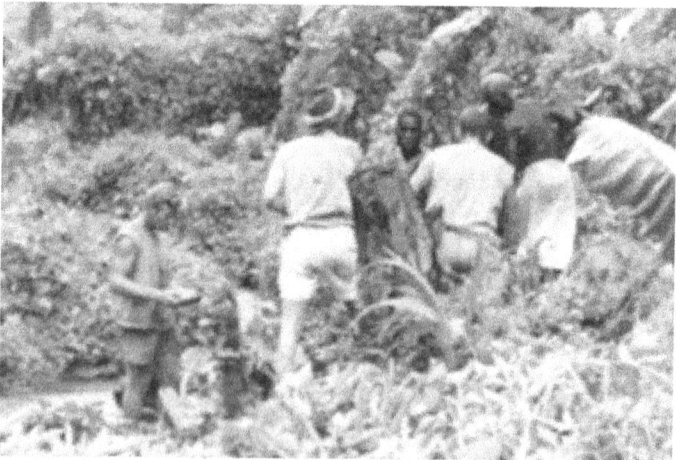

Villagers near Panguna copper deposit in Bougainville Island try to stop CRAE geologists Geoff Scott (left) and Warren Atkinson (second from right) from collecting a stream-sediment sample. The guy in the Akubra hat is the Kiap.

SOUTH EAST ASIA

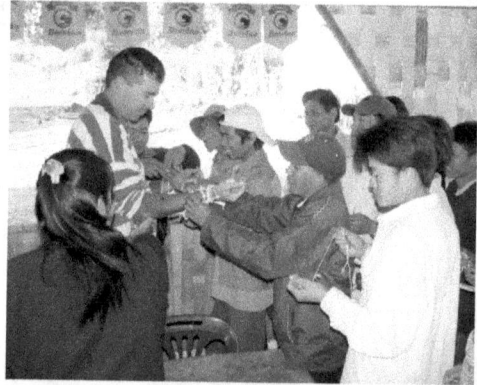

Clockwise from top left: A wooden chute built to load zinc oxide into trucks at the top of the mountain in Laos; The Original camp during the diamond drilling period along the track up the mountain; Our workers tying string to my wrists and attaching money for good luck; A Lao worker and his hard hat taking a break; Our site "office"; The road down to the drilling site at Nam Yen

7

Lost in Translation

Various Asian, 1982-2010

Anthony Williamson

Anthony Williamson has over 40 years' experience as a geologist in Australia, Asia and PNG, in both the government and private sectors. Taxi driving wasn't his most challenging or rewarding experience, but being the Director of Mining for the PNG Government was. Helping to grow their multi-billion dollar industry by developing a globally competitive fiscal regime, by maintaining investor confidence and by promoting the country, was stressful but satisfying. He has been credited with pivotal roles in numerous successful private ventures in PNG and South East Asia, none of which returned him a windfall profit or even early retirement. In summary, honest and equitable dealings with beneficial outcomes are important to Anthony, and his career and life reflect these values.

The Pre-start

I was brought up to respect old-fashioned values, so it's not surprising that truth and honesty, and equitable dealings with beneficial outcomes to all parties are important to me. My career and life repeatedly reflect these values. Nevertheless, I should add that I was a bit rough from the get-go, and later as a teen in the 70s I enjoyed all the bad things in life prompted by music that's since stood the test of time. Life turned out to be quite a heady time post-high school but, without the school regimen, came no drive and no plans. However, on a positive note, the chance came to work as a film librarian at Mr Packer's Channel 9 TV station. There were opportunities and challenges in that workplace but I was not only drained by school, but numbed. Fortunately, one of the mature female co-workers said: "You will be bored here by February, why not go to uni; go on, give it a go." She was correct: I did get bored, but she had enrolled me in the applied geology course at NSWIT and, lo and behold,

I was accepted, so off I went to uni. Thank you, Dear. Luckily for me, my extracurricular activities were restrained enough to permit me to take it on. I recently found out that some of my friends from that time didn't make it out of the post-school days intact.

My fresh-start first year of uni was disappointing. It felt like I was back to the same school-like routine. It wore thin quickly, trudging with the herd down Sydney's George St to uni, then the plodding through the same ol', same ol', subject matter. Study, books (pre-internet days), quiet places and yes, the things of the past, libraries. It was so easy to do a turn and drop into one or two of the many 'early opener' bars on the way to uni: sobriety was not an issue for me at that stage in my life. And I did drop in. But slowly and not so surely, my girlfriend, peers and the creeping challenge of the work at the uni knowingly, or otherwise, brought me closer to reality, dampened my wild spirits and encouraged me to raise the intellect bar. It worked: like a good horse, I was blinkered, winkered and encouraged by colleagues, bad guys and the constabulary, all the way until I finished my studies. The last two years seemed easy, because I realised that I liked being challenged by the science of geology, and by the lecturers. After finishing my last exam in the late 70s, I went to the pub with a few close friends, drank too much as usual, went home, crashed and made it to Kingsford-Smith airport for the first plane out of Sydney to start fieldwork the next morning in Far North Queensland with Newmont. I was proud to be a geologist and have never looked back, but sometimes I couldn't avoid questioning what was really going on in my life. My first job as a professional for Newmont was very similar to the experience I had had as a student geologist with Jododex (a Phelps Dodge – St Joe Lead JV) and also Kennecott, possibly because they were US companies. Although all companies are different in their own way, it became clear to me I wasn't destined to be a 'company man'. So, after a couple of years with Newmont, it was clearly time to move on and get down to some serious geology.

Papua New Guinea

In 1982, I threw out some job queries and an opportunity came up to do something completely different: my former lecturer, Dr Richard (Rick) Rogerson, offered me a chance at my first offshore job – as a regional geologist with the Papua New Guinea Geological Survey (GSPNG). The recruitment process took several months and when I finally got accepted by the PNG government, I flew straight from a cold windy Sydney winter's day to overnight in balmy coastal Port Moresby, and then early the following morning on to the foggy chill of PNG's Eastern Highlands. The Highlands' unforgettable grass-covered hills, clear skies and smoky smells accompanied us on the drive from Goroka down the Highlands Highway to Kainantu and on to our digs at Norikori coffee plantation, for a few late afternoon – pre-bedtime beers on the veranda overlooking the plantation.

On only my third day in PNG, I had to stop and take stock of what I was now involved in: I was walking up a stream in canvas jungle boots, behind a couple of local blokes carrying bows, arrows and bush knives, naked other than long leaves draped over a string around their midriff covering their arse and penis and, shivering in the cold rain, thought: "What am I doing here?" That first thought seems to have been a recurring question throughout my career. To tell all the experiences I had in PNG; its highland areas, seas, swamps, wet tropics and peoples, many now close friends, would fill a storybook in itself, but I will share one PNG story here. Not long after my arrival there, our small team of a few geos and about 20 carriers were mapping eastern Misima Island on foot. Late in the day and a few hundred metres short of our first base camp, we were walking along on a narrow track cut in the side of a steep hill on our right, with about a 6m drop to our left; I put my right foot on a wet root and 'whoosh' away went my balance and me, and the next thing I knew I was standing upright on the mud flats. It took me a few seconds to realise that I was standing in-between spikes, about 1m apart and shoulder high. Each spike had been a tree that had been just cut by a bush knife to a razor-sharp point. How I stayed upright was incredible in itself, but how I missed all the cut trees is beyond me. It's obvious that the PNG story didn't end there, and neither did the incidents, but suffice to say that I had a fantastically rewarding four or so years doing geological mapping and studies in PNG, but I still wanted more challenges.

I wanted to get back into exploration, but also to experience different peoples and cultures. I opted to take up an offer from well-known geologist Peter Macnab to work in Indonesia, so I wound up my PNG government job at the Department of Minerals & Energy and attempted to straighten out my messy personal affairs. My first child was born a year earlier to a Manus girl called Piuru. I thought I had explained everything about my work to Piuru, and the need for me to go where the work was, but apparently this was only one side of the story and I soon found out that I hadn't listened to what she had had to say about the matter. I had my bags packed, went to Moresby's Jacksons International Airport early, checked in, and did the usual waiting around until boarding was announced. I was about the last to board and just settling into in my seat when two big blokes wandered down the aisle looking straight at me and flashed their Royal Papua New Guinea Constabulary badges. "Mr Anthony Williamson?" they enquired. I was smart enough to realise this wasn't something that could be denied as the leader added: "get your belongings and come with us to the station." At the station I learned that Piuru was not happy about my departure and had sought assistance to bring me to task. About a week later, after many meetings, I managed to re-assure her that this was not the last she would see or hear of me, and she finally give me the OK to leave PNG. Contact between us continued but, sadly, she passed away some two years later. Only twenty-six years old! She was strong to the end, and at her behest, our strong headed son joined me.

Indonesia, The Philippines and Malaysia

Peter engaged me from the GSPNG to come to Jakarta to assemble and lead the original teams for the Contract of Work areas (exploration tenements) that he had selected for the 'umbrella' company, Teweti Limited. This included geologists Andy Kemp and later Ken McKillop, (killed a few years later during field work in the NT in a car crash). Teweti was later taken over by Kerry Packer's Muswellbrook Energy and Minerals that was in turn taken over by Aurora in about 1990. As it turned out, my Indonesian move proved to last less than a year.

I then moved to The Philippines and the business side of my life in the late 80's centred around setting up Banahaw Mining in that country and, as a Board member, recognising the then Co-O prospect for the potential it held, and moving the project towards production (Refer to Anthony's story, "Streaming Greed", in *Rocks In Our Heads* (Ed.)).

I won't discuss Banahaw here but will say that, well after the first gold was poured at Co-O, and while having a few very late-night San Miguel beers and vodkas with Vietnam veteran and good Texan mate, geologist Mike Spadafora, on his fourth-floor penthouse apartment's patio overlooking the Green Hills Shopping Centre's massive empty concrete carpark, he asked me: "What next Mr Tony?" I thought my marriage to a Thai woman at that time seemed to be going OK, so that couldn't have been the point of his question, so I assumed he must mean what does a geo do after bringing on the Co-O prospect to a mining reality?

I must have given him the mistaken impression over the past months that I was being a bit too laid back and hadn't given the next stage of my career too much thought. But, like Mike, I was a geologist: I was programmed to question, discover, explore and create. In fact, I had been using my Philippine base to do fortnightly, USD350 ASEAN Fare round trips to Malaysia, Thailand and back to The Philippines. I had also been through Japan, Taiwan and some other ASEAN countries in my search for worthwhile prospects. Funding my ASEAN project would be an expensive exploration exercise in the long term and certainly well beyond my budget, even with the cheap baht, peso and ringgit, house rents and labour costs in the region, but luck was on my side. I managed to fund it for long enough to garner some reasonable properties and set up staffed offices in the Petaling Jaya suburb of Kuala Lumpur, at Lampang in Thailand; and to set up another company, Pacific Phoenix, in The Philippines, to serve their respective exploration tenements. I had exploration geos on the ground in Sarawak, Sabah, Peninsular Malaysia, Thailand and a few places in The Philippines. So, I brought dear Mike up to date on my ASEAN circuit. I needn't have bothered, the voddies he was drinking kicked in too soon, and I found out the next day over breakfast at the local Jollibee burger outlet that he couldn't remember a thing we spoke about. Nevertheless, it was worthwhile for me to vocalise what I was doing. And so, life went on for Mike and me as usual, catching up from time to time between field trips and shooting pool over a few beers in Makati, until his death in 2012.

On the business side of life, I was fortunate that there was a ready partner to buy out the entire South East Asian package I had assembled. Mr Edwin Stoyle of Pacarc NL had recently done a deal with Packer's

Muswellbrook Energy and Minerals, and that left Pacarc without an exploration portfolio, and a restriction on its activities in Indonesia and The Philippines. So, I was free to roam again, and to assess potential acquisitions. This resulted in a sustained, focussed, but widespread search by boat, bus, chopper, train and plane starting from the busy cities of Asia and extending to the countryside: I encountered its wonderfully varied peoples and foods ranging from the street borne hepatitis and typhoid carts to the unhygienic restaurants that served up fantastic tasting food, but with the element of risk. I took to Asia like a duck to water.

I couldn't say which of the late 80's 'rising tiger' centres of Jakarta, Manila, Kuala Lumpur, Bangkok, Hong Kong, or Singapore had the least road traffic risk. There was just a slightly different flow to it in each place. I guess if I were pushed to answer, I'd happily hand over the steering wheel in Bangkok but the weirdest driving experience had to be Myanmar because the steering wheel was on the right side of the car, but the traffic also drove on the right-hand side of the road. That made both overtaking and taking the intersections thought-provoking moments!

My work involved a lot of night &/or day travelling back and forth on the section of the Trans Sumatran Highway between Lampung and Jakarta. The term Highway might sound grandiose, but the 80's highway in Lampung province was not much wider than the single tar strip as we know it in Oz, except the poor road base meant it was a generally undulating surface, and often pot-holed. Vehicles had to come off the sealed section of road to pass each other, so the roads' edges were often scalloped. You couldn't go far off the sealed section or you would hit vendors' stalls or kids walking to/from school. Even on the road one had to be careful not to run over beans and other produce laid out to dry on mats. Any drive was a frightening experience. Daytime driving was a constant honking of the horn to warn pedestrians and other cars of your presence. I couldn't understand why people couldn't see or hear a large vehicle approaching. The night drive proved to be most nerve- wracking. As vehicles approach each other at night, at a distance of anywhere from 10-50m from each other, they turn their lights off, pass, and then turn their lights on again. I found a frequent variation of this in Thailand where the car approaching would turn on high beam a few metres in front of you.

Despite the obvious challenges to safety, I preferred to start my travels late in the day so I could get through as much field or office work as possible. Faxes and poor communications dominated the internet-free 80's and, being in developing countries, those communication problems were exacerbated by language issues, inefficiency and plain old misunderstanding. Once, I had been doing fieldwork in the Bicol area of Luzon, in The Philippines, and needed to catch up on our Malaysian exploration program, in particular the drilling progress at Sarawak. The phones in the small barrio I was currently in were not working, so I drove to Naga City, found a public phone, and asked my secretary in Manila to book me a flight from Manila to Kuala Lumpur. She said she'd check, and "would I please call back, Sir?" All foreigners were called "Sir" in the Philippines. I can only guess it came from the American presence there during and after World War II. Anyway, calling back was not a simple task, I had to go to the end of the queue of people waiting to use the phone, and when it was my turn again, it was a roll of the dice as to whether or not the operator could get through to Manila.

When I finally did make contact again, she told me that I could get from Manila to Kota Kinabalu (KK) in Sabah but the flights were fully booked from there to Kuala Lumpur (KL). I had to go, so I asked her to get the travel agent to please get whatever ticket she could. I remind you this was pre-Flight Centre days and no online booking existed, lol. My secretary finally got me the 'last first-class ticket' on a special flight out of Narita, via Manila through KK, and on to KL with Malaysian Airlines. She sent the driver to get the ticket from our usual travel agent and I continued on from Naga City back to home in Manila, showered, packed, rushed to the airport, checked in and went straight to the departure area. I was travel pumped, but I had to admit that the boarding gate seemed very quiet even though it was open.

I was a bit late, so assumed everyone had already boarded. Anyway, I boarded and parked myself at the pointy end of the plane and "Yes thanks, I'll have that Bloody Mary before take-off." It didn't take an observant geo to notice there was nobody else travelling in First. Once airborne, I walked back and opened the curtains between First, business and cattle class and, bugger me, it was also totally empty. I sat back down and wondered what was this all about. A couple more drinks helped bring me to the idea that the plane would fill up with Japanese tourists being lifted from KK to KL. Not true: we landed at KK; DC-10's are thirsty so we refuelled, and nobody but me and some of the crew boarded there, and off we went again. It's the first and last time I've ever been the only passenger on a commercial flight in Asia. The DC-10 had a bit of a dodgy safety reputation in those days but, after we hit cruising altitude, I took comfort in the knowledge that we could probably glide to KL if needed. So, never having been one to let a chance go by, I relaxed, put on the fancy slippers that came in a nice little bag the stewardesses gave me, spilt some more drinks down my throat to wash down the canapes, and got into a bit of Bahasa Malay with several of the more friendly crew members the rest of the way to KL. Yummy.

Japan

Back to property acquisitions, the Australian guy I had in charge of the KL office had good connections to a very wealthy Tokyo businessman who held title over a closed gold mine on the Izu Peninsula in Japan. When this came to light, we established contact with the businessman regarding mineral exploration over his property (what entrepreneur isn't always on the lookout for an opportunity?). Fortunately for us, our first contact with the Japanese businessman came back positive, so it appeared we had something worthwhile to follow up on. Off to Japan! I had been offshore in dysfunctional countries for a few years by this stage so Japan was difficult for me to grasp. It was full of people with an attention to minutiae, in a country that worked, with trains running on time and taxi drivers with spotless cars, white gloves and no English. What could go wrong? Absolutely nothing, I naively thought, because I was with my Philippine business partner, Bob Hisshion, who was a fluent Japanese (or Nihongo) speaker. Nothing went wrong, except the deal.

We arrived at Narita late at night from Manila, bumbled around customs and immigration for a while and trained it to Tokyo, checked into the hotel and had a couple of Kirin's and decided it was smart to hit the futons early. The next morning, we phoned the office of the Japanese businessman that we came to meet, Mr Basu (a pseudonym), and he assigned some of his senior staff to pick us up at our hotel at 8am the following morning. Bob had other business to do in Tokyo so, later that day, we briefly checked out Tokyo's hospitality, and yet again drifted off early to bed. The next morning, we were waiting in the pick-up/drop off area outside the hotel lobby, me in my jeans and shirt, as opposed to my usual field gear of stubbies (classic Australian shorts), floral shirt (hi-vis 80's equivalent) and Blundstone boots) alongside a street thick with Tokyo traffic, and thinking that this will be something different – a first try at field work in Japan.

It sure enough was, because spot on 8am, a sleek black stretch Cadillac with huge horns from Longhorn cattle on the bonnet, pulled up at the lobby entrance. I turned to look at Bob and his smiling face said our field vehicle had arrived. I sometimes wondered about the truth behind his one blind eye and 10% vision in the other. Anyway, in we piled, slid across the plush seats to the tune of konnichiwa from the guy in a suit watching TV and a beaming old Japanese guy in khakis. I ignored the TV, I couldn't pick the ads from the programs, but noticed the mini bar after the guy pointed his g-pick at it and said: "we drink gin and tonic, we find gold." I'm still not sure if it was the bar or the g-pick which gave him away as the geo, but there was not much lost in translation! Anyway, I tucked in all the way out of Tokyo and down the highway through Yokohama and off to the Izu Peninsula with Fujiyama in the distance on our right. I held off a bit on the gin as I was going over the surface map with their geo and Bob translating. After all, I was still expecting mountains to climb, rivers to cross, etc, etc.

We eventually pulled up in a large car park at the foot of a hill. The old mine was not quite what I expected. It was now a well-manicured park with Shinto shrines here and there, trimmed foliage and neat swept paths, and cute bridges over little ponds. Reverential, calm, almost otherworldly, but here's their guy in khakis, with a huge smile, leading us with his g-pick, bashing at any rock that had the strength to crop out. Apparently, he was one of the last mine employees and I had to light-heartedly ask him to stop cracking the rocks. I felt it was inappropriate and had become embarrassing. Is this what members of the anti-mining lobby feel like? We soon abandoned the park as it was never going to turn into an ore discovery on a rugged mountainside, as I had envisaged, so we took the path out and walked down to have lunch. At the base of the hill, we were shown some above ground concrete tanks holding fish.

The suit told us this was the "$10,000 tank" only it wasn't one tank but about five tanks, each teeming with carp. Just to clarify, he meant that each carp in that particular group of tanks sold for $10,000! We continued walking towards to the restaurant, past the $20 thou, $30 thou and finally the $50,000 groups of tanks, sitting out in the open, right beside Mr Basu's factory that manufactures components for Toyota! I was humbled by the serious money involved with Mr Basu.

We had lunch in a large restaurant in the Japanese tradition of a shoes at the door and sit down on mats-type of place while our Japanese hosts ordered. I had been into but a few Japanese restaurants so this was relatively new to me, but we all have to start somewhere, but what I haven't seen since was the manner in which the fish was presented. One of the first courses was sashimi, the fish were about 15cm in size, served whole, standing upright on their ventral fins with a skewer along the axis of the fish behind the gills and through a point just before their tail fin, and their fillets flayed on either side, all accompanied by the usual wasabi, soy sauce, shredded white radish, ginger, etc. Bob and I love sashimi so we got into the fillets until I noticed Bob had this strange look on his face and seemed preoccupied with one particular fish in front of him. He leaned over and pointing to the fish said ……."that fish moved." I first gave him a quick look, and then the fish. Bugger me, its mouth opened and then slowly closed. At that point, the wall-mounted and singing big mouth billy bass came to mind, and almost brought on a giggle. But no, this was real, we were in a formal restaurant on the Izu peninsula, with very formal people, eating very fresh food. We both slowed down on eating the fish, thinking it all a bit too traditional.

The light lunch finished, we went back to the Caddy for the drive home to Tokyo where we would finally meet Mr Basu for a late afternoon meal. Dinner turned out to be at a privately booked restaurant on the 42nd floor of a hotel in Tokyo where he had personally and carefully selected our truly haute cuisine food. Baby eels appetizer from Spain, etc, etc, complemented by some of the best red wine from around the world. Amazingly, there were no other patrons present and the front door was closed, only the staff serving us. I had a sudden déjà vu moment and couldn't help thinking of my recent Manila to Kuala Lumpur flight full of non-existent Japanese passengers. Bob was enjoying himself talking with Mr Basu's entourage on what we would like to see in a deal and what we could bring to the party. Even though Bob had told them he was fluent in Japanese, he told me later that they were openly discussing business tactics in front of him. I couldn't verify this, as I don't speak Japanese, but I believed him as I had seen it before. Yep, we got the message, mining was in Mr Basu's commercial past, but I liked the way he let his old faithful geologist employee bash a few rocks for old times' sake. When the old guy passed our few meagre samples over for me to cart back to the Philippines for assay, both he and I knew what had happened; we were on the same page, it was just another day out.

Indonesia and The Philippines

My language deficiency in Japan reminded me of my early Indonesian days before I had learned their language, or had a hint of their culture. Wisely, I usually ventured out with a fluent English-speaking Indonesian friend and geologist, Ari Mahide, or Australian geologist, Mark Gillam, who was a long-term, fluent, Indonesian-speaking resident. The company had a large Contract of Work in Lampung and we needed to check the logistics, so off Mark and I went on an extensive road trip that entailed driving many hours over almost inaccessible tracks fanning out of Bandar Lampung to get as close to the sample sites as possible.

On one occasion, we had driven about five hours only to reach a rickety bridge that was blocked by a truck whose load had shifted, and both the truck and cargo were perilously close to sliding sideways off it. I wasn't going to turn around, so I suggested we help pull it off. All the truck crew were OK with that idea, so we hooked up the truck to the front of our 4WD and I slowly reversed. Almost immediately the truck's loadmaster, who was standing next to my car door, started screaming at me in words I couldn't understand and pulled out his gun, waved it and put the barrel on my temple. Mark was also standing next to him yelling at me to stop. OK, I get it, I stopped. I couldn't see that the truck's rear wheels had slid off the planks of the bridge and it was resting on its axle with one wheel in space.

Mark quickly spoke with the loadmaster asking him to put the gun down and offering our help to unload the truck so it could proceed. The truckie agreed and the unloading / reloading took all night, plus we had to pay the locals hanging around to help. Mark explained that their reasoning behind this was that, because I had helped pull the truck, any and all the problems from that decision point became my fault. I have found this to be frequently the case in Asia and the Pacific. In retrospect, I should have turned the 4WD around or parked up till they solved their own problems. As an aside, one of the first words that I learned in Indonesia was *berhenti* (stop), and I think the second word was *berapa* (how much)! A couple of years later Mark was killed while he was being lowered during a helicopter line drop in the then Irian Jaya (now called Papua, or Papua Barat to differentiate it from Papua in Papua New Guinea) when the chopper lurched forward and dragged him through the trees. Vale Mark, a truly nice guy, son of a Sydney cabbie and then partner to Anna, the Acehnese princess (Mark's words).

My Indonesian language skills were gradually increasing, but not fast enough, so I made a determined effort to speed it up. It was really useful and rewarding and, to this day, hasn't gone away, just gotten a bit rusty. Knowing a bit of the language gave me quite a bit of confidence to wander around. I soon discovered the night food markets or the vendors with their food carts that, in essence, fed the majority of Indonesians. Most people were unable to afford white goods, such as fridges, so food needed to be brought into the cities and cooked reasonably quickly. I became a regular to the *pasar kaget,* or night markets, that sprung up along the roads when night fell. I ask you to just digest what was happening; there was no running water, water came in buckets, and the food was stacked on paper or in containers. The ingredients, especially the meats, were not the freshest, and had been around for some time prior to being cooked. Needless to say, I became quite ill after a few weeks of 'eating out'. I learned a few new words including *segar dan sehat* (fresh and healthy) but I had yet to find the same in Jakarta's night markets.

You would think that I should have learned a lesson by now about keeping myself to myself, yet here I was again, not two years later, in my car stopped at the entrance gates to my housing compound in The Philippines with another gun pointed at my head. This time it was an irate policeman banging on in Tagalog (the lingua franca of the country) accusing me of abusing Filipinas! This all came about after my house staff woke me up because they heard a lot of scuffling and noises one morning emanating from outside our house. There was a woman looking a bit worse for wear and in a bit of a fluster, so I offered to give her a lift to the compound gates. I went back inside, grabbed my keys, got the car out and dropped her with the security guards at the front gate to the housing estate, and drove back home thinking of my good deed for the day.

About twenty minutes later, I got a phone call from the security guards asking me to come to the gate because the woman was still there. I went but couldn't figure out why they had called me. What they didn't say was that there were some angry cops there too. She hadn't properly explained to the police, or the guards, that I had nothing to do with her distressed state and that I only gave her a lift to the gate. I knew I was in a bad situation because usually when a weapon is brought to bear in the Philippines, it is almost incumbent on the shooter to pull the trigger. It's a man thing. So, here's innocent me, pistol to my left temple this time, talking English at the speed of light to a cop with a poor grasp of the global language. Fortunately, the woman was quick to react and pulled his arm back and started yelling *Hindi 'po* (No sir), which defused the moment. He was about to snuff me, his face clearly showed he had no issue in doing the job, but her words slowed down the moment and he slowly withdrew his weapon. The outcome of my good deed was that it cost me a few hundred pesos for the cops' time to come around to defend their citizens' rights, take her home in their police jeep, and almost shoot me in the process. Another good one, Tone! It didn't take much encouragement for me to find another compound to live in after that episode.

By July 1990 my time as a resident in the Philippines was nearing an end. It had been quite a hectic few months both at work and in my personal life. I was then living in a nice upmarket three-story townhouse in Firefly St, Valle Verde 6, located just down the road from our office in Pasig. My mother had passed away about a year earlier and that left my father hanging around the family home on the NSW Central Coast. He was not good at this sitting around thing, so he took himself off and visited their past mutual friends around the world. I could have told him he was wasting his time: too many years had passed, people move on so, not surprisingly, he seemed in a bit of a funk when I called him on his return home. I gave his predicament some thought and suggested he come up to the Philippines to see his two grandsons. He didn't take much convincing because he arrived late on a Friday about three days later.

We all did the family thing which gave him the opportunity catch up with his grandkids, swimming, barbeques, and playing. A couple of days later on the Sunday, I took him downtown to old Manila Intramuros and we walked around the historic Spanish churches, forts and buildings. Despite being in awe of the cultural buildings, we began feeling the weight of the long-dead Spaniards. One can only take in so many historic monuments and buildings in one hit and, after about three hours, I suggested it was enough for us so I proposed that we go and have a beer. Naively, he asked: "It's Sunday, is anything open?" Well, on Sunday arvo the travel time from the 400 year-old Spanish era Intramuros to the 40 year-old Hollywood bars on MH Del Pilar was near instantaneous. I parked the car outside a bar and managed to give him a quick word of advice – that if everyone is not at Mass and someone happens to be in the bar, remember to only buy one girl one drink. He totally ignored my words and within five minutes all I heard was laughing and giggling behind me. He was surrounded by bikini-clad girls, all with drinks and smiles. Just what he needed. Talk about a kid in a candy store and, as the saying goes, Manila does it to them all, whether you're 16 or 60.

Dad came to Manila at a time when we both needed it. I had been away from Australia for about six years straight and needed to bond again with my culture and the man I grew up with. He enjoyed what The Philippines had to offer in the relaxed fun atmosphere that can be found there for the *touristas*. He took in the wet markets, used the local transport, tried to make phone calls, and visited some government offices with me. He could also see that the place was barely functional. One morning, he rang me while I was at my office and asked if I was expecting any important letters. I was busy, and abruptly asked: "What are you calling me about letters for?" "Well," he replied: "your postie is lying dead in the gutter outside your house because the security guard bashed his head in with his weights, shot him about six times, and then he ran away." Dad had just realised that the candy store had a sour element.

Talking about The Philippine postal system and delayed mail: not long after my father returned to Australia, I received a death threat letter. I had been threatened with death in the past by the NPA (New People's Army), but had never received a death threat letter before. When I received it, the letter was dated almost 12 months earlier, so I was totally flummoxed surrounding the intent/etiquette of such a letter. Did this mean that someone had waited all this time to let me know they were about to kill me, because they couldn't do it until I received the letter? Or did the letter just take that long to get to me? Later in the evening of the same day someone let off a spray of bullets from an armalite into the air over our house. Last warning, or just a coincidence? I took it seriously enough and decided it wasn't worth taking the risk, so I packed a suitcase and left my kids, wife, car and rest of my belongings and flew KLM to Thailand the next afternoon. My family joined me a day later with a suitcase each. That pretty much ended my time in The Philippines. I have copied the letter below in case you haven't seen what a death threat letter looks like. Sections of the letter are omitted because there are people still living who could be seriously impacted by the contents of this letter.

---

4 July 1989

Mr. ANTHONY WILLIAMSON

No. 16 Kalayaan St.

Kawilihan Village,

Pasig, Metro Manila

Dear Mr. Williamson,

First, I would like to inform you that I am a relative of----------. At present, I am connected with the Intelligence Community of the Armed Forces of the Philippines. Of course, I don't have to inform you of my identity for security reasons. But I would like to inform you further, that I have with me the contract of the house that you are renting for my relative to include the receipts of your payments for the said apartment. Also, I had reproduced several copies of your picture and I had already given a copy of the said picture to a close friend and associate connected with the NBI and to our relative----of-------------. Upon my advise, ----------------.

I had tried to talk to you but my attempts proved negative because------- would not talk to me.--------------.

I would like to inform you also that of this writing, I had already posted several of my "assets" to monitor your daily movements. I don't mean to threaten you but I don't appreciate the way you------------. You should understand that here in our country, we value------as we value our lives. I had already killed a lot of people and I would never hesitate to do it to you.

Lastly, I would assure you that you could never leave our country alive.

Signed

SSS

Sadly or not, it was goodbye to the dominantly Catholic Philippines for me and, via Thailand, it was on to my old mate Graeme Fleming's newly rented house in South Jakarta. His place was situated just far enough away from the local mosque to reduce any newcomer's surprise at the pre-dawn call to the faithful for morning prayers. Graeme's last couple of jobs had come from Indonesia so it was the logical place to be. Initially, there wasn't much work in Indonesia for me, but at least it was safe and there were some opportunities. Graeme and I were cautiously optimistic about future work because we had been in Aceh a few months earlier for PT Krueng Mesen Minerals carrying out a regional recce which yielded very positive results at Miwah in Aceh Province.

That six-week field trip was about the most exciting and challenging bit of fieldwork I have ever done. We did a week or so in the Jakarta office planning the trip. Our start base was at the Woyla River alluvial mining operation in Aceh Province at the northern tip of Sumatra. The treatment process was yielding at least two different types of gold and large boulders of altered rocks were a common component of the gravels. We had a few hours' use of a chopper to plan our staging points and check out the few hundred square kilometres of prospective ground upstream from the mining area. We removed the rear door behind the pilot and flew at a low-level along drainages, following the altered boulder train upstream. Low enough along the creek that we once snuck in below the overhanging trees to follow the alteration. Yes, crazy. I was the spotter, Graeme the navigator. I strapped myself in the back seat behind the pilot, turned outwards and put my feet on the skids and off we went. I suffer motion sickness so it wasn't long before continuously looking down brought on the dry heaves. Just as well because it kept my mind off the very real possibility that the standard lap seat belt I was wearing might flick open and I would fall out. Anyway, up and down each stream we went until we could no longer see any altered boulders in the them and either one of us would yell out "they've gone" or some such equivalent, and we'd return to follow up the next confluence. We were following one boulder-choked stream and I'd just about had enough when Graeme said "look at that." A kilometre or so in front of us was the massive altered bluff which we named the Miwah Hill. What a feeling: we knew we had nailed something significant. Subsequent exploration turned up about 3Moz Au in resources at Miwah and is now a matter of public record, if one knows where to look!

At the Miwah Hill, the chopper couldn't put down due to dense vegetation and uneven and boggy ground, so we asked pilot Martinus to hover, drop us, go away and shut down for an hour or so and then come back for us. He hovered as close to the boggy ground as possible and we jumped out, laid low till the chopper went, and then scrambled up the hill grabbing samples as we went. We also slashed a makeshift chopper pad for our pick up and later to return with food and supplies to set up a camp. Not a bad day's work: we had found at least one prospect and selected two other food drop sites for the recce trip, so we got to sorting out the food and people at the base camp for the next days' drops. That was the start of our six-week bush trek.

There were moments of risk on that trip that still stop me. If I'm asleep, I will involuntarily spring wide awake or, if awake, freeze and take a breath like a punch to the stomach, or a tug on the balls. Involuntary and instantaneous. I'll get to a couple of those moments later, but first on with the trek. One of the bravest, or most stupid, choices was when we had to face our first crossing of the Woyla River, which just happened to be after a bit of rain. Prior our trip, wet weather had resulted in three Filipino geologists trying a similar crossing of the Woyla but being swept away downstream. Never to be seen again.

Anyway, the river current was a bit too fast for us to cross and some of the team couldn't swim, but we had to go on. So, we tossed a coin to see who would swim across, Graeme or me. In the end he volunteered because he was the stronger swimmer. Whew! The river was only about 30m wide but the central current was strong. Graeme tied a rope around his arm and he walked a couple of metres slowly out to near the edge of the rapid current, then dived and off he went, sideways, thrashing his best freestyle but was propelled about 10m downstream in a heartbeat. The tethered arm was rising and falling more and more slowly with each second. I could see he was tiring but luck was with him and he got footing on the shoulder of the far bank and scrambled ashore. I honestly can't say I was deeply concerned; all the troops were laughing at his dangerous crossing, myself included, because we thought we could pull him, or more likely at least an arm, back to our side of the river. Anyway, we made the river crossing and on we went, and found a jade boulder, and another porphyry or two on the way.

We were a couple of weeks into our trip and our immediate survival target was the next food drop at the 'Junction' of the Sipopok, Leuping and Woyla rivers. Almost all of our 30 strong support team had bailed out on us when we reached the Junction after the first week of the trip. I can't blame them; carrying heavy loads while negotiating the steep banks of a deep, fast-flowing river was a hard and dangerous walk. We were left with about six carriers so we stayed at the Junction a few days to think through what to do, while checking out what we thought was a low priority area. We had quite a few canned goods at the Junction drop site that we knew wouldn't be needed, so we bartered with the local people that drifted through the area from time to time for fresh fish or smoked venison.

We were literally in tiger country. A local gold panner or '*dulanger*' was taken by a tiger just a few days before we arrived. A *dulang* is a locally made, broad conical wooden dish used for panning gold. Due to the loss of our crew we basically had to ration work to the guys that remained with us so as not to tire them out. The outcome was that many of our traverses were done alone, and I can assuredly say that wandering around in tiger-infested karst terrain, especially near sunset, is exceptionally nerve-wracking, despite the fact that we drew eyes on the back of our raincoat hoods! On one traverse we came across tiger prints in the sand, I laid down my g-pick for scale and the prints were a tad shorter than the length of the pick. We walked a little bit further and we saw a wet paw print on a rock with a 20cm dribble of water running down the rock face. The animal was just ahead but moving away from us, or more likely watching us. Obviously having seen this before, the two local guys we had with us fled back downstream yelling, "*keluar se cepat mungkin, pak*" (get out as fast as possible, sir). We had intended to go round the next bend in the creek, but Graeme and I looked at each other and backed out almost as fast as the locals went forward in their retreat. The area had low potential so we decided to pull out and called in the chopper to take us direct to Miwah.

When we resumed work at Miwah a few years after the first reconnaissance and discovery, the pilot Martinus was killed while working for us there. At the time, we thought that the machine going down was a bit odd because we had just completed a safety audit by a Perth-based Australian outfit and carried out the recommended repairs. I was later reliably informed that the chopper was shot down, possibly by anti-government rebels operating in the area. Fortunately, there were two survivors from the crash, and one very grateful Tony who had missed the chopper's departure from Medan due to a delayed incoming Garuda flight.

The Miwah stories are many and varied, but I will share just one here and that one is about a wasp. We were about five weeks through the tour. Nothing was dry, it was raining most of the time and there isn't much sun in thick primary jungle, every scratch was turning septic, and our supply of antibiotics was nearly finished and the pills weren't really working anyway. Damn near everything was always wet and we were working at about 1200m altitude, and nights in the rain had become a bit miserable, so we cherished early finishes to the days' traverses to sit down in the late afternoons against a setting sun, a coffee kampung, a kretek (clove cigarette) and a well-earned rest. Late one afternoon on the scarce flat parts of the slopes of Miwah in just such a moment, I was sitting down cross legged talking with the crew when this bug about 3cm in size landed on my leg. I didn't react fast because it came in slow and just landed behind my knee (semitendinosus muscle); no circling, no buzzing or threatening moves, not a green hornet moment at all, but just quietly set itself down. Graeme said: "there's that wasp again, be careful", I was having a good look at it and replied "it's got a round bum, so I don't think it's got a stinger". Well, I just got that out and that was it, 'Zing'. Instantly the most intense local pain I have felt for years, and there it was the little fucker wiggling its arse trying to get the stinger in deeper, regardless of my flicking and flapping at it. I regrouped and grabbed the little bugger, pulled it off and flung it wherever. I was left with a bulls-eyed shaped sting site, and couldn't walk for a day or two. The worst sting of all was Mr Graeme's comment "that's the fastest I've ever seen you proven wrong". Bless him.

While I'm on fauna, there is another tiger story that is worth mentioning here. An Indonesian guy came to my office in Jakarta offering a KP (Mining Lease) over an old Dutch gold mining area some distance away from Padang, the provincial capital of West Sumatra. Being quite an underdeveloped area, it is also one of the last refuges of the Sumatran Tiger. Graeme drove us as far as we could go and then we did the usual geo thing after that, worked our way up the hills. Our steady upward walk was a few kilometres long, hot and humid as usual, but not steep. We were happy for small mercies. The area was easy to find and identify from the old Dutch maps which is not what usually happened in the pre-GPS days. Anyway, there were a few narrow veins around the area but there seemed to be a bit too much of Sumatra between the veins to make it commercial. To confirm what we were looking at, we grabbed a few samples to send off to the lab in Jakarta for analysis.

The main workings were near the ridge line so we decided to go into an adit to get some samples of the main vein. We only had a little torch with a dodgy battery, so we walked slowly in with our eyes closed the first few metres to get as much night vision as possible. The adit did a left turn and we shuffled along a few more metres. Then I spotted these glowing marble-sized spheres, down near the ground in the flashlight beam. At first, I thought damp spider webs but, as quick as a flash, I knew that I was looking at eyes, quite a few in fact. I backed up and said to Graeme: "Hold this torch, I am out of here." Naturally enough he took the torch and, while I was backing out, I said: "Can you see what looks like silver tiger eyes at the end there? Well, I think that's what they are!

Get out now!!" Both of us shuffling out backwards asap. We grabbed our gear we had left at the mouth of the adit and only then did we notice a lot of tiger prints on the spoil dump all around the adit's entrance. We had completely missed seeing them when we were preparing to go in to it. We grabbed our samples and backpacks and scrambled back down the hill, threw them into the car and went back to the hotel. Our post-field trip analysis brief, cobbled together over a few beers, included the words "poor potential, negative, close proximity to established farms and structures, report forthcoming", while we were mumbling about threats to fucking life 'n limb, the heat, stupid Dutch prospectors, mumble-mumble, steep hills, lucky the mother wasn't there, mumble, mumble.

Still on wildlife, the largest snake I have ever seen in either the zoo or in the wild was in North Sulawesi (Sulut). My team of about 12 guides (yes, 12, it's sometimes hard to tell them they aren't needed but, at a few bucks extra, it isn't worth the hassle to say no). We were trudging upstream through the jungle west of Gorontalo, and I was about to round the corner when all the labourers came running back past me yelling "*ular, ular*" (snake). I let them whip past me and then walked up to the riverbend to have a look at what was going on, and there it was, a reticulated python and it came as no surprise to me later to find out that it had taken several lives in that area. It was at <u>least</u> 10 m long, (please, this is a snake story, not a fish story) spanning the rocky river showing absolutely no interest in me whatsoever. Not like the slim, but deadly, tiger snakes, black or brown snakes I knew from my time as a fieldie in the Bathurst to Queanbeyan Lachlan Fold Belt of NSW, but something else, something more, something much more substantial.

Thailand and Burma

It's awkward for a person like me to bring the story about my Thai exploration effort to you because, to be truthful, it was unsuccessful, but this was not due to lack of effort. I was accompanied on most of my trips through Thailand by Thai geologist Khun Metha Amornsirinukroh, who was very experienced and helpful. When I first went there to look for projects, there were only a couple of exploration tenements throughout the country. So, I bought as many technical books as I could and went on a drive around the country with Khun Metha sampling altered rocks that we spotted on the roadside. A few weeks later we returned to Krung Thep (Bangkok) and packed up the samples for the lab in The Philippines. One interesting anomaly came from my reading and that was the reported antimony-gold production in central Thailand. There were hardly any producing antimony mines in Thailand at that time but there was substantial production recorded in the Thai export statistics. Something didn't add up. We went and visited many of the mines and most weren't producing ore, but were processing it, so we asked what the deal was. The lease owners openly told us that they were buying ore from Burma, which was illegal. I don't agree with smuggling, it's counterproductive except for lazy short term-minded, greedy people, but nevertheless decided we should go and follow the money trail. We had already hired a couple of 4WDs so we went back to town and bought supplies for our trip that took us down a bush road heading west towards Burma because, apparently, that's where the ore was coming from.

Well, it took us a couple of days of driving through pouring rain on boggy westbound overgrown tracks with numerous creek crossings before we finally broke down. While our guys were trying to fix the vehicle an old military truck came nosing east along the track. On the back were a bunch of armed men, not in uniform and not threatening, but I was told they were with the Karen Army. They had a quick look at what was going on, quizzed us all, then offered to help fix the car and towed us back over the border into Thailand. Happy chaps were we. They took us to a village that was set in hilly terrain studded by karst peaks, wispy clouds, and bamboo, just like one of those cheap paintings of Asia you can see on at least one wall of Chinese restaurants the world over. While the vehicle was being repaired it gave me the opportunity to ask them about the source of the ore and, yes, it was in Burma. I decided that I would save the knowledge until I could come at the source from inside Burma.

Back to Thailand, the rest of the anomalies I had identified on the previous swan through the country were frozen for applications and the Government opened a tender process. Not being one to buy at auction, I opted for finding ground outside the reserved areas. My company did find a couple of areas in Thailand that were worth an exploration dollar but, unfortunately, they weren't the stars we were after. I guess it's fair to say that the ground I selected turned out to be as equally disappointing as the Thai woman I chose to marry. The upside of my relationship with Thailand was that I learned to speak Thai, could read a bit of the Thai script, enjoyed the food and culture, and the fact that a wonderful child came from our relationship. He has stood the test of time and is a worthy person.

By the mid 90s I knew I had been in Indonesia too long, the job of exploration manager for Highlands Gold had lost its lustre and I had lost any creative element that I may have had. My need to get back in the game was insistent as a twisted nerve: I needed to make that discovery, the fix, and things were moving too slowly around me, so I made a few calls to old friends to see if there was anything interesting elsewhere. My old mate, Mr Kerry Doble, the former Chief Government Geologist for the PNG Geological Survey, came back to me with a job offer running a team in Burma with the Brisbane-based company Pacarc NL. The downside was the exploration areas at Mt Popa and north of Mandalay were already selected and acquired, so it was a "set up a team and manage it" kind of job. The upside was a creative opportunity to work in good company and put a team together in a country just opening up to the west. Hey, I couldn't get there fast enough and, when I arrived, I knew I had made the right decision. Everything about Burma seemed rudimentary – probably because the country had been closed for business since the early 1960's and only began re-engaging with the world in the late 1980's. It's now the 2020's and the country is still on a rocky road to global re-engagement, but I can't see that happening: more likely there will be years of conflict, and a breakdown into ethnic groupings, albeit loosely aligned, if at all.

I was lucky that Kerry already had Burmese manager U Than Shwe; a really nice guy, geologist, with a good command of the English language. Even in the 90's, the military wanted to get involved in everything. I recall watching night-time TV when this bemedalled General came on saying they had set up a new government water bottling company, "and today we open the company that can produce water pH 7, but this is the start of something big; next year we will do pH 8 and do better each year …". I thought at the time politicians are truly self-serving, show ponies no less, and the hard work is done at the Departmental level. Over thirty years' later, and nothing has changed my 1990's observation of politicians there or elsewhere.

Despite the opening up of Burma and the accompanying march of development and construction, there were still many 19th century British colonial-style buildings in Yangon (Rangoon). Downtown was dominated by warehouses and single to three story high, square, thick walled, small-windowed buildings with sufficient ornateness to differentiate them from the scattered heavy squat soviet-style architecture, but with no less gravity. The people had not fully embraced a western style of fashion; jeans were rare, all the men wore a *longyi* with a collarless shirt and the women an *eingyi* shirt top and most of them had *thanaka* paste applied quite stylishly to their cheeks. Walking around the old buildings and waterfront of the city always gave me a surreal anachronistic feeling. People from all over the country came to Yangon seeking work, but the Bamar, Shan, Mon and Karen peoples dominated, and their Theravada Buddhist beliefs and pagodas added depth to the visual experience of the city. The people's charm and easy approach to life so influenced me that I converted to Buddhism in Yangon in the 90's. I must confess that I am a poor adherent.

The country had only recently opened to the West so their commercial skills were still transforming and, despite the banks having the appearance of solid institutions, which indeed they were, they followed no procedural resemblance to that we, now, in the West associate with banks. A simple withdrawal was not that, it was a laborious task that could end up with the teller not being satisfied with your identity and refusing to give you any money from your own account! A transaction with the bank followed a routine that must have originated in the early colonial days, and stuck. To make a withdrawal one had to sit down with all the 20-odd other poor sods awaiting their turn to be called, and then be escorted through waist-high swinging doors and directed to sit in a chair in front of a clerk manning one of the innumerable desks. The stoic clerk would ask that you sign the back of the cheque, they would countersign, and then you would get escorted back to the waiting chairs, unless everything was working that day and you would go to the next clerk who would do the exact same thing but – note this – the pen colour was different. This process could be repeated, but I could never figure out why. Anyway, the final phase was sitting down in front of a desk manned by another clerk, with sleeve garters, leaning over a thick wood-bound ledger about half a square metre in size. He would open this big ledger, enter all the data on a line, both of us would sign the back of the check again, and he would then direct us back to the waiting chairs to be called by the teller to collect the money. It was not unusual for this process to take a couple of hours, or longer, with frequent misunderstandings due to language issues.

The Burma I recall is of a land only partially released from its colonial past. Steam trains were still a viable means of transport: elephants worked late nights near Mandalay loading teak logs into heavy goods trains at swirling smoky foggy sidings on the pagoda-studded banks of the Irrawaddy River.

Mongolia

As it turned out, the Pacarc NL Burmese job led into a Mongolian one. It's worth recounting how such transitions can come about. Kerry suggested we should wave the Burma tenements story and the corporate flag at the annual Prospectors and Developers Association of Canada ('PDAC') conference in Toronto. There, we bumped into a couple of my old mates who told us a positive story about Mongolian geology and its favourable attitude to foreign investment. That was all Kerry needed; we flew to Mongolia via China at the end of the conference.

The Russians had only recently pulled out of that country and had left it with no monetary unit, a dysfunctional city-based society, a decimated Mahayana Buddhist structure, and an inexperienced government. The Russian expatriates were ordered to get a bag packed and head for the airport or train station and leave. So sudden was the departure in the east of the country, that I witnessed many apartments in multi-story buildings with teapots and cups and saucers left abandoned on the kitchen tables. Eerie. Airports had hangar dugouts to shelter the MiGs and, on the southern boundaries, tanks were still in place with their turrets pointing towards the Chinese enemy.

After the departure of the Russians, many workers were gifted part of the factory or machine that they had previously worked on for their soviet masters. They were told they could earn their own money the capitalist way but those that remained in the city didn't do so for long. There was no money, no economy, no demand and most Mongolians returned to the countryside to take up their generations-old nomadic farming lifestyle.

So, here we were in Ulaan Baatar (or UB as it's locally known) looking for business. Fortunately, we fell on our feet and set up a company with arguably one of the most distinguished and westernised persons in the country at that time, the Former Foreign Minister Mr Batbold Sukhbaatar. Over the coming months, we pegged an old soviet porphyry copper target in eastern Mongolia, put a team together, bought some cars, *gers* and camping gear and off we went. A ger is a 'house' built from a circular expanding wood-framed lattice covered by about one-inch-thick horsehair and canvas. The door of all gers face south to avoid the cold northerly winter winds. Inside are beds and a little pot belly stove. Gers are supposed to be warm but, for me, in the shoulders of winter, I couldn't get warm no matter how hard I tried. The ger life in summer was quite different, hot and stuffy, and after a few weeks, the heavy odour of hanging mutton, kept inside to avoid it becoming flyblown, became quite nauseating I still only rarely eat ovine. Mongolia is vast: there were no corner stores, nor fences then; even in the towns it was difficult to buy provisions, so it was often easier to run a supply line from UB several hundred kilometres away.

To this day it's difficult for me to grasp why I hold such fond memories of Mongolia. The sub-zero temperatures and the summer bugs certainly didn't do it, not to mention the mediocre cuisine. It must have been the people. The population was predominantly young and enthusiastic and the pop music, fashion trends and cars originated from Europe. I was able to bring the first album by the band Aqua to Australia before it was even on the Australian radio. For once, I contributed to the kids' education and had them learn the Barbie song by Aqua which they sang at school on show and tell before it was released in Australia. Speaking of which, the mention of Mongolia makes me think, "if only I could turn back time ... if only I could".

Many a good night was had in UB. My good friend from my previous time in Indonesia was well known prospector, entrepreneur and

Canadian lovable rogue M. Armand Beaudoin: he had been there for some time doing business. At a very high-profile business function he funded, he once brazenly announced to all that he was so happy to be here because "the food was great but the girls were ugly." Unfortunately, it wasn't long after his business was looking up that he died from a heart attack. Sad, but I was soon after informed that he had died 'with his boots on'.

Mongolians love their wrestling and like most places the combination of girls, bars and macho-men often leads to skirmishes. Bar fights in the late 90's usually occurred between Mongolians and generally didn't involve foreigners. One night my good friend Mr Doug McGay and I were having a few drinks when an all-in brawl erupted at a very popular nightspot called the Matisse Bar. We saw it developing but were a bit late to react which is a bit like being the third dog out of the inside box in a 286m greyhound race. Within the first few seconds, lefts and rights were flying and blokes were falling about and wrestling. We jumped up and headed for the only exit, and our female companions, Oggie and Moggie, were ducking and weaving onward and up the stairs out of the melee and clear of the front door. My dear Moggie, at that point wailed: "OOOOOOOOOOh I forgot my bag, it's still down there." "Well, fuck me, aren't they wonderful creatures", I thought. Anyway, I'd had enough vodka by that stage that all my better judgement was voided, so played prince valiant and said no worries love, I'll get it, be back in a minute, and raced back into the shitfight, down the stairs, across the main drinking area, found the bag next to an upended chair, grabbed it, raced back upstairs dodging a few elbows and bodies and came triumphantly out the front door holding her bag in the air, and asked where to now? Outside, it was about minus twelve degrees which, together with enough alcohol and excitement, prompted agreement all round that it was time to go back to our respective apartments. Wise move.

The Mongolia – Burma story went well but ended fairly rapidly under a general global economic downturn in the late 90's aggravated by the insistence of the IMF that the Mongolian government impose a 10% sales tax on gold. You have to know when you are not backing a winner. Anyway, Batbold was a good friend and business partner and later become the Prime Minister of Mongolia. There was not much I could do by staying in Mongolia, I was instructed to sell all assets and leave no problems behind. I did the best I could but it did cause significant problems for our Mongolian partners and to this day the friendship is lost.

The loop again

I haven't mentioned redundancy or unemployment in my yarn thus far, but what's the point? All geologists can reasonably expect to be out of work for one global economic reason or another in cycles of about five years. I scuttled out of the Mongolian cold and the Burmese heat and told stories to people as a taxi driver in Cairns, went through a divorce, and finally landed work back in PNG again with the government. That job finished after five years and was extremely rewarding which I hope I get the opportunity to discuss elsewhere on another day. Oddly enough, after PNG I toddled off to Indonesia, again. This time as a consultant to Archipelago Resources at the Toka Tindung Project in North Sulawesi – ground originally pegged by Peter and worked on by myself back in 1986. Since then, the Aurora geologists had done most of the work at Toka. Abelle merged with Aurora in about 2002 and Harmony bought Abelle. Prior to the Harmony takeover, Archipelago Resources picked up Toka and then listed on the AIM in about 2003. There had been almost no serious geological work done since the Aurora geologists left and the site went onto care and maintenance in about early 2000. It wasn't until the PT Rajawali Corporation, led by Indonesian billionaire Peter Sondak, became involved with Archipelago that some serious geological work restarted. I had the opportunity in 2010 to work for PT Meares Soputan Mining (MSM) at Toka during its construction stage.

These takeover and mergers highlighted to me the irregular and discontinuous nature of resource ownership, inherent corporate redundancies, and related loss of geological knowledge. It's no wonder that many exploration geologists have only a shallow corporate commitment; but there are, and always will be, those who are willing to continue to take the corporate compliance pills. I know firsthand that the 2018 failed drilling program by Rio geologists of the Manus Island lithocap in PNG is a good example of geologists compromising drilling success due to enacting unrealistic corporate health, safety and environmental policies, and that was all compounded by a failure to communicate. In my opinion, Manus Island probably has one of the best undrilled porphyry targets in the world.

Working on a mine development site as an exploration geologist was a first for me. But I love the rigs turning, and the discipline required to achieve objectives. Maybe because that was how I started my career in 1979: running round as a junior geologist chasing a handful of near-sober drillers on as many rigs in a Queensland tin project drillout phase. I would counsel all wannabe geos to do the same after completing their degree. Anyway, Graeme Fleming was in charge at Toka and we hadn't worked together for near on 20 years but we met up again there and were tasked with meeting the management request to increase the mineral resource by half a million ounces of gold, in under a year. What was also needed was an updated Joint Ore Reserves Committee ('JORC') – compliant mineral resources estimate and Perth-based Stuart Masters guided that job. As a professional geologist I have mixed feelings about JORC, it ranges from dislike (on a par with the Miwah wasp) through to guarded acceptance because I've come to realise it's a safeguard protecting the investors from the many rogues in our industry. Really no surprise there because our industry always has had, and always will have, the rogues. Another plus is that the smart developer/promoter can use JORC to their advantage.

Geologists' attitudes play a significant role in successful exploration and one of the first things I noticed about Toka is that it did not nurture an enquiring attitude. In addition, other than Graeme and myself, they didn't have an exploration geologist but, in all fairness, that's why we were hired. In retrospect, the geological information to find the needed ounces was there in all the maps, etc. The logical way forward was to validate all the previous work completed by the Aurora geologists some seven-odd years earlier. I use the term validate here to mean getting out in the field and remapping and then rechecking drillholes and cross sections. At the end of the first week on site I sensed something was wrong about the geological interpretation because the mineralised area was much larger than the resource estimate indicated. Alarm bells should also have been ringing in the minds of the mine and grade control geologists. They certainly started ringing for me when the mine staff told me it was such a difficult project that they couldn't predict what the grade would be in the blast holes. The senior operations management were oblivious to the problem because nobody dared to say they had a problem. That was all I needed: I now doubted the computerised drill intercepts, so I went back to basics, pulled out the old paper maps and sections and went to work. It took a few weeks, but the layout of the Toka mineralisation revealed itself. The previously interpreted distribution of mineralisation about vertical veins were true in the underlying andesite, but the veins were dissipated in the overlying agglomerate and the mineralisation had an overall flat- lying sinusoidal shape. On scant evidence, Dr Greg Corbett had proposed a similar geological model some years earlier. I'd seen this geology before, there were similar controls in play at Toka to those at the Co-O orebody in the southern Philippines. The drilling from that point was focussed and we rapidly met and exceeded the management resource target.

The wrap-up

I feel like I've laboured my South East Asian experiences, so I'll close down after a few more short comments. Working in the high-risk exploration business in developing countries creates vivid memories. Some are good, but others I would rather forget. Those bad memories can involuntarily spring into my head at any time, and I mean anytime; asleep or awake, and force a shudder, a few quick gasps, or a random spasm, all in a millisecond. Worse than heat on a bimetallic strip. None of those memories could be even remotely construed as fun, but they certainly were character building. Fortunately, those moments are not as frequent as they used to be but, like the pain of the loss of a loved one, they are never ending and just become muted over time.

Expatriates tend to blame themselves for the mistakes that happen at work. For those that understand working in a foreign country, I guess this may stem from the introduction of western business practices and technologies into older cultures in developing nations or, maybe, it's just a plain old failure to communicate. On this note I offer my last tales. Graeme and I were connecting up his stereo system at his house in Jakarta and were missing about 4m of cable. We had a 20cm sample of cable and gave it to the maid and asked her to go to the shops to "buy some cable like this." She came back after about five hours and said nobody would sell her such a short piece of cable. We forgot to mention we needed 4m! In my view, this last story typifies the lack of control we actually do have over our own lives. Lastly, I chartered a small aircraft from Port Moresby to Gurney in Milne Bay and flew the entire journey in cloud. After a bumpy couple of hours of zero visibility, the pilot said to me that we should be over the Bay by now, so we'll slowly circle down to have a look. I wasn't really paying attention until he said: "have a look out the window and tell me if you can see any trees." It's a continuum between failing, getting the job done in difficult circumstances, and a disaster. As soon as I digested the implications of his words, I shot back: "Pull up now and land where we have visuals!" I heard the pilot died in a crash a few years later. Departures are not generally significant for geologists, but there's no need to be silly about them. There are still many sad and humorous yarns from my work-related travels that I have not included here. These I shall keep. At least for now.

8

A Zinc Oxide Adventure in Laos

Laos, 2004-2009

Ian Mulholland

Ian Mulholland graduated with an Honours degree in geology from Sydney University in 1979, and completed a Master's degree in Exploration and Mining Geology at James Cook University of North Queensland in 1987. He has over 40 years' experience in mineral exploration in Australia, New Zealand, Indonesia, South Africa and Laos, and worked for a number of well-known companies, including Western Mining Corporation, Esso Minerals, Otter Exploration, Aurora Gold, Archaean Gold, Summit Resources, Anaconda Nickel, and Conquest Mining. In 2004 he formed and floated Rox Resources, and retired from that company in 2019 after 15 years as Managing Director.

During his time with Rox he participated in the discovery of the Fisher East Nickel sulphide deposits and the Teena Zinc-Lead deposit. Ian and two JV company colleagues were awarded the prestigious AMEC Prospector of the Year in 2016 for the Teena discovery. In the early years at Rox, Ian shared his time almost equally between his home base of Perth, South Africa (where the company was exploring for diamonds) and Laos, until the Global Financial Crisis caused the company to drop the overseas projects and concentrate on Australia, where Ian was able to acquire the Teena and Fisher East projects. Ian is now semi-retired, combining consulting work with a number of hobbies.

Sometime in late 2004 we were told about some mineral exploration and mining opportunities in Laos. I knew little about the country, but a couple of other Australian companies, Oxiana and Pan Aust, had been successful there with copper and gold projects. Culturally it was like Thailand, and the language, writing and religion were similar; but politically it was like Vietnam, being a one-party, socialist state. We were told there was a Lao entrepreneur who had several promising projects and we should check them out: there were gold, copper, zinc, lead, iron ore and coal ones, and all held by a company called First Pacific Mining. So, I made a trip there in the company of our Australian contact, Campbell Baird, in mid-December 2004, to meet the Lao fellow, Vongvichit Nitiboun. Thus began an almost five-year association with Laos – the country, and its people, entrepreneurs and politicians.

I had floated Rox Resources and we listed in April 2004. We had started corporate life with a suite of tenements at the old gold mining centre of Menzies, about 110km north of Kalgoorlie in Western Australia. The idea was that we could explore further cutbacks of open pits that had been dug in the 1980s and, on the way down, pick up high-grade veins and shoots of gold left behind by the miners from the early 1900s. I had some experience in pulling old mining information together, so it sounded like a good opportunity for the company, especially with the increase in the gold price from 1987 to 2004.

To cut a long story short, after six months of fairly intense effort, we realised that the old-time miners didn't always mark mined areas on their plans and a lot of the hoped-for high-grade veins in the Menzies goldfield had actually been mined. In addition, there weren't a lot of extra tonnes available from the pit cutbacks and the whole thing then looked marginal at best.

Consequently, the company decided to change direction and look for other projects to get involved in. You need to do this in a small exploration outfit, where capital and other resources are scarce. So, the contact with the Lao entrepreneur sounded encouraging. At the same time, we also found out about some promising alluvial diamond prospects in South Africa, so for a year (2005), as MD of the company, I divided my time between Laos, South Africa and Australia. The plan was to see which of the projects looked the best by the end of 2005 and pursue that one.

This story is about my time in Laos. The South African story, involving gun-toting diamond dealers, suspicious Afrikaaner farmers, corrupt government officials, and robbery in the street, remains to be told another time.

I travelled to the capital of Laos, Vientiane, to meet Vongvichit. I had never been to Laos before, and the flight was via Singapore, Bangkok and then into Vientiane. We would get tourist visas at the airport when we arrived. As we circled to land, I wondered just what Laos would be like. I had heard it was still a communist country and very poor. Certainly, from the air, the small villages and rice paddies suggested a mainly rural economy. I had seen no evidence of any substantial town even.

We arrived and were unloaded downstairs onto the runway at the Vientiane International Airport. It was a single concrete runway surrounded by forest and rice paddies. Everything looked underdeveloped. I figured there were at most only a couple of international flights a day, namely to Bangkok and to Hanoi, but I noticed a grey concrete domestic terminal nearby with some old-looking Russian-type planes, so presumably there were internal flights as well. There were also a couple of Air Force fighter jets and three or four large military helicopters. Background information on Laos can be gained from Wikipedia.

It took a about an hour to get through immigration. Only the Lao people could go straight through, everyone else had to line up for their visa on entry. This involved filling in several forms and paying US$50 cash. You shuffled along a series of counters in line, passed your forms and passport in through one window and then moved to the next where you waited until they had been stamped (several times usually). Then you proceeded to the final window and paid your money and received your passport back with a very official looking visa now attached to one of the pages. It was a nice little earner of foreign currency for the government. Luckily, I had opted last time I renewed my passport to get the 68-page version (rather than the 34-page option). I could see a lot of pages taken up with visas if I were to travel regularly here. Later on, I would collect blank forms, fill them out on the plane and then bolt to the visa line to get through as quickly as possible. Later even still, I had a Lao business visa which got me through in the locals' line.

When we got into the terminal it was bedlam! There were people everywhere and Campbell said there should be someone to meet us. Well, we waited about an hour and saw no-one. Eventually we spotted a minibus outside with the name of our hotel on it, so approached them for a ride. We weren't on their list, but a bit of fast talking saw us allocated a seat and we set off for the hotel. Thus began an ongoing experience of communication difficulties and misunderstandings. No-one at the hotel had heard of us either, but they were able to allocate us rooms.

Apparently, Campbell thought Vongvichit would book the rooms, but obviously Vongvichit thought Campbell would book the rooms. As I was to find out, it was always better to book everything yourself and thus avoid the inevitable communication breakdowns.

Until sometime in the 18th century, Lao was only a spoken language, but then a written version was developed. While the language itself sounds like Thai, and Thais can understand Lao and vice versa, the written versions are quite different (even though they both involve similar looking scripts). During the French colonization they developed a 'transliteration' of Lao words into characters of the Latin Alphabet (which is what the English, French, German languages use). So, you could read these words and they would allow you to pronounce the Lao word. The trick was, as I found out later, that because they had been developed by the French, you had to pronounce the words and interpret the phonetic sounds with a French accent! Most street and road signs had both the Lao script and the transliteration script, but that didn't help if you didn't understand the Lao word in the first place.

Vientiane is a smallish city of about 500,000 inhabitants, and is situated on the banks of the Mekong River. During the dry season the riverbed consists mainly of sand with a meandering stream flowing through. It is a typical 'braided stream' in geographer's terms but, in the wet season, it is many metres deep (perhaps 6-10 metres) and extends from one side of the river bank to the other, perhaps 100-200 metres. Very much like a lot of North Queensland rivers, such as the Burdekin.

Along the Mekong River bank there are a number of restaurants and bars, and right opposite the main city block there is an area with deck chairs. Food and drink vendors own the chairs, and if you choose to sit at one vendor's location, then you buy your food and drinks from them. They seem to have their territory mapped out and agreed, and I never saw an altercation between vendors over it. The price of a large, 750 ml bottle of Beer Lao was 8,000 kip, about US$0.80. Really cheap, and an excellent beer. The food, mostly snacks – barbequed pork and things like that, also cheap. A very pleasant evening could be spent on the riverbank, just sitting there and taking in the river and the atmosphere.

A block back from the riverfront was a very European plaza, called

Opera Square, which had a number of very nice restaurants, including one or two outstanding French restaurants, and a steakhouse to be proud of. I would often entertain visitors from Australia there (or other Lao business contacts), and I usually would have a meal with Vongvichit and his wife there each trip, since they seemed to enjoy French food. I found out later that she had lived in France as a child.

Eventually, through the hotel staff, we managed to make contact with Vongvichit and First Pacific Mining, and a driver was duly dispatched to pick us up from the hotel and take us to meet with him.

Now Vongvichit Nitiboun is a relatively short name for a Lao person. Most have names at least twice as long as that, usually shortened to a one or two syllable nickname, and everyone seems to know them by this – even in official circles. His was still Vongvichit – or "Vong" as his wife and family called him – and she was "Pok". I never did find out what her real name was. They had different surnames (presumably denoting their family lineage), and hers was at least 15 characters long! I always called him Vongvichit (the long version), but called her Pok.

Vongvichit was younger than I and was "on his way up in the world". He was a businessman and was willing to take on anything that offered a profit. It turned out he had a furniture manufacturing business as well, but he was more interested in mining, since it was something relatively new for the country. He was quite well educated but, unlike Pok who spoke French, Vongvichit spoke English as his second language.

The Australian companies that were already mining in Laos were making good money, so Vongvichit's enthusiasm to get into mining was understandable. The problem was that he had almost no knowledge of the mining business. The other issue was that things were being done the 'Lao way', that is with very little money and an even more scant regard for health and safety. That probably explained why he never booked us the rooms at the hotel.

His projects were all north of Vientiane, mainly near the town of Vangvieng, situated on the banks of the picturesque Nam (River) Song, and surrounded by scenic soaring limestone kaarst mountains. Vangvieng was a popular backpacker destination with young adults 'getting away from it all' in South East Asia. It was very cheap and reasonably accessible by bus from Vientiane. Tubing down the river was a very popular pastime, although I suspect there were probably lots of germs in the water due to the poor sanitation in the villages along its banks.

The 'third lane'

I had heard about the 'third lane' from other people who'd been to South East Asia, but hadn't realised what it really meant until we were on our way north from Vientiane to Vangvieng the next day in Vongvichit's Toyota LandCruiser. The roads in Laos are quite undeveloped in general, very bumpy, narrow, full of potholes, and characterized by slow, diesel fume-belching trucks and recklessly ridden motor bikes and scooters. Bitumen was a rarity. More often it was dirt, although the main north-south highway through the country was sealed, as were a lot of the streets in Vientiane and Vangvieng. Road works were a constant sight, to fill in the potholes created by the large Chinese- and Russian-made trucks, and to repair the probably poor road construction in the first place.

Driving there, as I started to do later, was not an easy task. There were so many things to watch out for, including the potholes, children playing on the road, water buffalo who would not move off the road, and motor bikes coming in from side roads at full speed without looking. Plus, of course, you drove on the right-hand side of the road, rather than the left as in Australia. This made the border crossing into Thailand, where they drive on the left-hand side, quite interesting. After you exit Laos, the road does a scissor and takes you over to the left-hand side. You needed to watch the "X" in the middle where the lanes cross! It all seemed to work, but I could envisage head-on collisions based on the way most people seemed to drive.

Consequently, there were a lot of road accidents in Laos, and the relatively unprotected motor cyclists often came off second best with a car or truck. Plus, the wealthy Lao drivers in plush European cars had little regard for anyone else on the road and drove way too fast for the road conditions with the result that braking was often extreme when slowing from 160 km/hour to zero to avoid a water buffalo. Children and motor bikes had less chance with these low flying vehicles.

Hence the development of the 'third lane' which straddled the white line down the middle of the road. A BMW or Mercedes (or Toyota Land-Cruiser) travelling at 120 km/hour had unofficial priority to take this path, and the slower traffic (often only moving at 40 km/hour) somehow knew to get out of the way. Most of the time I just closed my eyes and prayed, except when I was the driver of course, when I prayed with my eyes open ; I must admit, there seemed to be an understanding from the other road users that this was necessary, and I often used it myself when driving to get around slower vehicles, they all seemed to move to the left to let you through, but you had to be confident when undertaking such a manoeuvre.

There seemed to be an elite class of Lao citizens who had European educations and wealth, and who were presumably politically well-connected, juxtaposed in a socialist/communist ideology against almost penniless proletariat peasants who represented the majority of the population, and who were totally subservient to the bourgeoisie.

Pha Luang and Ban Phatang

On our first trip to his projects, we were picked up at our hotel (the top-rated Lao Plaza Hotel) by Vongvichit and his driver for the 150km drive north. To my surprise, Vongvichit was driving, and his driver was sitting in the back. I got to sit up front and wished I hadn't, as we streaked down the third lane, swerving between buses and trucks, and braking suddenly for motor bikes and children.

Thankfully arriving in Vangvieng in one piece, we were taken to a very nice hotel on the banks of the Nam Song. We each had our own small bungalow, and there was a restaurant and bar which overlooked the river. A beautiful setting and I took my family back there a few years later for a holiday. I got to know the inside of my bungalow very well a couple of days' later.

The first project we visited, and the "jewel in the crown", so to speak, was a zinc-lead project called Pha Luang (the "h" is silent). In Lao, "Pha" means mountain, and "Luang" means large. It certainly was that. Vongvichit had built a road up the mountain, which I can only describe as the scariest drive of my life (although, again, in due course, and after appropriate instruction, I did drive it myself), but the effort behind building this road was simply amazing! It had largely been achieved using old-style hand-held rock drills, explosives, an excavator and bulldozer, and a few trucks. Apparently, it had taken two years to build.

The overall slope of the mountain was about 45 degrees. That's really steep. Most cars can't go up any slope more than about 10 degrees. Vongvichit and his workers had built a three-kilometre-long track that ascended nearly 1.6 km in altitude to the top. It had 25 switchbacks (hairpin bends), and you needed to be in four-wheel drive, low range, first gear. You literally crawled up the mountain, and you could probably walk faster (if you had the energy and stamina). It rose from an altitude of about 200m ASL in the valley to almost 1,800m ASL at the top.

The view from the top of the mountain was spectacular. The road went up to some outcrops of mineralisation grading an incredible 50% zinc. The mineral was a curious white colour and extremely light in density. We figured out it was an oxide mineral called hydrozincite (which we later confirmed by XRD). The advantage was you could mine this stuff and send it straight to a smelter. It didn't need any extra processing – although you had to get it down the mountain, which was the challenge.

There were several occurrences of this zinc mineralisation along the mountain range over about a 10 km distance, so the potential was amazing. To say I was excited would be an understatement. Vongvichit was mining the zinc oxide, trucking it down the mountain, and then driving it up to China where he sold it to a processing plant. But the tonnages were very small, and he wasn't making much money (if at all, after the costs of constructing the road).

He was after a capital injection so that he could expand operations and then start to develop some of his other projects (I mentioned coal, iron ore, copper, lead and gold).

Campbell, being an experienced mining engineer, suggested to me that a cableway (like a chairlift) would be an efficient way to get the ore down. He'd seen this done in Europe, and other places. We may have mentioned this to Vongvichit, not knowing that he would take us at our word and attempt to construct one in the "Lao way". That happened a bit later and I'll get back to it.

There were also some outcrops of almost pure lead (in a mineral called galena) lower down the mountain. These seemed to be much more limited in extent, but it confirmed my thinking that we were dealing with a Carbonate-Hosted Zinc-Lead type of mineralisation (similar to the Lennard Shelf in Australia, or the Tennessee deposits in the USA). I could see the potential of discovering more of this mineralisation at Pha Luang and turning it all into a decent-sized mining operation. The trick would be finding the sulphide host to the mineralisation (this was the exploration model), since the surface stuff, especially the high-grade zinc oxide, was remobilised from presumably a sulphide source at depth and seemed limited in extent (but impressive in outcrop just the same).

After our visit to the mountain, we stopped at a village, Ban Phatang, on the way back to the hotel for lunch. First Pacific Mining had a field office there. I had read that it was impolite not to eat what was put in front of you if you wanted to do business in Laos, so I ate a bit of everything. I think this is common around the world, especially in Asia, where the people, often poor, will put on a huge spread to impress potential business partners.

We then spent the afternoon looking at maps and generally getting ourselves excited about the exploration and mining opportunity. Campbell had lots of good ideas about how to mine and transport the ore, and I could see a whole new province of zinc-lead mineralisation being defined. A celebratory dinner was had, with copious Beer Lao and pizza. Yes, Vangvieng, being a 'Mecca' for back-packing European tourists had a couple of pizza restaurants that weren't half bad!

In the early hours of the next morning my body woke me to what I knew was going to be a terrible 24 hours. I literally ran to the toilet and it all came out – both ends. I felt terrible! I tried to drink as much water as I could; and I took my anti-gastric and diarrhoea pills, but to no avail. When the knock came to go for breakfast, I was curled up in the foetal position where I basically remained the whole day and until the next morning. I figured it was the chicken's feet that I'd eaten at lunch in the village.

Consequently, I missed out on visiting the iron ore and gold projects that day, but they seemed minor compared to the zinc project. I did manage to catch up with Campbell that night, but could not face eating anything. The next day we were to go back to Vientiane, Vongvichit having gone ahead, saying he would send his driver to get us. That's when I learned about "Lao time".

Lao time is basically any time you want it to be. If you think it's an hour, it is probably three, and can often be a day. Our driver, due to arrive (we were told confidently by the hotel manager) at 10am, eventually arrived at 4pm. We sat and cooled our heels all day, not knowing whether we would be leaving or not, having minimal grasp of the local language (phrase books are useless in these circumstances), and not being able to find anyone who could interpret for us. Our trip back to Vientiane was even scarier than the drive out, being at supersonic speeds in the dark.

Striking a deal

I told Vongvichit I would go back to Australia and talk to my Board, and that we were very interested in getting involved in his project, especially the zinc one. We had a sumptuous French cuisine celebration dinner in Opera Square with Vongvichit and Pok, washed down with a couple of bottles of French wine. The price for four people, dinner $US20, wine US$30. What a bargain!

It was early March 2005 before the company Board met and gave me the green light to go back to Laos and try and negotiate a deal. Vongvichit was waiting for us, and a deal involving a six-month option to conduct due diligence, and then to enter a joint venture, was duly struck. That's when the fun really began.

My first task was to find an 'in country' manager, someone I could trust to promote and protect the company's interests. I was lucky to find an expatriate geologist, James Venables, who was living in Laos, and who was married to a local woman, Kat. They were an invaluable pair for the company, James managing the technical work, and Kat managing our Lao workers. I trusted them implicitly and they never let me down.

James had worked for Phu Bia Mining (the local subsidiary of Pan Aust) at the Phu Bia gold-copper mine near Vangvieng, but was looking for something different and he heard about Rox on the local grapevine. His call was very welcome, and I hired him on the spot.

James set about setting up our operations in Laos. It was now April, and the field season ended in May for the wet season which ran from June to October. Our task within the first six months (by September) was to drill some holes to test the theory that there was sulphide mineralisation at depth below the rich oxides at surface. Being so late in the field season, and not having any roads established, we opted for a 'man-portable' drill rig, which we basically pulled around with a bulldozer. The wet season had hit, and each day there was a torrential downpour regularly at 3pm. Some days it just rained all day through, and driving up the mountain was quite tortuous. As it turned out, we were able to construct a crude track through the jungle and pull the rig into position with the bulldozer. The first two holes drilled beneath the zinc oxide, didn't hit anything, and we scratched our heads a bit. Where did all that zinc come from? The third hole, at a different prospect, did hit some lead mineralisation, but it was all oxide. Nevertheless, we had established a thickness of about 10 metres, and we confirmed the potential enough to exercise the option.

Ban Na camp

With the weather mainly inclement during July to September, we needed a camp on the mountain, near the drill rig, since the drive up and down the mountain each day was very dangerous, and some days impossible. You tended to slip and slide around, and it was possible that an uncontrolled slide could take you over the side of the mountain. James designed and built some bamboo huts along the track, which sort of hugged the side of the mountain, so you had to be careful, especially at night, that you didn't disappear down the steep slopes when you stepped out to relieve yourself. On my visits I would stay up at the camp for a few days to inspect the progress of drilling. The mud was thick, and going in and out of doors was an exacting process of carefully removing your muddy boots, cleaning them off, and then putting on a pair of "inside" shoes.

The toilet was an old tin upturned with the bottom cut out, placed over a hole in the ground. You "flushed" it by emptying some lime down the hole to neutralise your deposits. This was fairly routine to us in Australia as a bush toilet, but the Lao workers just went anywhere they wanted. We had to threaten them with the sack to get them to use our toilet and maintain some decent hygiene. Every few days we filled up one hole, and dug another.

We lasted about eight weeks up there until the drilling was finished and we demobilised back to Vientiane, where we leased a house and set up an office. Our plan for the next field season was to build a larger camp at the bottom of the mountain, and operate only in the dry season (November to May), so that we could drive up and down the mountain safely without the heavy chains needed in the wet.

Under our agreement with Vongvichit, he was allowed to continue mining the surface zinc oxides and sell them as he saw fit. We were exploring for the deeper zinc and lead sulphide mineralisation, and most of the time we didn't get in each other's way. We set up separate camps at the bottom of the mountain beside a beautiful stream (the Nam Na), which picked its way along the valley floor. You could almost hear it talking to you as it bubbled away, and the setting, complete with fireflies at night, was idyllic.

As we built up our staff of local workers, we added more huts in the camp. At one stage we had 80 locals working for us, and over the four years of our operations in Laos I do believe we made a positive impact on the local economy, providing employment, regular wages and training. I was very proud of that, although it seemed that government and wealthy Lao people didn't really care about it. As I found out, there was a strong vein of corruption and nepotism in Lao society, which is common in Asia, and which will hold those countries back in their development.

Mealtimes were conducted in two or three sittings, to feed our 80 workers, although we had a big mess hut. There were four Lao women who did the cooking, and breakfast consisted of sticky rice wrapped up in bamboo leaves. Sometimes it was coloured purple. Kat taught the Lao women to cook some western meals for James and me. We had one other expatriate working there, Pat, but he was just as happy eating Lao food, so often it was just a separate meal for James and me – we just couldn't get used to the hot Lao spices in the food.

The cooks would get up at 4am to have breakfast cooked and ready for everyone at 6 am. The workers would take lunch out with them, so there was a lot of food prepared in the early morning. Dinner was usually something containing beef or chicken, and I heard that jobs with the company were highly sought after because of the good meals and accommodation we provided.

Our Lao workers were each issued with a pair of work boots, since most of them had turned up for work in thongs, or at best sandshoes. Vongvichit's people continued to work in thongs (and I hate to think of the accidents they probably had) but, being an Australian company, we required a much higher standard of Occupational Health and Safety (OH and S). James made sure each employee signed for them, since they were worth several week's wages.

The next day most of the workers turned up in thongs. We said: "You must wear the boots that were issued to you yesterday". Glum faces were our response. "We sold them", one worker sheepishly replied. "Well, you'd better get them back", we said. No boots, no work. I think most of them had to buy them back at a higher price than they had sold them, but they learned a lesson that we were serious about safety and wouldn't compromise.

We had similar issues getting our workers to wear hard hats when it was required. They just didn't have this work safety culture in Laos, and thought we were quite mad when we insisted that they wore personal protective equipment (PPE). In the end, I think they were quite proud that we treated them the same as we would any worker, whether a local or an expatriate.

Pay day was an occasion worthy of mention. We paid each labourer 35,000 kip per day – which was the going rate, about US$3.50, – the equivalent of four and a half large bottles of Beer Lao. Believe it or not, they seemed very happy with that rate of pay, since the cost of living in a village in Laos was quite low. At that time, the largest denomination note in Laos was 10,000 kip, so each worker was paid in 10,000 and 5,000 kip notes. For a week that was 245,000 kip, a small wad of 10,000 and 5,000 kip notes, multiplied by 40-50 people, that was about 10 million kip for the whole payroll – which was about a suitcase worth. James and Kat would go to the bank in Vientiane and withdraw the cash, put it in the suitcase and then drive up to site and distribute the money.

Later on, the government introduced 20,000 and 50,000 kip notes, and our workers got appropriate pay rises to keep up with inflation and work productivity, but they still preferred being paid in the smaller denomination notes, so the suitcase got well used and travelled.

I used to worry about James and Kat being robbed on their way from Vientiane to Pha Luang with the suitcase full of wages. There was a story from the mid-1990s when a bus load of tourists (mainly locals) was ambushed on the road between Vanvieng and Pha Luang by a group of Hmong rebels. Three people were killed. We used to drive past the spot on our way to our camp. We had heard stories of the rebels raiding villages to steal food. James, Pat and I, as expatriates, were obvious targets for kidnapping, but James and Pat seemed less concerned than I was. James did a deal with the local militia (there were both army and police stationed in the Nam Na valley), where they would provide us with security. We paid them wages "on the side" to what they supposedly got paid from the government (but hardly ever received). We also provided them with accommodation and meals, so they jumped at the chance. I think the whole valley was very secure from the bandit rebels because of this.

James said he had run a quick 'economic survey' of the valley. There were now at least a dozen motor bikes (prior to our arrival there were none), plus people had been able to upgrade their tractors and the huts in the villages were all a lot better kept. I think we made a strong contribution to the well-being of the locals.

Pat was spending four-week stints at the camp, and he decided that he would buy a satellite dish so that he could tune in and watch Australian TV, particularly the State of Origin Rugby League series, being a staunch Queenslander. The satellite dish duly arrived and, after some mucking around by Pat, it was installed, but all he could get were Thai TV stations. Pat went without his rugby league, but became a hero to the 80 Lao staff who just loved to watch the Thai soap operas and movies in the evenings.

I also spent a lot of time on 'government relations' in Vientiane and seemed to have endless meetings with various officials regarding our 'foreign investment licence', something we needed to get before we could own an exploration tenement or mining rights in Laos. The rights we had were contractual, through a joint venture contract with Vongvichit's company, which owned the mining and exploration rights. We really wanted to be able to own these rights in our own name, but the government seemed to make this very difficult. We would have numerous meetings where they would say, "yes, the foreign investment licence is under consideration, just be patient". I relayed this to our investors in Australia, but concern started to build about the transparency of the whole process.

The blessing ceremony

A Lao tradition is the blessing of the workers and the boss, usually held at the beginning of the year. James' wife Kat ensured that we had a memorable ceremony each year to bless us all and wish us good luck in our endeavours. It all started early one crisp January morning. A mat was laid out facing the mountain and several shrines, food, drink and sets of flowers were set up as offerings to Buddha. Kat and the head of the local village said a few words in Lao and the assembled workers all chanted their replies, and then I was asked to say a few words, which were duly translated to Lao for our workers to understand.

Then the most incredible thing happened. I was mobbed by the workers. They wanted to touch me and tie string around my wrists. To the string they attached money. Apparently, this is to wish that I became rich, because if I became rich, then they would also become rich being in my employ. It was really quite touching to see the genuine look on their faces as they tied the string on my wrist, attached their money and said a blessing to me.

I was feeling quite elated at the end of it all, and really felt that we were going to find a mine here and establish the prosperity of this valley for a long time to come. I remarked to Kat and James that I was touched that the workers had given me their money. Kat smiled and nodded. James told me later that she'd given them the money to tie to my wrist, since most of them couldn't afford to give me their own: still, it was a touching moment. I bundled up the money and asked James to distribute it back to the workers somehow. His solution was to throw a big feast and put on some free Beer Lao for them all.

The flying fox

One day, when I was meeting with Vongvichit in Vientiane, he said: "Ian, you need to see the new cableway I've built up at Pha Luang. I can now get the zinc oxide ore down the mountain quickly and cheaply, and won't be needing to spend so much money on maintaining the road." I was keen to see what he'd built, imagining something like a ski chairlift construction.

I was undoubtedly optimistic in my expectation and totally underwhelmed by what I saw. Essentially what he'd constructed was more like a flying fox or cableway. There was a taut wire supported by poles, and the idea was that bags of zinc oxide (each one weighing about 50kg) would be sent down the wire from the top of the mountain to the bottom where they would be unloaded. Suspending the wire down the mountain would not have been an easy task, since on the steep slope it would have all needed to be done manually, and Vongvichit and his workers' endeavour was impressive, if not ultimately misguided.

What he hadn't understood was the effect of gravity and other physical effects when a bag weighing 50kg is let go down a wire sloping at about 45 degrees for a distance of one kilometre. By the time the bag got to the bottom, if it hadn't derailed on the way down, or hit a supporting pole and split and spilled its contents down the mountainside as it swung from side to side, it was travelling bloody fast. Too fast to be stopped safely, and most of the bags in the first trial hit the last pole at the bottom and split open as Vongvichit's workers scrambled to safety, creating a large area of fine white zinc-rich powder covering just about everything. I struggled to keep a straight face as Vongvichit showed me the results of his efforts and explained how he was planning to fix it. I realised then that he was no engineer.

After several deep breaths to regain my composure, I suggested that he needed some way to control the speed of the bags, and to stabilise them as they travelled down the wire. This then led to a series of attempts, involving a second thinner wire that was attached to each bag and unwound from a pulley at the top of the mountain by an operator. What hadn't been anticipated was that the weight of 1km of this wire to let the bag down the mountain was too great for his operator to wind back up again, and the whole process was agonisingly slow. They tried attaching several bags at a time, but again, the overall weight of everything was just too great for a human to manage. I dreaded the inevitable attempt to motorise this process, which I could foresee would lead to fingers and hands being caught up and severed. I was therefore suitably impressed when he came up with the idea of wrapping the wire around a truck wheel and controlling the speed of the bags down the mountain, but this required a lot of coordination between the top and bottom of the mountain, and ultimately proved to be too slow. Eventually he abandoned the whole idea and went back to driving trucks up and down the mountain, which was dangerous enough in itself.

16th century technology

I mentioned that Vongvichit had other prospects that contained coal, iron ore, copper and gold. He hadn't done much with them, because he was concentrating on trying to make some money from the zinc oxide mining. However, when in China one time, he had seen a processing system that was in use there, and he decided he would build one in Laos. I hadn't seen one in action before (because the technology is about 200 years old) but had read about a very rudimentary process to liberate zinc from oxide ores, called a Waelz kiln (probably originally described by Agricola in his 16th century work, *De Re Metallica).*

The Waelz kiln is technology developed in the 1920s based on a rudimentary chemical extraction process. What Vongvichit built was a variation on it, which involved heating up the zinc oxide in a clay vessel with some coal. The coal, which is mainly carbon, reacts with the oxygen in the zinc oxide ore ideally producing carbon dioxide and zinc metal. Vongvichit started mining coal (which was of very low quality) and had a big crew of workers making long clay pots to hold the ore/coal mixture. These were placed in rows in the kiln and the whole thing heated with wood and coal. On a small scale his initial efforts were successful, and he managed to produce a zinc metal button about 1cm thick and 10cm round.

The plan was to build larger kilns holding around 200 pots at a time, and Vongvichit had built about 10 kilns when I visited his plant site. Trials with the first kiln had produced some zinc metal but, in a lot of pots, the reaction hadn't proceeded to completion. This was probably because of the impurity of the coal used, and because the mixture was required to be heated to about 800 degrees Celsius, which required a lot of coal and wood, and that temperature hadn't been achieved.

The other thing was that a couple of his workers had been severely affected by gases that were given off during the process and had been taken to hospital. There was talk of a 'blue gas' that had been given off. I put my chemistry training to use and deduced that it was probably a potent mixture of carbon monoxide (which would also be produced by the reaction), plus some other "nasties" that would be released from the ore, like mercury, arsenic and cadmium, all very poisonous materials. No wonder his workers had gotten gravely ill.

I was appalled at this apparent lack of forethought and concern for his workers' safety but, looking back on it, what was going on was a reflection of the sort of industrial practice that had also existed in Britain and Australia, before we had developed expertise and understanding of what we were doing. Doing things the 'Lao way' meant cutting corners and copying from China, where some of this 200 year old technology was still in use. Anyway, in due course, Vongvichit was visited by some government health officials and the whole operation was shut down and abandoned.

Later, he visited China again and saw a more modern version of the Waelz Kiln and tried to build one of these. I called this the Zinc Oxide Plant. It required heating to 1,200 degrees Celsius, which is incredibly dangerous and difficult to do with wood and coal, and again quite a bit of money was fruitlessly spent trying to implement it. This one involved an oven at one end, with large pellets of pressed coal used for heating, and also a source of CO_2 (when coal burns CO_2 is given off). The heat and CO_2 were passed through a series of steel tubes which undulated like a large snake, along a distance of about 30 metres. The lower parts of the steel tube were heated to maintain heat all the way along.

Smaller pellets of zinc oxide ore were placed in the steel tube at various places. The idea being that the CO_2 reacted with the impure zinc oxide ore to produce fine particles of pure zinc oxide, which were blown along the pipe by the heat and some bellows. At the end, the output was passed through some fine filters hanging vertically, where the zinc oxide dust was collected. He showed me some pure zinc oxide, but I later found out this was a 'sample' from China, not one from the new plant.

I doubt that the temperature got much above 500 degrees Celsius, and so the whole exercise was a waste of time and money. It was frustrating to watch this happen, but Vongvichit simply wasn't interested in, and probably couldn't afford to do things the proper western way, despite my efforts to advise him. I guess he wasn't getting the technology that he was trying to copy from China right, and it would have been wiser to buy in this knowledge and expertise. After that, he stopped trying to process the zinc oxide ore in Laos, and just went back to trucking it up to China.

Pha Sod

Apart from the prospects we had at the eastern end of the Pha Luang Range (which included Pha Daeng, Nam Yen and Bon Noi), there was one other prospect we were told about by Vongvichit at the western end called Pha Sod. It was close to an area, off our mining lease, that had been mined by the Thai company, Paedang Zinc, so was considered very prospective. To get there we had to drive all the way around the mountain and then up from the other end. I negotiated with Vongvichit and he agreed to build us a road up there (it was a goat track at best). After about six months, he advised the road was ready and we eagerly drove up there.

The prospect area covered about 2 x 2 km at about 1,500m above sea level. There was a flat area and then a cliff where you could see sparkles from outcropping lead sulphide and also lots of white zinc oxide. The flat area was covered with small pits where Vongvichit and his men had been digging in the past (walking the material out in buckets carried over the shoulder). We used a portable X-Ray Fluorescence (pXRF) instrument to measure the lead and zinc content of the soil samples we collected. We calibrated these readings with about a hundred random samples we sent to the assay laboratory and got an excellent correlation. We had a cheap and quick way to sample and evaluate prospects without having to wait for the assays to come back from the laboratory, and without that expense (although the portable XRF instrument wasn't cheap).

I refused to let the pXRF into the field, since it was too delicate and too expensive. It was kept in camp. The field teams would collect their samples in calico bags and bring them back to camp. It was hard work scrambling over the outcrops and cliffs, so they had Sunday off in camp, where we showed them how to use the pXRF to analyse a sample (taking several readings from each and averaging them in a spreadsheet). They loved this aspect of the work, and they excitedly went about this task every Sunday. By that night we had a valuable addition to our soil sampling database, and we could plan the next week's work using the newly acquired data.

In the end, we defined a soil anomaly at Pha Sod covering about 1 x 2 km, and dug some trenches. The results were amazing. The Zn+Pb values were about 2% in the soils and, after we dug a few pits and trenches ourselves, I estimated there was probably something like 100,000 tonnes of zinc and lead just in the soil. That's a lot of metal. We undertook a bit of a study to see whether this was a viable mining option, but the figures just didn't add up unfortunately; the biggest issue was how to process the material from the oxide form to elemental form cheaply and efficiently given the low grades.

Nam Yen

Our most successful project was called Nam Yen, after the stream that ran beside the prospect and then over the side of the mountain to the village of Ban Nam Yen some 1,000m vertically below. The prospect had a small area of outcropping lead sulphide (galena) which assayed an impressive 60% Pb. Our first drill hole there during the due diligence period had intersected about 10m of lead oxide mineralisation so, after we exercised the option, we went back in the dry season, built a proper road to the prospect; then took in an RC rig and drilled some deeper holes, which were stepped back further from the outcrop. In the third or fourth hole, we hit 33m of 11% Zn+Pb, which was enough to send the company's share price soaring, from 15 cents to about 65 cents, and we took the opportunity to raise some money to fund further drilling.

Eventually at Nam Yen, I reckon we defined a resource of about

500,000 tonnes grading 6.5% Zn+Pb (in about equal proportions of each metal). It didn't meet our minimum criteria of 5 million tonnes at that grade, or better, but it was pretty much all we were able to define with drilling. The literature talks about the close-spaced drilling that is required to define resources at these Carbonate-Hosted (or Mississippi Valley Type (MVT) Zinc-Lead deposits, but that was nigh on impossible in the terrain on the mountain at Pha Luang with average slopes of 45 degrees to contend with.

To drill out the deposit, we built tracks/roads that followed the hill contours along for about 200 metres each, and then stepped down the hill 50 metres and built another. We had four crews of four men each, each with an air compressor and rock drill. They would drill and blast the road out of the limestone. Each 200m contour track took the four crews about a month to build, so it was quite a long and laborious task. We would then prepare a drill pad every 50 metres along and set the rig up to drill an array of holes at different azimuths and dips to cover all the angles we needed to cover. Then we'd move the rig along to the next pad and do the same. Occasionally we'd come to a cliff that we couldn't get past, so we'd have to find another way around to the other side and then back up to where we wanted to drill.

Because it was known at the time of doing our deal, Vongvichit had kept the mineralisation at Nam Yen as his, rather than a JV, asset. I wasn't too concerned, because it was sub-economic to us and we certainly wouldn't mine it, and I couldn't see Vongvichit mining it underground. He surprised me therefore one day by informing me that he had hired a Chinese mining company to take over all mining operations at Pha Luang. He would continue to transport and sell the ore to China, but the Chinese company would do the mining and be paid a share of profits.

On my next site visit I was keen to see their mining operations. Again, to say I was underwhelmed would be an overstatement. Their open pit operations were the most unsafe, sloppy, dangerous and environmentally messy I had ever seen. I banned our staff from going anywhere near them, and I told Vongvichit what I thought. He shrugged his shoulders in typical style, smiled ruefully and said, "they're making me money."

While they were mining the higher altitude zinc oxide deposits I wasn't too concerned, but then they started underground mining the lead and zinc sulphide at Nam Yen. In one way I was a little peeved, because they were mining out material that we had drilled out, but it did help us understand the highly poddy and discontinuous nature of the mineralisation, and we realised that our company couldn't possibly have mined it and made money, so our decision on whether to keep going or pull out of the project became clear.

Not quite a nightmare in Laos

Sadly, we didn't discover the sulphide source for the zinc and lead mineralisation, and while we looked at some other projects in Laos, we didn't really find anything that would allow the company to grow, so we pulled out of that country. It was difficult to negotiate realistic deals with local partners, and you needed a foreign investment for every project you wanted to get involved in. It became obvious that this was a ploy to play us along and extract various bribes, which were subtly revealed to us. The goal posts kept changing and you never knew where you stood. I remember taking some Australian investors up to Laos and Vongvichit told them face to face that our foreign investment licence was about to be granted – but it never was. I don't know that he really knew either. I think if we had indicated that we wanted to invest a lot more money and develop the mine we would have gotten our licence, but not just for exploration apparently.

One day I was approached by a Thai businessman who asked if I would meet with him next time I was passing through Bangkok. A meeting was duly arranged and, when I arrived, he informed me that he had secured some exploration licences in Laos adjacent to our project and he enquired whether I would like to buy them from him. I looked at his map and saw that these were areas that we had applied for 12 months earlier. I asked how he had procured them when we were the first applicant. He just smiled and asked, "did you pay the facilitation fee?" I knew exactly what he meant, and said no. He just shrugged and said, "you'll never do business in Laos then".

Eventually I knew the time had come for us to cut our losses and leave Laos. We left 80 people and their families out of work, and the local officials never batted an eyelid. Despite all the meetings and platitudes, the Lao government did nothing to help us. We had spent about US$4.5 million there, a lot of which had benefitted the local community, but that was never appreciated because no-one in officialdom had benefitted – we had been very careful with how we spent our money, and we bribed no-one.

Along the way however, I had found out just how subtly corrupt the government and officials were in Laos. Perhaps I was naive, and now I would not recommend doing business up there unless you are prepared to pay bribes and make under the counter payments demanded – all of which of course are highly illegal under Australian law.

We also had to compete with the Chinese who started to invest heavily in Laos to exploit its natural resources, and who were prepared and able to feed the corruption cycle, to our commercial disadvantage. I remember one day visiting the Office of Foreign Investment in the company of the Australian Ambassador to Laos, to see the Minister about our foreign investment licence – which was always promised but never delivered, just as six brand new Mercedes Benz cars were being dropped off by the Chinese as a 'gift'. The Australian Ambassador just shook his head and gave an ironic laugh with a comment, "there goes your foreign investment licence, Ian."

Laos is a beautiful country, and its people are largely friendly and generous, however, like most Asian countries, it is plagued by widespread corruption of power and promotion of individual self-interest. Anyone who challenges the accepted way soon learns to their detriment not to rock the boat. There is a book titled *Nightmare in Laos* that I would encourage anyone to read. A sobering story of how an Australian couple were unjustly incarcerated in Laos for over 10 months for unwisely trying to set up a business in competition with the local power brokers.

Laos was an adventure, and I saw and experienced some amazing things there – both good and bad. Chinese interests have now invested heavily in the country and I assume there are benefits flowing from that, but I can't help but think how better off the country would be if it could shed the traditions of corruption and nepotism and embrace a more egalitarian outlook. I doubt that will ever happen.

9

You're the Piano Man

Indonesia, 1997

Jeff Rayner

Not knowing what he wanted to do in life, Jeff Rayner attended a Geology 101 lecture out of curiosity at Melbourne University and emerged with a major in geology in 1986. He worked for the Big Australian at Coronation Hill and two years later valuable lessons were learnt as a mine geologist at Ora Banda, Kalgoorlie. BHP Gold and Newmont Australia merged as Newcrest Mining and Jeff continued in the exploration role in eastern Australia and Indonesia. In 1998, he was unfortunate enough to have landed a job in Sardinia, Italy with Gold Mines of Sardinia, and was based in Italy for 20 years. A leap of faith by seed investors saw him become MD of KEFI Minerals in 2007, a junior explorer listed on the LSE. Exploration projects were spread out in Africa, Middle East and Turkey. Since 2016, Jeff has worked as a consultant for gold and base-metal miners and explorers in Australia, Ecuador, Asia and the Middle East.

I'm sailing away from an active volcano in the Pacific Ring of Fire, which is one of the Maluku Islands, called Ternate, and its only city Kuta Ternate; sailing to the larger island to the east, Halmahera. The Maluku Islands were once a hub for merchant sailors on tall ships and for international trade. Today it is largely forgotten, set in an idyllic tropical paradise, where existence is low stress, the modern world yet to impose its calamity on the docile Indonesian village way of life. The tourist trade is off the beaten track, backpackers mostly, looking for untouched civilisations and wilderness on 10 dollars a day. Ternate was once a major world producer of cloves and was governed in an endless tug of war between the Portuguese in the 16th century, the Spaniards and Dutch in the 17th century, Islamic Sultanates and the British in the 18th and 19th centuries, and finally the Japanese in World War II. The spice trade has been replaced with a new commodity, gold. The quest for instant wealth has brought an influx of hungry miners, mostly from Australia and the Americas.

It's a very exciting and inspiring moment for me. A new jungle adventure. I romanticise being on board the SS *Venture*, battling through thick fog, in angry seas between a treacherous narrow rocky cliff opening to *Skull Island*, going to meet *Kong*. OK, maybe I am a dreamer and I'm exaggerating just a little bit. The vessel is just a clapped-out, rust bucket, overcrowded, Indonesian ferry; there is no blinding fog, there are no deadly cliffs and *Kong* swam to Borneo centuries ago before he made himself extinct. There is no wretched blonde beauty to save, no epic story of beauty and the beast but, nevertheless, I'm here to map some trenches at a rich gold find, deep in the jungle, and no doubt a very dangerous place to be.

The destination is an exploration drilling camp somewhere in the thick, unpopulated jungle of the Halmaheras. There was an easier option to get to the camp from Ternate; a 35-minute flight in a chopper, but that's an expensive 35 minutes, even for a major mining company. So, the romantic cheaper way is a two-hour ferry ride, a bumpy dusty two-hour 4WD ride, and a two-hour trek through the dense jungle to the camp, carrying a box of beer on my shoulder. I worry immensely that this box won't last the duration of a six-week swing.

The Indonesians did a remarkable job building the camp five years previously. With slippers on, and no eye or ear protection, they chainsawed wood planks ½ inch thick from felled rainforest trees and knocked up a decent camp which, unfortunately, was located in a steep, forested valley that trends east-west and apart for two hours a day, never sees the sun. I find out later what a worthy medium that tropical heat, soaking humidity and jungle-filtered light is for fungi and bacteria to colonise a suitcase of jocks and socks.

Diamond drilling rigs had done their job: kilometres of sticks of core have been logged, cut and sampled. The ore deposit has been modelled, but the geological model doesn't quite gel. The job in hand for me is to work out why the seams of gold are so rich and what are the little structures which trapped them so. This means re-logging drill core and re-mapping trenches for the lads in Perth to unravel the secrets and so make money for the punters.

I am bedded down in a row of half-walled wooden huts, inducted to where the lone western dunny is, where the mess is, where trapped invading rats are fed to captured pythons in the square-dug pits, where the First Aid is, and, most importantly, where the beer fridge is. The camp was built right on the banks of a narrow creek but, don't be fooled, it can be a raging torrent of white foamy surf, tossing two metre-diameter boulders around from an unseen, local distant tropical downpour further upstream: you hear it before you see it. Now I am settled into the camp, I am briefed on what is needed from me on the morrow and, after a Bintang or two, retire under a thick mosquito net, the vital protection that keeps the malaria parasite from dwelling in and defiling my liver.

The speckled light of dawn awakens the jungle and it's time to go to work; it's seven o'clock breakfast on a Saturday morning and the regular crowd shuffles in. Indonesian drilling camps, Halmaheras, you know how it goes. There's an old Indonesian geologist sitting next to me, "He's quick with a joke, or to light up your smoke, but there's some place that he'd rather be". He says, "Son, can you go sample those trenches up the hill? I could do it when I wore a younger man's clothes".

The trenches lay idle, five years have passed since they were dug by hand: one-ounce dirt uncovered in thick jungle forest. A small but rich epithermal vein had punched its way up from the Molucca Sea Plate about 2.8Myrs ago. Now drilled like a Swiss cheese and about to be DFS'd, crushed and sprinkled with cyanide.

Dead leaf litter carpeted the trench floors and rough lichen painted the walls vivid green. The Indonesian field assistants had spindly shovels and stopped digging when the floor was impenetrable: that happened to be the top of the mineralized, flat-dipping, Hanging Wall vein, so it wasn't sampled: you know how it goes.

Well, a few weeks later, there were just two metres left in the last trench to re-log. I was mapping alone, the large leaf litter started to move like a wave and rustle about seemingly on its own. It caught my attention and there, calmly gazing at my face through venomous eyes, was a black snake almost as thick as a man's wrist. I was shocked, paralysed, momentarily shit-scared. I know snakes can't talk, but I'm pretty sure it said to me: "Man, what are you doing here?" I jumped out, I leaped out. One more step in that trench and there would have been a decaying mess of necrotic jellied tissue for the paramedic to scoop out. No anti-venom, no snake bandages and no safety-buddy. That was Health, Safety and Environment in the 90s: now look at HSE, almost as toxic as a black snake trapped in a trench.

A few months passed; the daily grinding tasks of logging drill core and trudging up and down jungle treks were paying off, progress was being made. My respite was a six- and two-week roster: back to Perth for a debrief, then home to Brisbane to sterilise my socks and jocks. It was an arduous journey: returning from camp-to-Ternate-to-Manado-to-Jakarta-to-Perth and to Brisbane; then shortly, all again in reverse.

Back at the camp on my third swing, I happened to be lunching in the mess and there seated near me was our regular contract chopper pilot, 'Fast' Eddie. Not long back from a drop to the Bre-X camp in Borneo. He kept mostly to himself, and sometimes a droopy tear leaked out of an eye, splashing on the recurring tin plate of kampong chicken and rice. Some traumatic horror in Vietnam must have revisited him. Fast Eddie opened up to me, and said that he was in the Bre-X camp at Busang when it burnt down. He filmed it. Separately four buildings caught fire in the dead of night. Strangely, no bodies could be seen panicking in the darkness trying to put the blazing inferno out. When dawn broke, a D8 dozer ploughed the ashes into the dirt. No evidence left. He panned out his lens and, there it was, a two metre-high fire hydrant stood proud 80m from the ashen camp and 80m back from a major river: pity no one thought to use it, you know how it goes. I may have been one of the few persons to ever see that footage: Fast Eddie plummeted to the ground around 18 months later.

Later in that month the Exploration Manager requested a geology update in Jakarta. I was fortunate that it coincided with the monthly drinks night sponsored by various generous foreign companies. In the smoke-filled, alcoholic pub haze, I spotted a handsome young lady wearing a yellow t-shirt with a Bre-X Share Certificate embossed on the front, lovely job, even had a hologram that glittered under the seedy karaoke lights. She turned around and on her back was printed 'Official Sampling Team'. I wanted that shirt, I needed that shirt, but no matter what I offered for it, she did not want to part with a piece of history.

I'm on my last swing, the job was done after nine months of hard yakka; even the geotech guys in Perth believed the geological and structural interpretations. Dinner time was up, my last camp dinner, kampong chicken and rice again. The morrow promised a cow, choppered in, inside a sling. The camp cook learnt from his first effort, this time to instead let the poor beast recover consciousness for three days before pointing its throat to Mecca. The first beast flown in was petrified, we ate thin slices of adrenalin-marinated steaks that time. But that wasn't enough persuasion for me to stay another day.

In Ternate to celebrate my last night in Indonesia, for everyone was glad to see me go, the evening entertainment began, and it was a pretty good crowd for a Saturday. The bar is open, and the manager gives me a smile. A local half-starved lass lazes her hand on my knee and shows off her English saying, "I can, I can". I shake my head and say: "I can't, I can't". In a dark corner is geo Dan, a seppo, "he's at the bar and is a friend of mine, he gets me my drinks for free". When not at work, he lives in places greater than 46°N or S of the Equator: no liver-resting parasites there, no biting bed bugs, no cows with slitted throats point Mecca-wards; snakes shiver in holes, and the beer is icy cold. We drink and drink and drink and reminisce good times and not so good times. Dan's a tough guy, he's roughed it in the jungle for five years, but tonight he falters and says: "Jeff, I believe this is killing me, if only I could get out of this place".

"Well Dan", I say, "I am glad I did this job, made some good friends, but I'm glad I'm going home, and it was a pity about missing out on *Kong*". You know how it goes.

(*Apologies to Mr Joel*)

Postscript: Over 20 years later, I did a small consulting job for a gold mine which was going bust in Sumatra. The entire mine workforce ate salt fish and rice for lunch and dinner, 13 days straight. A couple of the Indonesians working there were previously mine geologists at the Halmahera gold mine. They showed me some cross-sections of the structural interpretation – to my surprise they were my drawings. I took the risk and asked timidly, "How did they work out?" My head swelled when the reply came, "We used these all through the mine operations". I had to walk out the door sideways.

From Michael Fellows' story, "Magadan on Mogodon" (page 171)

A gulag gold miner who presumably died on the job

CHINA, RUSSIA AND THE FORMER USSR

10

Chinese Burns
China, 2009

Ian Mulholland

The Chinese MOU

Before the Global Financial Crisis (GFC) hit, and certainly after it, a lot of Australian companies sought Chinese investment in their projects. The Chinese had paid quite high prices before the GFC and were seen by many Australian CEOs as a way of attracting funding when it was no longer available from the Australian capital markets, and possibly as 'easy pickings'. I can tell you they weren't.

We were approached by several 'investment consultants' who claimed they could broker us a great deal with a Chinese company. The problem was they wanted to be paid a retainer up front, plus a commission on any deal done. None of them had track records of doing deals, so it was difficult sorting out the genuine ones from the 'wannabes'. Eventually we were introduced to a guy who did have a track record and didn't want any payments up front. He was Chinese (as all the investment consultants were, and probably needed to be), but he had lived and worked in Australia for some time, so was not only familiar with how Chinese companies worked, but also with how Australian companies operated.

He proposed lining up a number of prospective investors (mainly private companies), but also a couple of State Owned Enterprises (SOE) and we would then travel to China to meet them. One company in particular was quite interested, and it happened that they were in Australia at that time looking at some other projects. After a quick meeting with them in Perth, they invited us to China to visit their head office there. We ended up signing a Memorandum of Understanding (MOU) with them, and although we signed it with great hoo-ha in China, it wasn't really worth the paper it was written on. This is the story.

Negotiating the MOU was an experience I will never forget. We first met the Chinese group in Perth, and they seemed very interested in our project. It was still at the exploration stage, but we had made a number of good drill intercepts of zinc-lead mineralisation and had defined an area of it at the Myrtle prospect, which is only 15km south of the giant McArthur River zinc-lead mine in the Northern Territory.

After the initial meeting in our office, we proposed that we take them out for dinner at the best Chinese restaurant in Perth. They were suitably impressed with the meal and, at the end of it, produced a couple of bottles of Chinese whiskey, called Moutai. They introduced us to the tradition of toasting your hosts and guests. I didn't much like the whiskey. It was white, about 53% alcohol, and really packed a punch. I noticed the Chairperson, Madame Zhang, didn't partake in the whiskey, but instead would toast with red wine. I thought perhaps this was a male-female thing, but I now don't think it was. Later, I used this ruse myself but, that first night, I was all about impressing the Chinese and hopefully doing a deal with them. Luckily, I only had a couple of Moutai toasts and then went home (since I was driving).

We told them that we planned to attend the China Mining Conference the next month, so we arranged to meet up there and discuss the project in more detail. This we duly did, but the conference was very busy, and I got the feeling that the Chinese company did not want its competitors to see them talking to us, so instead we met off-site. It was a quick meeting, since we were keen to get back to the conference and explore other opportunities, but no-one else seemed to be interested in us. That should have rung alarm bells for me but, being relatively naïve in dealing with the Chinese, it didn't. We were invited to return to China at a later date at our convenience, and to negotiate the details of a deal.

We flew back to China a few weeks later, to one of the more than twenty-plus Chinese cities of more than ten million people, and were picked up by a group of four of our hosts at the airport. One was to be our main contact. He was the 'manager' of the sub-company of the SOE that we would be doing the deal with. There was also his assistant/secretary, who would look after all of our logistical needs, a driver, and the last was introduced as the "party representative". I have to admit I was surprised that the Chinese Communist Party would need to send a representative to monitor and supervise the details of our deal, but it turned out he spoke perfect English (better than the others), so he was certainly handy to have around. It turned out he was permanently assigned to this SOE and was involved in all of their negotiations with foreign companies.

As we drove into the city from the airport, I noticed the large number of cranes on the skyline. There appeared to be building going on everywhere, and there were countless tall apartment buildings along the route, either in construction, or ready for occupation. We were accommodated in the SOE's own hotel in the centre of the city. It was quite luxurious, consisting of about 12 floors with maybe 50 rooms per floor, and seemed to have other people staying in it, but I don't think there was anyone else on our floor except the three of us. Each night we were entertained by their staff over lavish banquets. Luckily, I love Chinese food, but I soon developed a severe aversion to Chinese whiskey!

The first night, we arrived and were taken to their restaurant. The food was amazing! The best Chinese I've ever eaten. I was seated next to the Chairperson again, Madam Zhang. She was a little younger than I, but obviously very successful. Her English was OK, but not great, but was a lot better than my Chinese! I made the mistake of asking her if she had children, and if so, how many. "Just one, of course!", was her answer, me forgetting the Chinese government one child policy! It was difficult to make small talk with the language barrier, but the party representative was seated next to her, so could translate. We managed to make some sort of polite conversation for about an hour, and then the manager stood up, glass in hand.

The tradition that I referred to before at these dinners was about to start. You had to drink shots of Chinese whiskey while saying "Gambei" – Chinese for "cheers". This particular night, the brand called Moutai, that I'd had previously, was not offered. Instead, it was another brand which tasted quite sickly, but still packed the same wallop at 53% alcohol! Previously, over our dinner in Australia, we drank maybe two shots of Moutai, which was plenty for me. However, I was now told that if someone toasted me, I had to toast them back. Now there were twelve people from the company, and they all dutifully stood up in turn and toasted me, so I was expected to toast them back. Of course, only myself and the toaster had to drink, everyone else just sat by.

I have to admit that after only two or three of these toasts with this sickly tasting Chinese whiskey, I was just about ready to throw up, but my glass kept being filled up. A couple of times I managed to surreptitiously pour the whiskey onto the floor and make like I was toasting, but with an empty glass, but my sleight of hand was poorly developed. The other two Aussies in our party had wisely only partaken in a couple of toasts, and in fact one of them had said he didn't drink alcohol, so was toasting with Coca-Cola!

I would say that by the end of the evening I ended up having about 8-10 shots of this stuff as I toasted them back, plus a couple of beers and glasses of wine. I do remember somewhere during the evening asking Madam Zhang whether she liked Chinese whiskey and her saying "No, I wouldn't touch the stuff, as you can see, I toast with red wine". I should have realised this was a survival mechanism that I should have also used, but it was too late. By the end of dinner I was quite jolly (to say the least).

The alcohol hadn't quite hit me though, so when it was suggested we go out to a night club, I was all for it. Another few beers and then I started to feel it. Walking back to the hotel, I needed some support, and I poured myself into bed fairly quickly, with the room spinning and I knew I was in trouble.

I woke up about 3am and rushed to the toilet and did what Bazza McKenzie called a "technicolour yawn". This continued until I had nothing left in me and, exhausted I crawled back into bed. In the morning I woke up feeling like I had been run over by a tractor. I looked in the mirror. Both my eyes were bloodshot red, I had a black eye and a split lip. I vaguely recalled head butting the toilet bowl sometime during the night. I had vomited so violently that the blood vessels in both eyes had ruptured. I looked (and felt) terrible.

There was a knock at the door. It was the Chinese manager's assistant/secretary. "Are you coming down to breakfast? We've organised a sightseeing tour today up to the nearby mountains, and then tomorrow we'll start negotiations". Through the crack in the door that I'd opened I said, "I'm very sorry, I am not feeling very well. I might come down later". She scurried away.

Soon after, another knock on my door. It was one of my Australian colleagues. "How are feeling mate, a bit under the weather?"

"I'm not going anywhere today mate, except back to bed", I whispered. "I'll be fine, I just need 24 hours to get over this alcoholic poisoning. Just leave me alone for the day".

"They certainly did a number on you last might mate", he said, "I'm not feeling the best myself, and I probably drank a quarter of what you did".

"Enjoy the trip up the mountain", I said, and went back to bed.

I think I slept most of the day and night, except for a phone call home to my wife looking for sympathy (I got none). The next morning, I could barely face eating anything, but managed to keep some fried rice and coffee down. We negotiated for 12 hours that day! I was exhausted. My colleague decided that it had been their plan all along to weaken me for negotiations, and he was probably right. The terms that were offered were OK, but nothing special, but we felt we had to go home with something, so eventually a deal was agreed. By the end of the day I was feeling a little better, and because I hadn't eaten anything for 24 hours, I was actually starting to feel quite hungry.

"Great", said our Chinese host, "Let's go to dinner and celebrate while the MOU is drafted and afterwards, we'll have a signing ceremony with the Party officials".

"I'm not drinking any Moutai tonight", I said.

"OK, you drink coke", was his reply.

Why hadn't I thought of that two nights ago I asked myself? Or at least red wine. Twelve shots of that would only amount to a couple of glasses, and I would have been fine. Damn you Madam Zhang, for keeping that little secret to yourself until it was too late!

The signing ceremony was scheduled for 8.30pm. At 8pm, as we were just finishing dinner, one of the Chinese staff entered the room looking very serious and whispered in Madam Zhang's ear. She looked shocked and turned to me. "We have a problem", she said. "The Party officials won't approve our deal. We will all lose face if we don't sign since the press and a lot of city officials will be there".

"What's the problem"? I asked.

It turned out that we had negotiated a significant up-front payment (US$250,000) to ensure that the Chinese were serious and not just tying up our project for six months (the term of the MOU) while they did due diligence. They would be parting with some 'hurt money' to make sure they didn't just muck us around. "We can't make the up-front payment" said Madam Zhang. I knew then that the whole thing had been a charade and this was all part of their negotiating tactics. Their ploy was to agree to the payment and then withdraw that at the last minute when we were committed to the deal. "I'll have to phone my Board", I said crossly and walked out of the room.

We didn't really have much option. If we walked out, the deal – worth more than ten million dollars – would be off. It wasn't worth doing that for $0.25 million, so we reluctantly agreed. We signed, but I knew the deal wouldn't go far, and I was right. For more than five months I heard nothing from them so, a week before the expiry, I phoned the party representative at their office in China and spoke to him. "How's your due diligence going?" (knowing that they hadn't conducted any, since we hadn't heard from them at all). "The MOU expires in a week, do you need any more information"?

"I'll talk to Madam Zhang and ring you back", he said. Four hours later he called to say "we want to do a site visit tomorrow". I told him that was impossible because it would take me (and them) at least two days to get to the project area. "I'll have our Australian representative get in touch", he said. Unknown to me, the Chinese company had established an office in Adelaide to review projects, staffed with a Chinese-Australian manager (who spoke fluent English and Mandarin, and understood how Australian business worked), and a Chinese geologist (who spoke little English).

The next day, a very well-spoken Chinese guy rang up from Adelaide. "I can be in Darwin tomorrow", he said. So, I dropped everything and organised myself to get out to the project area. I phoned James (our Exploration Manager from Laos), who was now living in Brisbane and working on this project for us, and arranged to meet him in Darwin. I then phoned our investment consultant in Melbourne and arranged for him to fly up to Darwin as well.

The arrangements were that James and I would drive out to Borroloola, while our investment consultant would meet the Chinese company representatives in Darwin and fly out to McArthur River on the Air North flight. We would pick them up at the airport the next day and take them to visit our site. Because there was only a flight every third day, we would all then drive back to Darwin together.

James and I drove from Darwin out to the project area. We had decided to stay in the small town of Borroloola, located about 100km inland from the Gulf of Carpentaria and about 70km from the project site, since we had packed up our camp for the season. A closer place to stay would have been the Heartbreak Hotel, the roadhouse at Cape Crawford, a locality where the Tablelands and Carpentaria Highways intersected but, at short notice, it was booked out by grey nomads.

We managed to book the Chinese group into the Borroloola Guesthouse, but they got the last rooms, so James and I had to stay across the road at the less than salubrious Borroloola Hotel. No problem, we'd stayed there before. It had donga-style rooms, but they were airconditioned and comfortable enough. The Hotel had lost its liquor licence a few years earlier when the owner at the time had been caught selling sly grog illegally to the local indigenous community. If you stayed at the hotel, you could be served alcohol with your meal, but you couldn't buy take-away liquor. For that you had to go to the General Store, where they recorded your driver's licence details to ensure that you only bought your 'quota' (which was one carton of mid-strength beer per day).

We had dropped in at the General Store to buy a carton because you were never quite sure whether the hotel would have any beer. It was approaching sundown and we'd settled into our adjacent donga rooms. James and I were sitting on the verandah outside watching the sun go down when, to our surprise, John Doyle, a.k.a. Roy Slaven of Roy and HG fame, arrived that afternoon with a film crew. We joined them for a hilarious evening, which somewhat offset the impending Chinese disappointment.

Where's the mine?

Next day, we drove down to McArthur River to pick up our Chinese guests at the airport. They wouldn't have understood any of the evening and morning's hilarity, so we just kept that all to ourselves. We took them to the exploration site and proudly showed them the drill core demonstrating our discovery. They were interested and the geologist asked some good questions.

We had organised a bush lunch, laid out on a couple of trestle tables from our camp, with some billy tea (they loved their hong cha). Over lunch, they asked, "Where's the mine?"

"Oh, it's about 15 kms that way", I said, pointing north towards McArthur River.

"No, the mine we are buying from you", they said.

"There is no mine yet. This is a new discovery and is still at the exploration stage. It's all in the data we showed and sent you. Do you think we'd be selling you a developed zinc-lead mine for only ten million dollars?" (It would be worth a lot more than that).

"Oh. We'll report that back to head office then", was their long-faced reply.

We drove the 900km back to Darwin, largely in silence. The communication gap was very annoying, and I knew we had been stitched up by the Chinese. There never was going to be a deal, and this 'field visit' by them was largely a face-saving exercise on their part. On the way back we had to stop overnight at the roadhouse at Daly Waters, on the corner of the Stuart and Carpentaria Highways. There were dongas out the back, and a bar and restaurant at the roadhouse to eat. We had an hour or so before sundown, so I suggested we go for a drive up the road to the old Daly Waters settlement.

I had passed through the settlement a couple of times. About 7kms off the main highway, it was an interesting detour. There was a historic World War II airstrip with an old hangar/terminal. On the wall there is a map and timetable of a flight that once existed from Perth to Daly Waters in the 1930's. The schedule showed it took three and a half days! There were two overnight stops in WA and one in the NT before you arrived at the thriving metropolis of Daly Waters, population 20. I wondered to myself who on earth would actually have taken that flight!

The village of Daly Waters itself consisted of about six old corrugated-iron houses, a caravan park, and a pub. The pub had old-style petrol pumps out the front, and was decorated inside with all sorts of number plates from all around the world. The town was chokka that night (being a Saturday). The caravan park was full of grey nomads, and the beer garden at the pub was standing room only. There was a guy singing and playing country songs on a guitar, and the crowd was lustily singing along with him. It was quite a party! The Chinese were amazed at the little town and these goings-on, so we decided to stay for dinner. If you ever do get to visit Daly Waters on a Saturday night during the grey nomad season, it's not to be missed.

I never heard from the Chinese again after this, but heard around the traps that this Chinese SOE company had signed about ten different MOU's, each tying up a project for six months while they evaluated what would be the best deal for them. Meanwhile we were hamstrung and couldn't really do anything. The other issue was once an MOU was signed and announced, no other Chinese company would deal with us. They had effectively gotten a free and exclusive six-month option over our project.

Eventually I heard that they did a deal which went sour and retreated back to China. It was a classic example of them not really knowing anything about the Australian mining industry, but having deep pockets. Unfortunately, there have been very few successful Chinese-Australian mining deals that have worked out well (for either party) without lengthy legal wrangling. Just ask Clive Palmer. The cultural gap is just too wide.

Epilogue

A few years later I went back to China to try and negotiate a deal on another project our company had. I remembered Madam Zhang's ploy of only drinking red wine in toasts. It worked a treat for me this time and, apart from a nasty bout of flu, I was not otherwise health-afflicted during that trip. Alas, a deal did not eventuate from that effort either, despite another attempt at signing an MOU representing a free option over our project, which I politely refused!

11

In Search of the Golden Fleece
Georgia, 1994

Bob Besley

Bob Besley graduated with a BSc (Honours) in Economic Geology from the University of Adelaide in 1966. He commenced his career as an exploration geologist in South Australia, soon followed by a series of international appointments before returning to Australia in 1980. During this time he worked on managing exploration programmes, project evaluation and feasibility studies on a variety of commodities in a range of global settings, and worked for large international mining and oil and gas corporations based out of the US. He then transferred to Sydney to head up operations in Australia and the Pacific.

The next stage of his career was associated with the junior mining sector looking at Australian and global exploration opportunities. He managed the creation, listing and operation of two successful mining companies; CBH Resources Limited, which he led as Managing Director from a small exploration company to Australia's fourth largest zinc producer; and Australmin Holdings Limited, which brought into production a gold mine in WA and a mineral sands mine in NSW. He was also a director of other Australian mining companies.

More recently he was Chairman of Silver City Minerals guiding that company through detailed exploration for a repeat of the famous Broken Hill orebody. He is currently Chairman of Image Resources, a successful mineral sands miner in WA. Bob has served in a number of Government advisory roles including several years as Deputy Chairman of the NSW Minerals Council.

Gorgeous Georgia is often described as the Switzerland of Eastern Europe. It has towering mountain ranges, great rivers and lakes, ample rainfall and amazing agricultural potential. It was the source of the great wines consumed in the Soviet Union. In fact the wine industry of Georgia goes back to Neolithic times and has been recognised with a UNESCO world heritage listing. Some of the world's oldest continuously inhabited cities are in Georgia. There is evidence that man's first agriculturally-based settlements, marking the transition from the nomadic Stone Age to community living, was around the present Black Sea – before the Mediterranean broke through the Bosporus, swamping the fertile Black Sea plains with salt water, and creating Noah's Flood. Ancient stories and legends abound, including that of Jason and the Argonauts collecting gold from the rivers of the Caucasus Mountains in the Golden Fleece.

With the breakup of the Soviet Union in the late 1980s, Georgia gained independence as a nation in 1991. Not long after, an old football acquaintance from Adelaide, Hank Lindner, showed up at the office with his latest entrepreneurial adventure. Hank and brother Don were outstanding Australian Rules Football players for the North Adelaide Club. Hank was known on occasion to have stopped by his local pub before a game to strengthen his resolve. The Lindners ran a successful scrap metal business and Hank was the roving initiator of the business. On this occasion Hank had seen the opportunities afforded by the Soviet breakup and, having developed contacts in a number of the resultant newly independent countries, believed this was the time to gain access to a range of mineral deposits in countries such as Georgia, Armenia and Azerbaijan that had previously been totally under the control of the Soviet Union system. At Redfire Resources, we took up the challenge on the basis there was an early mover advantage – despite undoubted problems.

The flight from Frankfurt to Tbilisi was on an old Aeroflot plane rebadged Air Georgia. I was sure the front door was not closed properly but, after a noisy, cold flight we landed safely at Tbilisi Airport and I headed into the mayhem that was the customs shed. I do not know what it is but, in such a system, I often start near the front of the line and somehow finish at the back. However I was rescued by a stern-looking Georgian man who took me by the elbow and, waving a bunch of papers guided me through the system, and into a waiting car, a large black Volvo! Somehow my luggage was in the boot. This was the start of my Georgian adventure.

Hank had procured a house in Tbilisi, set up as an office, accommodation, dining facilities and bar. At this point it became obvious that the driver, interpreter and a couple of assistants all carried revolvers in their belts. It was explained that, to get anywhere in Georgia, we needed to align with one of the families that lie behind most activities in that country and have done so for a very long time. Hank 'had found just the right family' to assist us with introductions, provide transport and deliver security (hence guns in belts). So we headed off to meet the family, and pulled up at a classical Soviet-era drab concrete block of apartments. I was starting to think of an exit strategy at this point. As it turned out, wealth was not to be displayed in Georgia and, to my surprise, we entered a private lift that opened into a huge living area connected to a series of apartments that took up the whole floor.

Inside there was not much in the way of luxurious appointments not on display. What followed was an hilarious night of introductions, eating and drinking that was constantly interrupted by toasts to everyone by everyone. Each toast was accompanied by a scull of vodka. As the night wore on the speeches became louder, longer and delivered with increasing arm waving and gusto. Apparently among the former Soviet Union nations, Georgians were considered the greatest toastmasters of all time. It was all on display that night. From the few words I could pick up and a smattering of translation, it seemed Georgia and Australia were to become brother nations with Australia helping Georgia to prosperity through its mining investment and development expertise.

With a sense of welcome, the next day we headed off to the Ministry of Many Things (a number of Ministries were still being formed) where we were introduced to the Head of Geology. It turned out that, under the Soviet system, exploration and mining were centralised, but in totally different structures, such that all exploration and development had no connection to mining. Once a feasibility was completed it, was handed to Mining: Exploration walked out, and Mining walked in. Therefore, the incentive for Exploration was not to complete a feasibility study. The other interesting revelation was that, under the Soviets, Tbilisi was the centre for geology and exploration for Georgia, Armenia and Azerbaijan but, after independence, what went where was somewhat of a mystery and information was not very forthcoming. We gained some verbal insight into the projects that were high on the list for development and one that gained my attention was a copper-molybdenum porphyry deposit in Armenia, not far south of the Georgian boarder. Despite my persistent questions on reports and data, we walked out empty-handed. As is often the case in many parts of the world, data may be the only thing of value they have and, in the case of Georgia, it was doubly true as the geology department personnel had not been paid for a year or more. As we walked down the street in the early evening fading light, I noticed what looked like a map roll sticking out of a garbage bin on the side of the road. My curiosity got the upper hand and, on close inspection, there were a whole bunch of map rolls and folded plans in the bin that had what looked like assay data for trenches and tunnels. Not knowing what this all meant, we bundled the pile up and headed to the office. It turned out this was data from that copper-moly deposit in Armenia and I had to think this was left for us to find. But why?

The next step was a visit to the Minister for Titles. It seemed he had all manner of titles under his administration and, in the course of a most entertaining morning, we were offered vineyards, a half-built brewery, a synthetic ruby plant and a small nuclear submarine – but not much in the way of mineral titles. At this point the conversation drifted off in the direction of the great mountain sheep of the Caucasus Mountains and how prized they were as a trophy for an avid hunter such as the jovial Minister. The catch was it needed a helicopter to take the hunters to the highest part of the mountains where they were assured of a successful shoot and endless stories to follow. Here I managed to steer the conversation around to gold, laying out the expertise that Australian miners had for rapid and efficient development of gold deposits, and it was our aim to bring that expertise to Georgia should we be granted titles to explore and develop for it. I even relayed the story of Jason and the Argonauts and the Golden Fleece that likely derived from ancient gold production from the rivers draining the Caucasus Mountains, and that we could test that myth and explore for the source of the gold. However, the interests of the meeting kept returning to helicopters, guns and mountain sheep – with the implication that one bagged sheep was worth a gold title. At this point, I had to break the news that we were not in a position to fund a helicopter trip into the mountains. Reality was approaching fast; we were too early for any sort of mineral foray into Georgia, as the legal and financial structures necessary to support a mining investment did not yet exist.

The next day we were asked to again attend a meeting with the Minister at which he expressed his disappointment that we were not able to join him on a trip to the heart of the mountains but, then with a flourish of the arms, he produced a title document in Georgian, complete with red and gold seals. The literal translation was the grant to Redfire Resources of all the alluvial gold in Georgia to the bowels of the earth. What to do next?

It seemed with such a piece of paper in hand, we should at least search out the potential source of the Golden Fleece. The Caucasus were reported to have ancient workings for gold, silver, copper and iron, but no active mines in recent times were known. We loaded up the black Volvo and headed for the ancient city of Kutaisi in western Georgia, not far from the Black Sea. Accompanying me was our interpreter, driver and an add-on whose role I never quite understood; all with revolvers in their belts.

Kutaisi was the ancient capital city of the region of Colchis and is regarded as one of the oldest continuously inhabited cities in the world. It is situated on the Rioni River that drains the Caucasus, and is known locally as the source of the Golden Fleece legend. Indeed it is present on the coat of arms and flags of the region. From Kutaisi we drove up that river towards the mountains, passing multiple gravel terraces, being evidence of active uplift and large-scale erosion. We were the subject of much local interest as we stopped along the way to sample and pan for heavy minerals in active streams. Interestingly, pyrite (iron sulphide) showed up in some of the panned concentrates, thus indicating very rapid erosion and nearby sources. Normally sulphides oxidise on exposure to erosion and are not preserved in stream sediments. However not a speck of gold showed up in the pan. I was beginning to wonder if we needed a sheep's skin.

After another riotous night in Kutaisi fuelled by vodka, which I managed to avoid by sticking to beer, we headed back to the mountains to look for hard-rock examples of alteration and sulphides that are the tell-tail signs of mineralising systems capable of hosting gold and metal deposits. We traversed a winding mountain road that crossed a range of geological terrains but none of these signs were evident. We did however come across a poly pipe sticking out of the side of the road in the middle of nowhere from which flowed a beautifully refreshing sparkling mineral water, a welcome stop. As we climbed further into the mountains, we came to a pass that looked down on a green valley extending to the northwest. I had noticed the Georgians were getting restless and nervous and, by the time we reached the pass, they were definitely agitated. They would go no further. What lay ahead was the disputed territory of South Ossetia where hostilities had recently increased and the Russian backed 'Rebels' apparently had artillery. We had come to the limit of our foray into the Caucasus.

One of the best-known economic mineral deposits in Georgia is the copper-gold Bolnisi mine, in southern Georgia near the Armenian boarder. This needed refinancing and refurbishing, but had been an important producer for a long time. The drive from Tbilisi was through beautiful rolling hills and mountain meadows with extensive fields of wild red poppies. An introduction got us in the front gate and a tour of the facilities, classical Soviet-style massive concrete structures that were surrounded by an environmental mess. Geologically it was very interesting. The mineralisation appeared to be epithermal in style; with copper-gold hosted in breccia pipes, and with lead-zinc massive-sulphide lenses peripheral to the breccias. Mining had focused on the massive-sulphides, leaving the copper-gold breccias still outcropping, but extensively weathered, producing an oxidised cap. This was said to have interesting gold values and inspection revealed a large area of gossanous (oxidised sulphides) outcrops.

Back in Tblisi, I was able to have discussions with emerging banks and government officials; and it became even clearer that a framework for foreign investment was still being established, and it would take a few years of persistence and presence to participate in Georgia's emerging mining industry – with no guarantees of a workable outcome.

A few years later, an Australian ASX-listed explorer with the backing of one of the Georgian families undertook a drilling programme on the oxidised capping of the Bolnisi deposit and confirmed a significant oxidised gold deposit. The company became Bolnisi Gold and proceeded to feasibility study and production. Oxidised gold deposits are usually low-cost to develop, have low-cost mining by open pit methods, and usually have high recoveries of gold. In addition, the ore was suitable to the low-cost heap leaching recovery method. Successful operations commenced in 1997 at a capacity of 70,000 ounces of gold per year. The wheels fell off a few years later when a plane load of gold heading for Switzerland never arrived, disappearing in thin air. Bolnisi exited with grace and, as far as I know, the mine continued to produce gold.

Such was the evolution of mining in Georgia in the 1990's.

From Ian Tedder's "An Infidel in Saudi Arabia" (page 189)

A sketch of buildings in Jeddah, Saudi Arabia

12

Getaway in Siberia
Russia, early 2000s

John Garlick

John Garlick graduated as a geologist from London University in 1969 and began his career in Southern Africa, exploring for diamonds. After assignments in West Africa and Australia, he completed an MSc in mineral exploration in 1974 at the Royal School of Mines, Imperial College, London. Mining industry engagements followed in Iran and the UK.

In 1978, John relocated to Perth to manage the mining consultancy Mackay and Schnellmann. In this role, he has consulted to numerous companies on assignments throughout Australia, Africa, Asia and North America.

Early in the 2000s, a Western Australian summer had just begun when a short assignment in the Russian Far East suddenly came across my desk. The initial destination was Vladivostok – the famous port city and Pacific terminus of the Trans-Siberian Railway.

My contact in Vladivostok was Dmitry – a large, gregarious Russian who took charge of my unfolding itinerary with the help of his secretary Katina. Dmitry was also the dealer leveraging his and others' interests in various mineral properties in Siberia to potential overseas investors. Initial impressions pointed to an interesting assignment with plenty of interaction with him and his entourage.

My job was to assess the prospectivity of what was on offer. This tale, however, covers the incidental events which played out beyond the professional brief that make travel as an itinerant geologist so interesting.

The afternoon before I left the city of Vladivostok for site visits in the countryside of Primorsky Krai, I slipped out of my hotel for a stroll to get a sense of my surroundings. The weather was cold, as expected for a December day in Siberia, but the sun shined weakly in a clear sky. Most roads from my hotel led downhill, so I followed one towards

Golden Horn Bay and the extensive infrastructure of the port city. From a high embankment looking across a stretch of water, I was surprised to see that ice covered the near shore. It must have been thick enough to support the weight of anglers, for some had pierced it in order to try their hand at catching a meal. This reminded me how cold it was in the wintery sunshine, probably around minus 5-10 degrees centigrade, so I decided to return to the comfort of my hotel. On the way back, I stumbled upon a corner shop. Inside was a glass-covered counter displaying a large rather soulful fresh fish. There was no need for refrigeration as it was below freezing both inside and outside the 'deli'. What was curious about the place was the complete lack of anything else in the food line for sale, apart from cartons of some imported orange juice. Other items to entice the locals into parting with their hard-earned included stacks of vodka – perhaps to go with the orange juice – and numerous varieties of cigarettes. One brand was sardonically named 'Black Death' and others, designed perhaps to attract the young, included 'Kiss' and 'Sweet Dreams'. I left the place with the matches I needed to keep my pipe going and a sense that offbeat, bordering on black, humour was perhaps essential in Vladivostok to get through the average day.

In the evening, travel to the Russian countryside began with a train journey from the architecturally impressive Vladivostok railway station. A truly imperial building with beautiful ceiling mosaics. Not far away outside, a statue of Lenin, with his right arm outstretched to rally the masses, provided an interesting counterpoint.

The subsequent train ride over several hours was comfortable, and when I arrived at my destination, with a place name that I forget, it was still dark, cold of course, and snow was falling. Outside the station, there was a horse-drawn sleigh which, in the swirl of falling snow, looked like a scene from Doctor Zhivago. My ride however was less romantic, but warmer, for a taxi took me to a hotel for some rest.

Some hours later, the prospect of breakfast looked promising as I joined the field party at a restaurant outside the hotel. Dmitry and his team had arrived ahead of me from Vladivostok by road. We sat at tables decked out with chequered tablecloths, condiments and small bowls filled with peanuts. Refreshment arrived in the form of vodka all round and my hope for a more nutritious start to the day was quickly dispelled.

The subsequent meeting considered logistics and planning for the field trip whilst I otherwise consoled myself by consuming peanuts.

In the following days, we travelled to various exploration camps to investigate the mineral properties on offer. All were remote from settlements and in mountainous terrain covered by forest. Commonly, large spruce and other fir trees hung heavy with snow and looked picturesque, but the snow-covered landscape made travel rather slow. Being winter, only a few sites were occupied with any personnel, apart from site caretakers who stayed close to their potbelly stoves in their meagre surroundings.

During this field time, I learnt from my hosts how they had managed to navigate the economic difficulties which began a decade or so earlier following the fall of the Soviet Union. Many had continued to work without pay, with food and other essentials being supplied by barter or other arrangements with others. Some schemes included: installing grandparents in country cottages (dachas) to grow food for their city-dwelling family; wealthy private individuals paying civil servants some salary in exchange for information and/or starting a Russian version of private/public partnerships; and even clandestine alluvial gold mining over the Chinese border in neighbouring Heilongjiang Province – formerly part of Manchuria.

The return to Vladivostok was by car with wheels clad in chains to keep traction on the ice and snow-encrusted road. Fortunately, I didn't have to drive in the slippery road conditions, but being a passenger also has its disadvantages when your Russian is poor. The journey was slow, and traffic was light. At one stage, we stopped at an isolated railway goods wagon surprisingly parked like the 'Tardis' on the roadside. It turned out to have been converted into a small home and food stop on the forest fringe. Here we purchased barbequed wild boar prepared on a homemade flame-fired grill. Mine host was a muscular weather-beaten elderly man who looked as though he would be the victor in any tussle with a quarrelsome wandering brown bear.

Travelling through Vladivostok later in the evening was a bit confronting when we were stopped at a roadblock manned by a group of men who were not in uniform. Our driver was engaged in some discussion for a while but eventually we were free to move on following an exchange of roubles. I fancied this extortion was a levy to pass from one local unofficial jurisdiction to another. The perpetrators realising, no doubt, that a bit of coercion beats labouring all day at a regular job.

The following days passed engaged in research at some institutions and the like. On the last evening in Vladivostok, I was invited by Dmitry to a venue down in the harbour precinct which I will call 'Blue Ice' – an appropriate name as night temperatures fell to around minus 18C. I was not sure what to expect at Blue Ice but, on arrival in the lobby, I was greeted by four men wearing camouflage uniforms and carrying Kalashnikov submachine guns. Fortunately, they recognised my driver, Mikhail, and we were waved through. Inside the venue was a ground floor dancing space bordered by an open curved staircase leading upstairs to a dining area. Following the stairs, we found Dmitry and his secretary Katina already seated and we joined them for a farewell seafood meal with all the trimmings. Sea cucumber was one of the delicacies consumed as Dmitry impressed with his knowledge of its medicinal virtues. Later on, descending the stairway, I made out the camouflage-bedecked security men we had met on arrival, this time without their firearms. They were on the dance floor with young ladies who had come along for a night out. So by then, I guess, all was well on the security front!

My return flight was through Seoul and Hong Kong and I made it back to Perth early one morning along with rock samples for quarantine clearance. When the sun was up, the temperature climbed to the mid-30 degrees centigrade and I was back in a Perth summer. I stowed my snow boots and rock pick in the garden shed and headed for the office to make a start on my report: for the trip was over and it was suddenly become just another day at the office; that is, until the next getaway assignment.

13

Magadan on Mogadon
Russia, 1994

Michael Fellows

Michael Fellows obtained a BSc with Honours and a Master's degree in Geology from the University of Arizona. After 12 years in mineral exploration in the USA with ExxonMobil, he and family migrated to Queensland in 1983. There he was involved in gold and base metal exploration for several foreign and domestic mining companies as a consultant. He finished his career working for the Queensland Department of Main Roads. Now retired, he lives in Castlemaine, Victoria.

In September 1994 I joined a Geological Association of Canada mineral deposits excursion to the Magadan region of the Russian Far East. This was my first trip to the former Soviet Union. Something I had dreamt of for a long time. From my high school days, I was, what a friend calls, a 'Russophile'. It was undoubtedly the Russian literature, particularly Dostoyevsky's works, which influenced me. Although Russian was taught at the high school I attended, I was taking Latin then and only took up Russian at university. It was an interest and not a passion thus I never became proficient in the language. However, I maintained my interest in things Russian by following the Soviet geological literature and purchasing a three-volume work on Soviet ore deposits. I probably could have gone to the country earlier had I made a concerted effort. But I hadn't, and now I grabbed at the chance.

The timing of the trip coincided with a visit by my mother-in-law. I got on well with her, but wags spread the rumour that I was going to extreme lengths, to Siberia in fact, to avoid her.

Two weeks looking at gold and silver deposits in the Kolyma region of the Soviet Far East was an experience. It was enjoyable despite the obvious lack of creature comforts which Westerners expect. The people we met were very hospitable. They were generous with what little they had to share. And they were anxious for help. The trip left me with the impression that, although there were exploration opportunities, the most pressing need was not for more discoveries (i.e. exploration geologists) but for massive injections of capital and modern management techniques to develop the known resources. Most of the operating mines (?) we saw were woefully inefficient. The technology was outdated and work practices debilitating. It was no surprise that the economy had ground to a halt. Wherever we went, it seemed that everything man-made had been crumbling even before the USSR breakup in 1991.

Although the geology and mineral deposits were interesting it was the landscape, the people and their history that made a lasting impression. Unfortunately, in the Russian Far East, it was all too easy to notice and harp on the deficiencies. This was somewhat unfair to the people who remained there after the collapse of the Soviet Union. But no matter how much help came from 'outside', they were going to have to rebuild the country themselves.

The trip was to start from Anchorage, Alaska. At the time Alaska Airlines operated daily flights to Magadan and Vladivostok. I travelled from Queensland to Anchorage via Los Angeles and Seattle. On the leg from Seattle I sat with a Russian from Chechnya. He was on his way to visit friends in Anchorage, home to a large group of Russian émigrés. I'll call him Sergei, because I have forgotten his name. He was quite interested in the reason for my trip but seemed overly preoccupied with the possibility that I might contract a venereal disease during my visit. He warned me that I "...must use protection". I told him that wouldn't be an issue, but he repeated it three or four times. I didn't press him as to why he was so concerned. (Perhaps from experience?)

When we arrived in Anchorage he was met by his friends, and I was invited to accompany them. I had arranged to meet a cousin whom I hadn't seen for over thirty years but had time available, so I joined Sergei. His host was an American woman who had lived briefly with Sergei's family in Russia. By an odd coincidence, she knew my cousin and insisted we make a brief visit to the school where she taught before doing anything else. Later I met some of Sergei's Russian friends who were residing in Alaska. I pretty much stayed on the sidelines, testing my Russian comprehension without much success. That evening I had a wonderful Alaskan salmon dinner with my cousin and her family. The following day was free, as my flight to Magadan departed in the evening. Sergei and I, along with his host, visited a museum for a special exhibit of Russian icons. Later we took a drive along Cook Inlet. I recall thinking that, several days before I had flown out over Magnetic Island, named by Captain James Cook in 1770, and here I was in Cook Inlet, which he entered eight years later!

Sergei was friendly and enjoyable company but even having spent two days with him, for the life of me, I can't remember anything he said other than beware of the whores of Magadan. He gave me a letter to post when I reached Magadan which I did.

The flight from Anchorage to Magadan was full. I surmised that one third of the passengers were businesspeople seeking opportunities in Russia, a third Christian missionaries intent on reclaiming the soul of the nation, and the remainder tourists and returning Russians. (I could have fitted into each of these groups. I carried with me several Bibles in Russian which my wife asked me to distribute to anyone interested.)

The tourist potential of the country was enormous for those willing to endure the Spartan conditions which would take time to improve. One target group already identified were big game hunters who were keen on bagging bears in Kamchatka. (Two weeks later our return flight to Anchorage was delayed in Vladivostok awaiting the remains of an American hunter who died of a heart attack.)

The flight arrived in Magadan in the morning. On the tarmac were dozens of planes. It was obvious they had been sitting for a while and unlikely to be going anywhere soon. The terminal was clean but stark. A kaleidoscope of mismatched marble. Stern looking guards and customs staff, seemingly peeved at having to get up so early, shepherded us through the arrival process.

Once clear, our Russian organizers bundled us into two large heavy-duty transports. They looked to me like old American army 'deuce-and-a-halfs' with a yellow school bus in back. These, however, were much larger, ten-ton, six-wheel drive vehicles, also ex-army, fitted out for passengers. They were ideal for the trip ahead. There were 18 of us. Six Canadians, five Americans, one Japanese, me the sole

Australian, and three Russian geologists with their two aides or 'fixers' who handled logistics.

The leaders were Sergei Srtujkov a research geologist from Moscow and Mari Gorodinsky, head of the local geological survey (Sevvost-geolcom). Another Russian geologist, an employee of Kubaka (Omo-lon) gold mine, operated by Cyprus-AMAX 700 kilometres northeast of Magadan, also accompanied us. These three served as translators. None of the rest of us were competent in the language. Magadan was 40 kilometres south of the airport. We however headed north along the Golden Circle road which connected several mining towns of the former Kolyma Gulag. The road was also known as the Road of Bones, the bones being those of the thousands of prisoners who died building it.

The sealed road was in reasonable condition, but lined with truck tyre carcasses kilometre after kilometre. A recycler could make a killing. The hills were covered with larch, spruce and birch – somewhat stunted in stature compared to the lofty varieties I was used to in North America. The leaves were turning and the landscape picturesque, but the sky overcast and gloomy. After several hours, the drivers suddenly turned off the road, drove down an embankment and into a modest-sized stream. Wanting to impress us with the vehicles' capabilities, they drove upstream several hundred meters along the gravel- and cobble-choked channel to a grassy meadow where we had lunch. This was the first real opportunity we had to get acquainted with each other. We exchanged business cards and chatted about our expectations for the trip. The group, all male, was a mix of industry, government, and academic geologists. (Actually, only one of the latter, S. Ishihara of ilmenite / magnetite-series granitic rock fame.)

First night was at what appeared to be an abandoned palace in the middle of nowhere. Neoclassical buildings one might expect in St. Petersburg or Berlin, but not in the taiga. In Soviet times it was a spa where industrious works were rewarded with a respite from their labour by a dip in the natural hot springs. From the number of ruined buildings in the vicinity, I surmised that it had once been the site of a gulag, but the history of the place was not mentioned. I later learned that there was still a prison in the town. The palace has been refurbished since our visit and is still a popular stop for the few adventurous tourists who are keen to see the Russian Far East. Accommodations were comfortable and the food, like all meals on the trip, were wholesome. After dinner one of the Russian organisers pulled me aside and queried me about my business card. He suggested that I not hand any more out during the trip. He said that the Russian translation of my surname was a slang term for a pig's penis. I had made the translation myself and used a wrong vowel! (I still have a stack of unused cards.)

The next day we crossed into the Kolyma River drainage basin proper and there was evidence of past placer operations on most streams. A few small alluvial plants were operating, some reworking earlier tails, others new ground. We stopped briefly to examine one such operation. The plant was a simple affair; a small bulldozer was pushing gravel as far as 400 metres to a grizzly feeding the rocker and riffles. Those familiar with similar operations in Alaska and the Yukon noted the lack of backhoes and the inefficient riffles. Despite the shortcomings, we could see the gold and they were obviously making a living. One of the trip participants, a Canadian based in Magadan, was trying to manufacture and sell modern placer plants of his own design. He had sold a prototype which impressed the local miners, but he was having trouble setting up a plant in Magadan to manufacture more of them. Finding a trustworthy joint venture partner, a suitable facility, and qualified workers were proving difficult. The placers operators were apologetic that they didn't have more modern equipment. They mentioned that during World War II their fathers were amazed at the capability of the bulldozers – Caterpillars – imported from America for the war effort. They also remembered fondly the wartime SPAM from the US.

We left the Golden Circle and took a gravel road 200 kilometres northeast to Omsukchan, the site of the Dukat silver mine. It was a large mine with good grades. Impressive colloform-banded, epithermal quartz veins, stockworks and breccias. However, neither the mine nor the concentrator was operating. The plant at Omsukchan produced a concentrate which, during Soviet times, was sent to Kazakhstan for smelting but was now not allowed to leave Russia. For a brief time, exports were permitted but stopped when the smelter failed to pay the mine. At the time of our visit a hydrometallurgy plant was under construction, but it didn't look like there had been much activity on the site for many months.

The memorable moment here was going to the loo. Like everywhere else, no toilet paper of course. No toilet seat either; being wooden, it had long since been used for firewood. If you were left-handed or an Arab, you were SOL (for non-Americans, Shit Out of Luck!) because the toilet was too close to the wall and your left side was hard up against it. The rim of the toilet bowel protruded about a centimetre into the plaster of the wall.

The meals however were great. Simple but tasty. Potatoes (boiled or fried), cucumbers, onions, cabbage, sausage, freshly baked bread, both beli (white) and chorni (dark). Sometimes fish and, to impress the visitors, caviar. For breakfast, coffee, kasha and kefir.

The people were extremely welcoming; the local geologists gave each of us several polished specimens of beautiful colloform-banded quartz veins. In bygone years, geologists were a well respected and privileged group. Along the highway and in town were monuments to individual geologists who had made significant discoveries. There was also a special club house in Omsukchan exclusively for geologists. It was now closed. With the breakdown of the USSR, many geologists, particularly the older ones, had fallen on hard times. Those few still working had set up a fund to help their less fortunate comrades (oops!) colleagues. Another memorable sight were rows of abandoned multi-story apartment buildings. It seems that rather than build structures capable of withstanding the permafrost terrain, here the Soviets just replaced damaged buildings by erecting one exactly like it next to it.

The Bibles I had brought with me were welcomed by a cook at Omsukchan but not as much appreciated, I think, as were trinkets I'd handed out with images of North Queensland saltwater crocodiles. I gave out all that I had, but mine staff would approach me and ask if anymore were available. They were a big hit because it reminded them of the Soviet-era satirical journal *Krokodil*. This rag published cartoons and articles that lampooned the Soviet system but never went so far as to bring down condemnation. It was as if the Commissars admitted that the system needed a comic relief outlet for the inanities of the system.

We spent several days at Dukat and took an adventurous trip to a prospect called Lunnoye (Moonlight) 60 kilometres north of Omsukchan just on the edge of the Arctic Circle. The first 20 kilometres were on a gravel road, but the rest was on rough bush tracks and driving along stream beds. I was surprised upon arriving at the prospect, to see a camp where 50-75 people were at work. Cabins were fashioned out of rough, freshly sawn timber. After a recent snow, the ground was wet and muddy, the air cold, hovering just above zero. Inside the buildings it was warm and comfortable, heat coming from wood burning stoves. We shared a simple but again wholesome meal with the workers and then examined the prospect. Impressive quartz-adularia veins with classic epithermal vein textures and high silver grades. The deposit had been explored by several drifts in Soviet times. Permafrost extended a considerable distance in from the tunnel entrances. The prospect was controlled by a regional 'oligarch', Valerii Korsh, a former government official who had obtained several 'state assets' for a song. He accompanied us on the visit, dressed in camouflage gear and with a hunting knife strapped to his belt.

That evening back at Omsukchan, the vodka came out at dinner (as at all the places we visited) and there were the obligatory toasts from host and guests. Invited to speak, I saluted Valerii for his initiative and for providing jobs and 'looking after' his workers at the camp. It appeared to me and others that the venture was well organised and efficiently managed. I was later chided by one of the Americans for my comments. He had formed the opinion that the workers were there under duress and were ill-treated, and not free to leave. The latter was probably true, because without transport provided by the company the walk out (if one knew the way) would have taken several days and, in winter, would have been perilous. The truth may lie somewhere in between. Several days later we visited an alluvial/lode prospect owned by Valerii. Impressive coarse, angular gold nuggets in the placer workings were similar to the gold in high grade shoots in the underlying quartz veins and stockworks. At this mine there was a government agent continually on site to ensure no gold walked off. The workers lived in the nearby village of Yagodnoye. If there was any coercion it was probably economic: people desperate for work; entrepreneurs keen to maximise productivity; local government desperate for income. The big sugar teat in Moscow had shrivelled up.

At another small prospect Valerii was developing, miners lived in canvas tents. These were only temporary during the field season but could not have been comfortable then because of recent snow.

There were obvious investment opportunities in Omsukchan that needed deep pockets. And unfortunately, most potential local joint venture partners appeared to be pick-pockets.

The Dukat mine was eventually reopened and Lunnoye developed by the Russian company Polymetal. Lunnoye is now one of Russia's largest silver producers. A photo of the mine and plant in the company's recent annual report is a far different scene from the rough wooden 'lageri' we visited in 1994. Ever suspicious of outside help, I suppose it inevitable that the Russians would be the ones to develop these prospects on their terms. Whether or not an effective 'rule of law' has been established in the country, the indigenous capitalists have succeeded. I am reminded of a Russian proverb: "When one lives among wolves one must act like one to survive".

There was a sense of desolation throughout the Kolyma region, and not just because of the stark landscape. The minimal activity, the vacant apartment buildings, the abandoned villages and farmhouses were evidence of the thrombosis that gripped the entire country. Those with roots elsewhere had departed, but those for whom the Kolyma was home had nowhere to go. On our journey back to the Road of Bones we stopped at a roadhouse in an otherwise abandoned village. The cafe catered to the infrequent truck drivers travelling to Omsukchan. The woman proprietor was pleased to see so many customers. The hearty borscht with fresh black bread was one of the best meals I've ever had.

Back on the main road we followed a river down to the Kolyma. Mile after mile of dredge tailings on the river and every side stream. After crossing the wide Kolyma River, we made a side trip to Sinegorye the location of a dam and hydroelectric plant on the Kolyma. It was the region's source of power. We were given a tour of the dam and I was reminded of the opening scenes of the David Lean film 'Dr. Zhivago'. At that time, the hydro plant was operating at 40% of its rated capacity because of the reduced industrial activity in the region. That night we lodged in a multi-story hotel, virtually vacant except for our group. The town around the hotel appeared deserted. Again, a night of excellent food and vodka with our entertaining host Gamlet (aka Hamlet). (There is no "h" in the Russian alphabet.) A week later I was reminded of Gamlet when one of our group unearthed and displayed a skull on our visit to Butugychag, a former labour camp.

Resuming our way on the Golden Circle, we moved westward towards the town of Susuman, here travelling upstream along a tributary of the Kolyma. The heaped debris left by the dredges gave the appearance of giant ploughed fields, kilometres long. Near Susuman several dredges were still working. We visited one which was producing about 300 kilograms gold per annum. We were allowed into the gold room and shown two large platters filled with nuggets. Most were pea-sized but the largest the size of a walnut. The fine gold was thought not worth our attention. The pay streak was a metre thick and averaged 10 grams per cubic metre. At one operation, hot water was being injected to thaw frozen gravel to make it easier to dredge. We were told the story of gulag inmates who came upon the body of a mammoth, intact but frozen in the gravel they were excavating. Before scientists could be summoned to examine the animal, the prisoners had eaten the flesh and were boiling the bones for soup. Large resources of placer gold remain but the depth of overburden hinders exploitation.

The scene at Susuman was again that of desolation. Large concrete apartment buildings half occupied. An abandoned prison surrounded by barbed wire fences, wooden guard towers falling apart. The land along the river chewed up by the dredges. The stone-faced hotel manager insisting we give her our passports as was required in Soviet times. A portrait of Lenin on the side of the building with the words "The name and work of Lenin will live forever". I bet they will.

The largest and most impressive deposit visited was at Natalka. The village itself was a pleasant township. The hillside setting, architecture, and the mining methods reminded me of places I had visited in Chihuahua, Mexico. The classic company mining town. Smartly dressed women a marked contrast to the generally slovenly men. Pyjama-like track-suit pants with stripes down the legs seemed to be the pinnacle of fashion for them. Here the accommodation was simple but comfortable. At the end of the day the obligatory trip to the communal bath house. (Separate ones for men and women.) At the 'banya' I was asked to stand up, stark naked, save a can of beer in my hand, and converse in halting Russian with the miners. What does one say in such a situation?

The Natalka gold deposit was a network of quartz veins in shales and sandstone. 75% of the gold free milling. It was an underground operation, but the equipment and mining methods were from another era. Mining was cut-and-fill in narrow stopes. Cable-drawn slushers pulled the broken ore to draw points where it fell into small ore carts. These were hoisted in winzes to adit levels and drawn out of the mine by small electric locomotives. The mine was operating in a desultory fashion; the ore stockpiled, the recovery plant was closed. The deposit was said to contain open pittable reserves of 100 million tonnes at 2 g/t gold. The mine owners were anxious to attract foreign investment. The need for capital was obvious. Twenty years on, it is now owned by Polyus, a major Russian mining company, and the mine is one of the country's largest gold producers. The village of Natalka has largely disappeared, replaced by neat rows of 'dongas' with blue roofs, the iconic colour used on trim in Russian houses.

We visited several other small mines, all geologically interesting, some with good potential for expansion. All being worked with outdated methods and equipment, the workers living in conditions only a step or two above subsistence level.

Our last field visit was to Butugychag, a tin-uranium deposit mined in the Stalinist era, now abandoned. At the camp were the remnants of the mine buildings and prisoner barracks, crumbling rock walls with barred windows. In one ruin on a window ledge were two plastic flowers with red and pink blossoms. A story behind them, no doubt, that we will never hear but can well imagine. Next to one ruin, hundreds of disintegrating remnants of shoes and rubber boots. What few roofs remained on the rock walls were caved in or burned; the damage said to have happened during a prisoner uprising towards the end of the gulag era.

The tin mines were in granite at the summit of a peak several kilometres away from the camp. The high grade cassiterite veins in greisen had been mined by hand in pits or small open stopes. It was a lonely eerie place made even more so by the cold and several inches of snow. We didn't visit the uranium deposits because of radiation danger. At the camp one of the party found a skull with the top sawn off. (It was explained that, looking for evidence of radiation damage, autopsies were routinely performed on dead prisoners.) The thought did cross my mind that the skull was a prop left to impress visitors. Although it was forbidden to remove anything from the place, I stole the rubber sole of a boot and a 20-centimetre-long piece of barbed wire. The latter I subsequently made into a holder for an Amnesty International candle.

That night we stayed in a hotel in Ust Omchug. A blur but for the lonely statue of Lenin in the town square, his hand outstretched. "Where did I go wrong?" he seemed to be saying, as he surveyed the forlorn rundown buildings.

Our last two days were in Magadan. A pleasant, bustling town but obviously struggling to survive. One or two modern buildings but most the standard Soviet beige- and white-trimmed pseudo-classical buildings and the ever-present, soul-killing multi-story concrete apartment buildings. There were fishing vessels in the harbour but with the depopulation of the region and distance from other markets, there was little activity. It was unlikely to improve until transport revived so that seafood could be shipped to where there was a demand. The economic foundations of the town and region were based principally on mining and there would be no recovery until it resumed. There was little other economic base, beyond fishing, mining and tourism, on which the region could build. Any potential oil and gas found north of Sakhalin could be more easily serviced from existing facilities further south.

Images from a stroll through the city: shops with some western goods on sale but too expensive for most of the population; busy open-air market with fresh produce; a well-stocked bookstore with many customers; in the seedier part of town several men, with jars in hand, lined up along a wall with a single window over which a sign says, "Piva" (beer).

A visit to the geological museum was interesting but I'd seen enough rocks and minerals during the previous ten days. Of more interest to me was the museum's guest register. I was surprised to see in it the names of tourists from Townsville as well as the names of geologists I knew (one my former boss at Exxon). Plenty of mining companies had obviously been 'kicking tyres' in the area for several years.

Customs agents had been organised to meet us at the museum to inspect the samples we were taking with us. I felt guilty (afraid of being caught is more likely) about smuggling a battered tin cup I'd pinched from Butugychag, so I hid it behind a display at the museum. All our samples were inspected and sealed by the authorities.

A visit to the Gulag History Museum was more sobering. It is dedicated to telling the story of the labour camp system and the prisoners. I was surprised at how openly the exhibits talked about the system of repression. I thought it brave of the local authorities to acknowledge their dark history in this way. Of course, many of the region's older residents had been prisoners, and the younger ones were descendants of prisoners. They wanted to ensure that their tragic history was remembered. I wondered whether, in future years, will these people's descendants boast of their gulag ancestry in the way Australian First Fleeters do? The pain of the gulags still lingered in Magadan and I suspect that, if it does happen, it will take a very long time. I recently heard disturbing news that Vladimir Putin ordered some records of the gulags destroyed in an attempt to 'rehabilitate' Stalin.

For lunch on our last day, the group gathered at Mari Gorodinsky's dacha on the outskirts of Magadan on a cliff overlooking the Sea of Okhotsk. The view out to sea reinforced the feeling of being in a very lonely and forgotten part of the world. Our lunch was simple but satisfying. That night at our hotel a final banquet with plenty of caviar and vodka. I retired early, anxious to leave in the morning. I had enjoyed the trip, however it was emotionally draining to see the hardships the people endured compared to the luxuries I enjoyed and took for granted in Australia. I hope in the subsequent passage of twenty eight years the fortunes of the people have turned around.

During the night I was roused by shouting and screaming in the hallway and banging on doors. Was it the whores of Magadan Sergei had warned me about?

14

Romanian Rhapsody
Romania, 2013
Martin Spence

Martin Spence arrived in Australia in 1972 from South Africa with the vague idea of training in TV production as, by then the Americans had put a man on the moon, but South Africa was yet to get TV. He studied Film & Television in Sydney but never worked at it, nor went back to South Africa to work or live.

Venturing around Australia in the latter 1970s, he discovered Perth and has made this his home ever since. A degree in geology from the University of WA gave him a wonderfully exciting life of hard work and constant learning. Working in gold, nickel, other base metals, diamonds, uranium, platinoids, rare earths, lithium and the many various types of occurrences thereof.

He has worked throughout Australia, most countries in Southern Africa, parts of Eastern Europe and is currently working back in Australia.

He considers it has been, and continues to be, a wonderful life.

After working in Australia and southern Africa for most of my geological career, an assignment in Romania exposed a bewildered Aussie/Yaapie to snow, mountains and fresh rocks, in a glorious country.

In 2013 a quick trip was undertaken to suss out a 10g/t gold concentrate left in a large gold mining stockpile by someone and to get a feel for the mining potential of Romania. We were based in a town called Baia Mare located in the foothills of the Carpathian Mountains in the state of Transylvania. It was late winter with a daily temperature of minus 2C. Covered in snow yet sunny enough for our nostrils to get suntanned.

Our manager there was a woman engineer, our interpreter was a woman engineer, our driver was a bloke lawyer and the fieldie was a local geologist. Much was yet to recover after the Commos got kicked out, unemployment was awful, the place was tatty, and infrastructure showed signs of cheap and shoddy work.

The language's closest relative is Portuguese, of which I speak a bit, but I did not find a single word I could recognize; but most of the minus 40-year-olds speak some or good English, so go young. The Romanians are a delightful, generous people. Eager to please, eager to learn, can be a bit pig-headed about doing mining things their way, but that's always negotiable.

Aussies at that stage were not that welcome as the tailings dam of an Aus mining company had recently leaked into the Danube. Needless to say, as soon as word of this got out, various wastes from Hungary to Bulgaria were emptied into the river while the poor Aussie copped all the blame.

So off we went in the snow-covered glorious mountains, looking at fresh volcanic rocks with wonderment, most of the outcrop showed signs of significant sulphide alteration, quartz veining and various other attractive mineralisation indicators. An explorer's dream, particularly as there appeared to be very little sign of previous working away from the historic lines of mining along what would have been the near surface, more easily mined, high grade trends.

The ex-Communist mine declines were remarkable works of engineering for their size and single-minded lack of attention to the environment.

"Comrade, we need a road to join mine A to mine B, 20km away!"

"Ja Comrade, a 20km straight underground road will do you, Comrade?"

"Ja! Bugger ventilation."

Our only navigation in snow-capped, fir-covered mountains was by using mobile phone satellite GPS; thankfully the EU had installed world class telecommunications in the 90s.

Romania historically produced a fifth of the world's gold, the Romans were amongst those who exploited it and their workings are as fresh as the day the last slave died. Crawl through small tunnels into large gashes that follow the quartz veins. The scorch marks from the fires that heated the rocks before being quenched by freezing water are still visible with shards of quartz strewn all over from the fracturing. We suggested to a ministerial type that these could be of historic or touristic significance but that was looked upon as a silly idea.

After examining the recent underground mines, we would lunch at a ski run on the surface; Mike skied like a champ while I floundered around with my dickey knee. Ski slopes were good, I am told, and lift pass prices were ridiculously low.

Pork, sauces, delicious pork and more pork dominate the menu (they say pork is their favourite vegetable). But God it was tasty. Forget staying slim, at least five courses at each sitting and you would be rude to refuse any of them. The supermarket products are familiar although the language is a huge puzzle, the open meat freezers are dominated by you guessed it, but the bread is preservative free, tomatoes taste like they used to and fruit and veg are seasonal, thus local.

Most locals have their own shack where they grow vines and fruit trees, keep a pig and brew moonshine called palinka from seasonal fruit, mostly plum, which was deceptively tasty.

A downside was that the Stasi policing equivalent are still in existence, two very urbane men caught me innocently measuring the pile of gold concentrate and took a dim view of my activities. I was told I could buy what information I wanted from the government printers at $10 a page after submitting my request through them. We suspected our geo/fieldie was an informer, employed to keep an eye on us.

The EU, in its wisdom, had decided that supporting Romania financially was an expensive exercise and told them to open up their mining borders, which they did without much practical experience and even less mining legislation, so all parties were winging it.

We mostly focused on the economic potential of old workings, some of which are operating on a small scale today, but the lack of infrastructure was going to mean a lot of fencing wire and chewing gum to keep things operational. There was nothing in our bailiwick that showed sign of large-scale production potential.

A Canadian company had spent millions getting a base metal mine to approval stage when it was closed down due to farmers' concerns; so there were sovereign rights issues for us to watch.

The ground water contains a large amount of arsenic (plus probably lead, copper, zinc and God knows what else). We were surprised to see a flooded mine dewatering into the Danube, chokkas with the above: when we asked how this was being filtered out, we were told by the engineer, standing in ankle deep snow, "evaporation". Hmmm, that wasn't going to work. Our friends had built a large base metal refinery in the centre of town; it has an 800-foot-tall chimney stack which spewed enough caustic material to melt, by lunchtime, any nylon stockings being worn: amazing how little they cared.

Travelling through the country appeared safe, and staying at various lodges was a treat, the wooden churches are built with love and panache, and we visited a delightful graveyard that wasn't shy about the reputations of the dead. A gravestone showed a lady angel in a red bikini with the description "Sofia loved the men", others about Slobo loving the bottle and someone getting knocked off by the bloody Hungarians. Hungarians cop a bit of a bashing, probably due to one of the recurrent wars historically fought in this area,

Mention of Dracula gets a "yeah, yeah" look while Vlad the Impaler is a part of history that is wished the West wouldn't gloat over. The priests are all required to be married prior to taking office and the national dress worn on celebratory occasions is a remarkably gorgeous mix of historic west and east.

Bucharest, the capital, has a lot of historically lovely buildings following a Parisian style of architecture, some ugly commo concrete blocks, while the more recent buildings copy most modern cities. The taxi driver will gleefully show you where the last dictator was put against an arc de something and shot, I think his wife may have joined him.

The outlying towns we went through generally offered a fabulous historical blend of baroque, gothic and renaissance architecture, now housing KFP, Maccapig etc. The countryside is lush, the scenery magnificent, friendly restaurants serve excellent food and some acceptable wine often in gorgeous settings.

Two mid-tier base metal mines have been commissioned by the company we worked for since our visit, but this has nothing to do with us apart from our enthusiasm for the potential of the country.

I make no apologies about this reading like a sponsored travelogue – what could be better? A geological excursion in a glorious country!

Surprise, surprise, the Chinese got the gold concentrate.

THE MUSLIM WORLD

A hard walk home if there is a breakdown

A Yemeni wedding party

15

An Infidel in Saudi Arabia
Saudi Arabia, 1979-1980

Ian Tedder

Ian Tedder grew up in the Solomon Islands, with secondary schooling as a boarder in Sydney before graduating as a geologist from the New England University in 1974. Vacation geological work in the Solomons led to a job in South Australia exploring for copper before wanderlust took over and he left to travel and get jobs in Europe, eventually ending up in Saudi Arabia.

The following 30 years he has worked predominantly in Australia seeking sedimentary and volcanic hosted gold and copper, mineral sands and lead-zinc deposits. He is now retired after the last 15 years of his career based at Orange, NSW, as part of the exploration team that discovered the Cadia porphyry system which hosts one of Australia's largest resources of copper and gold.

Why Saudi Arabia?

At the time that I started to think about writing up some of my diaries in 2015, Saudi Arabia was in the news as it waged a bombing campaign against the Houthi rebels in Yemen. The news photos brought back a flood of memories of southern Saudi Arabia where I worked for 13 months from early 1979. The exploration area was centred on the town of Muhayil, considered part of Yemen by some.

So, what got me into Saudi Arabia?

Towards the end of 1978 I was in Europe, exploring for uranium in Bavaria. I was enjoying geology again after an 18-month break, pacing across beautiful green pastures, collecting soil samples in sometimes admittedly freezing weather and surrounded by pine and beech forests. Just before the short-term contract ran out, I was to receive a telephone call from my old boss in South Australia, Nigel, from whom I had not heard for almost two years. He was in Australia guiding Utah Development Company's exploration for copper in sedimentary rocks – known as stratiform copper deposits. He asked if I would like to work for Utah again but this time overseas in Saudi Arabia. I hesitated. Utah had previously started an exploration project in Saudi around 1976 but it only lasted about a year when the geologists decided the rocks they were looking at did not fit the bill. Another attempt was to be made though, this time further south, closer to the Yemen border. The way he induced me to join, was to indicate it would be a short campaign and that I would be able to continue travelling because of the lengthy 10 day leave breaks. I accepted and in December was back in Australia to do a four week 'crash' course in Arabic, in Canberra. Sounded good, though I knew languages were not my speciality and was soon able to confirm this!

So I re-joined Utah Development Company on the second of January 1979, after almost a two year break, and by the third I was ensconced in Toad Hall at the ANU, Canberra, ready to do my best at learning Arabic, run by a Lebanese lady with a wry sense of humour. It was not a success. On one occasion I attempted three times in the class to repeat after her a particular Arabic word and each time, I pronounced it differently to her. Eventually with a smile, she threw her hands in the air and said: "Oh well, have it your way". The class had a good laugh at my expense. The course was a Michigan University four-year course condensed into four weeks! All the words we were learning were things about "university, lecturer, rooms", etc, but nothing like: "water, water please!"; "why is it so hot?"; "where is the nearest town?"; or "is that a real dagger?"

Jeddah

By 4 February, I and a new colleague were on our way to Saudi Arabia via a three-day stopover in Bangkok in order to pick up the necessary entry visas. On the 7th we flew to Bahrain in the Arabian, or Persian, Gulf for an overnight stay before catching a Gulf Air 737 to Jeddah on the Red Sea coastline of Saudi Arabia. We arrived on a Friday, the Muslim day of rest, and were whisked off down some new freeway to a sun-blasted sandy cricket ground at '29 Palms' on a bleak flat Red Sea coast to watch our Chief Geologist play a game of cricket.

To set the scene: Saudi Arabia was then headed by Khalid bin Abdulaziz Al Saud who was using the massive revenue from the post-1973 oil shock crisis to fund rapid development of infrastructure, particularly roads and modern buildings, especially evident in Jeddah on arrival. New six-lane highways slashed through the edges of the tall old Jeddah buildings with their paint-peeled wooden lattice shuttered windows overlooking dusty side roads lined with abandoned 'Yank Tanks'. Along with the rapid material development though, unknown to us, disquiet was simmering away in places such as in Medina where Salafism, an offshoot of the Wahhabi movement promoting rejection of Western imperialism, was gaining adherents. The year-long Iranian revolution was ending, with the return of Ayatollah Ruhollah Khomeini after the Shah baled out into exile. Jimmy Carter was the President of the USA, Leonid Brezhnev ruled the USSR, and Australia had Malcolm Fraser guiding the way.

The first full day in Jeddah was spent in a rented villa that was to be the short-term Jeddah base. The villa had a high wall surrounding it, but was not part of an expatriate compound as was common for housing most Westerners. It was a two-storey concrete construction, had really large high-ceilinged rooms and cool marble-tiled floors everywhere. The view was out over dusty asphalted streets lined with similar villas up and down them. There were no house address numbers, so no delivery of mail by the Saudi Arabia postal service: you had to pick it up at the post office. Mail took two to four weeks to get to Jeddah from Australia. At Christmas time, so much mail comes in that the authorities feel overwhelmed by the avalanche of Christmas cards (and probably are), so they take the bags of mail outside and burn them. We also had to start working with two calendars – the Western one and the Arabic one. The year 1979 was actually 1399 on the Hegira calendar, at least till the 20th of November, when the Arabic year of 1400 started. This turned out to be a significant date.

The early days were spent preparing and acquiring equipment and supplies for field work. The field area was some 550 km southeast of Jeddah down the Red Sea coast, so we needed to be as self-sufficient as possible while in the field area for six to eight weeks at the time. Our 'work horses' were to be short-wheel-based Toyota LandCruisers. They had been used in the previous program so needed cleaning, tools bought and stored, and batteries fixed up. Lists of supplies and equipment required in the field were compiled and hand tools sorted out for setting up a camp etc. The large canvas accommodation tents arrived, and one was set up in the courtyard for a practice run.

Other jobs over the next week included designing plumbing and electrical supply circuits, then purchasing the required items, including a 3.5kVa generator. Bank accounts needed to be opened, Saudi driver's licences acquired, and medical check-ups conducted. Supplies could be obtained at the modern shopping centres and souks. All this required an amazing amount of time, with the language barrier not helping but, fortunately, many of the Saudis in the modern shops spoke some English.

Shopping hours were different to Australia in that they always had a big lunch break during the heat of the day, closing at 12 or 1 pm then not re-opening again until 4 in the afternoon. They stayed open to well after dark, especially in the souks, which was 9 pm or later.

The souks were everything you read about. Crooked narrow alleys overhung by tall old buildings lined both sides by what look like open car garages, roller doors and all. Packed into each one and spilling out onto the pavement, so there was barely room to pass, were the goods for sale, varying from the typical cheap Chinese trinkets, to toys, kitchen utensils, Persian rugs, clothes, leather goods and shoes, right through to electronics. A dusty spicy smell thickly permeated the air, along with the babble of customers and sellers hard at it – haggling. Music, as bootleg tapes (no DVDs or CDs in those days) of modern Western pop as well as traditional Arabic chants and dirges. Usually, different laneways specialised in a specific range of goods – so food down that lane, gold jewellery down this one. The gold jewellery and stacks of gold ingots made for glowing displays cascading out of the stores, even invading the pavement. We often found ourselves in the money changers' alley as we needed to change to or from Saudi riyals before and after our leave breaks. The usually grubby, greasy bank notes were piled in great stacks of various currencies and denominations held together by rubber bands, all around the squatting money changer who often used an abacus with practised skills to work out the amounts to be exchanged. There was apparently no security – who would steal money in a country that chops off your hands when you are caught!

There were religious police patrolling souks and streets. They were ensuring everyone was abiding by Sharia law, checking that women in public covered up correctly and were accompanied by a male guardian, that shops closed during prayer times, and that there was the appropriate segregation of sexes. Pairs of submachine gun-toting regular police could be found, usually more obvious outside modern supermarkets for some reason, which always made me more alert. One of our party used to do a bit of early morning jogging around the block to keep fit. A police truck carrying 'illegal immigrants' pulled up alongside and they asked to see his passport. All he had to show them was his wallet and, after a bit of animated discussion, each in their own languages, he just set off again and luckily was not pursued. There was a regular round-up and deportation of unregistered foreign workers.

Getting the Saudi driver's licence simply involved us showing our International and Australian state driver's licences to the motor registry people but that involved hours of waiting in various queues for multiple form filling, then handing these in to be stamped, signed etc. Visitor's visas were a bit more complicated. To quote from a memo to our new incoming administrator:

> At the present time since Utah Saudi Arabia Inc. is not yet registered the procedure for obtaining visitor's visa is as follows: proposed visitor sends information on full name, date and place of birth, citizenship, passport number, date and place of issue, validity, religion, and place coming from (where visa is to be issued) to the Jeddah office. The Jeddah office prepares a letter in Arabic and English addressed to Mr. Ab...l R... A... R... of the DGMR requesting issuance of a visitor's visa to Mr. 'X' and giving his full particulars. If hand-carried (as it should be in the interest of time) this letter is endorsed by Mr. Ab...l R... A... R... (or his deputy Mr. M... B...) and given to Mr. A...z B... The DGMR (B...'s office) prepares a letter to the foreign ministry requesting issuance of a visitor's visa. Utah office expediter then takes this letter to the foreign ministry (or B...'s office can do it, but it takes longer) where a telegram to the consulate is prepared...This telegram is then given to Utah's expediter (or DGMR representative) and returned to DGMR to be registered. The telegram is then taken ...

I think you have the picture.

We all were required to visit a doctor who did a check-up of our general health, took blood for tests then, later, X-rays in an old section of Jeddah with many crumbling, four storey traditional buildings. He outlined some pretty horrible things that could happen to you. For example, he warned us not to dip our hands into, or walk in any water (except sea water). The place was rife with Bilharzia (technically Schistosomiasis), which is a parasite that has a life cycle involving water snails. If you touch the 'fresh' water, the parasite penetrates your skin and plays havoc with your intestines, liver, bladder etc. We were also warned about rabies, which is in the dog population. Dogs are everywhere on the edges of villages as they are the sanitary workers after the locals have been to the toilet.

The doctor also made some recommendations about taking salt tablets that some felt might be necessary due to the extreme heat we would be facing in the field. We would be better off eating normally and letting the body balance the salt needs by retaining or expelling salts as required. Certainly, in the first week in the field, I noticed in the afternoons a salty crust left by evaporated perspiration built up on my brow. Yet after that first week, the body adjusted and I expired much less salt even though, through this time, my salt diet was not altered.

In the second week, I had a minor complaint that required an overnight stay in one of the brand new modern private hospitals in Jeddah. It was very clean, efficient and staffed by a multitude of well- trained nurses and doctors from all over the world. I gathered from the nurses, mostly from Singapore and Malaysia, that they had a restricted lifestyle in Jeddah – confined to compounds when not working and being bused to shops, all the while being carefully guarded. My work colleagues graciously paid me a visit to wish me well but seemed more interested in these lovely nurses than in my health!

We got a 'pep' talk from an American manager of the Australasian side of things aimed at inspiring us for the coming field work. However, we were not that impressed as he was responsible for completion of a proposed contract between Utah International and the Saudi Arabian government which would allow us to work legally in Saudi Arabia, but it was still outstanding. The agreement was to be the first between a private company and the Saudi government allowing a private company to explore in Saudi Arabia. The continued failure to complete this agreement was to cause us a lot of grief at police gates as we could not acquire our 'Ngama' (residential permit) and it also caused delays to us receiving our entry/exit visas. Our draft working contract required us to work seven days a week, for six weeks, then a paid airfare to the 'nearest civilised city' – which was to be Beirut – but, due to the civil war there, was changed to Athens. However, the schedules never worked out exactly as planned due to various problems with entry/exit visas.

The initial team that assembled in Jeddah consisted of one American (as administrator) who remained in Jeddah and was helped by a Saudi 'fixer'. There were two English – Dick, who was appointed the geological field manager due to his greater experience, and Chris, and two Australian geologists – Genesio and myself – with two to three years' geological experience each. The field support team, built up over a couple of weeks, were Somalis. There was Mohammed, dark skinned, tall and serious – the 'preacher': he became the spokesperson for the fieldies (field assistants). Ibrahim, slightly lighter-skinned and shorter, the oldest fieldy in the group who had seen it all before and was great because of his relaxed attitude. Yusef was the bright young man in the group, a quick learner, hard worker and didn't have any 'chips on his shoulder'. Musa was a dour fellow who was designated driver for the camp, making trips every week or so to Jeddah to get supplies. As such, it was hard to get him to do anything else around the camp. Even his fellow workers were unimpressed with him, saying he was spending too much time in Jeddah while they 'slaved' in the field. The cook, Abdullah, was a tall, darker skinned fellow – a nicer man you could not hope to meet. Relatively quiet, but with a quick smile and we never got sick from his cooking despite his operating in quite difficult conditions. The overall exploration responsibility was held by an Australian manager – Nigel, who was there at the start and who made two/three trips into the field from Australia in 1979. In the field, we all worked fairly independently in typical exploration style, with a team of two geologists and their support crew always in the field camp at any one time. Chris and I shared a similar timetable of scheduled leave breaks and field work stints.

Friday, 16 February, being equivalent to a Muslim Sunday, was a
day off. We abandoned field preparations for the day and took our new
field manager, Dick, on a three-hour drive to the highland city of Taif.
It took us through the more modern outer suburbs of Jeddah, which
consisted of newish boxy concrete block houses enclosed behind high
walls with crude iron gates and trims painted in bright hues. Then into
sandy dune country spiked by ragged pyramidal shaped hills, brown
hack-saw rocky ridges sparsely decorated with a dry scattering of dull
low vegetation, all veiled by shimmering heat and dust. We saw nothing of Mecca as the by-pass gives it wide berth. The drive up the Red
Sea escarpment itself is spectacular with views back over the hot, hazy
coastal plain sliced by dry wadis. At Taif we had rotisseried chicken
and chai (tea) for lunch and, while waiting for a tyre to be repaired,
we smoked a hookah (flavoured tobacco, not hashish!). The only thing
I noted in my diary was that the mosques were better looking, that
I counted around 50 apparently new lime green fire engines parked
outside a fire station and that there were a lot of armed police patrolling around.

Then back to the frustrations at Jeddah. One of my problems was
trying to get my suitcase (unaccompanied baggage) out of customs
at the airport. I was gradually getting used to driving around Jeddah
which, to the uninitiated in those days, was chaotic. Driving was on
the right-hand side of the road. Jeddah was crisscrossed by some very
fancy freeways and lots of quite well-maintained major roads with traffic lights, lane markings, etc, but very few road signs and no street signs.
However, Saudis tended to obey traffic signals in a lacklustre way. So,
going through red lights was not uncommon. A four-lane freeway with
one left hand turn lane would sometimes have three lanes of cars wanting to turn left – blocking through-going traffic. At intersections where
there were no lights, it could be horrendous with five-car-wide traffic
jams on a road marked with three lanes, everyone blowing their horns
and yelling Arabic at each other, arms waving out windows. You had to
be aggressive to get through these situations – pushing forward to make
your way through the morass of vehicles – otherwise you would not get
anywhere. I was a passenger with Chris driving one evening, returning from some errand, when we got caught in a traffic snarl. He was

in tune with the Arabic way of driving, but we ended up grinding our vehicle against some other car. The Saudi driver angrily hopped out of his car to abuse us and tried to snatch the ignition key from through the open window. However, Chris gave back as good as he got (in English at least) and, luckily, we were able to get away quickly as traffic cleared at that moment. It was a dangerous situation for us. We had already heard enough stories of the consequences of having a serious accident in Saudi Arabia. The injured are taken to hospital and the uninjured driver is put in jail until any dispute with regards to rights and wrongs is settled and compensation is paid. The company administration had an unwritten but strict policy that employees were not to stop at the scenes of accidents they may have witnessed.

The Saudis often overloaded their vehicles. For example, we saw one Hilux-style ute filled with a heap of sand/gravel collapsed at one Jeddah intersection, its rear wheels splayed out just like you see in cartoons. It was not uncommon to see camels crouched in these utes too, which looked far too small to carry such a large beast. Many owners appeared to have no interest in servicing vehicles. The Saudi people got a lot of oil money in the form of subsidies from the government and could buy vehicles very cheaply. However, when they ran out of oil or failed to have them serviced and they broke down, they were dumped in the city streets and a new one obtained. The problem of dumped vehicles became so great in Jeddah that an Australian company won the contract to clear the streets of Jeddah's abandoned cars – mostly those 'Yank Tanks'.

The Saudi government was making some effort to more fairly distribute the proceeds from its bounteous oil revenue. It did this in the rural areas by handing out subsidies to 'farmers' and shepherds. They received cash for every sheep and goat they kept and for each block of land they tilled. For example, in 1979 this amounted to 30 riyals per goat per year, 50 riyals per camel per year. This allowed some at least to buy new vehicles when their current one broke down. The trick of the better-off shepherds was to know when the government agent was coming to count one's sheep and, before the latter arrived, to race over to the neighbour and borrow his sheep for a few days to double one's income.

In the field

By the third week of February, with time slipping away, the idea of a 10 day orientation trip was cancelled and now we were simply going to go to the field area and set up camp near a village called Muhayil.

I probably should outline what we were doing in Saudi Arabia. It had been identified from literature that the Late Proterozoic (550 to 1,600 million years) metamorphosed sediments of the Ablah Graben in Saudi Arabia may be prospective for stratiform copper. That is, copper mineralisation confined to specific sedimentary layers, in fact, muddy-sandy and calcareous beds that had been deposited in very shallow, intertidal to subtidal zones. In order to explore for this type of copper deposit, a very systematic field logging process had been developed that involved traversing across the sedimentary units, identifying rock types, thickness of units, alteration, mineralisation and sedimentary structures that could aid interpretation of the palaeogeography and record the collection of rock samples for analysis. The process was actually based on some French geological work on similar copper deposits in what is now the Democratic Republic of Congo and in Zambia. The benefit of the rigorously defined logging process was that different geologists with varying skills would be obliged to follow the same steps, thereby hopefully minimising variance of observation and interpretation between traverses done by different people. We also cross checked each other's work occasionally.

In early March we were finally ready and left for our field area down south with a convoy of five four-wheel drive Toyotas packed to the gunnels with everything we thought we would need to build a tent camp, with supplies and spares to sustain ourselves for months. A truck with more equipment was to follow us down two days later. The route was out to the Mecca by-pass, through a police gate, then south down the coastal plain road to Al Lith, Al Qufudhah, then east up into the hills, circle past Jabal Thaban ('Jabal' – the Arabic word for mountain) then south again to Muhayil – some 550km from Jeddah.

The aim was to arrive, seek out suitable camp sites (preferably away from the locals so our infidel ways would not offend), then start field work. It took us all day to get within 100km of Muhayil and we camped under the stars that first night. Next morning, we rose before the sun was up, had a quick breakfast, packed and proceeded on gravel roads to Muhayil, a town of possibly 1,500-2,000 people on the northern bank of the large dry Wadi Tayyah (or Taiah). The town had a shambled layout consisting mostly of the standard concrete brick, block-shaped houses with flat roofs; a few of mud brick or stone – all very dusty; while wheeling overhead, hundreds of scavenging kites and vultures. Stone houses were more common in the surrounding rural areas. The local men, in four- wheel drives or wandering the irregular streets, were dressed in off-colour thobes and a variety of kuffiyehs (head scarves) which contrasted strongly with the Jeddah men who seemed to only dress in pure white thobes with red and white patterned kuffiyehs. Disconcertingly, many were armed with holstered handguns. Also, in contrast to Jeddah, women were walking the streets and running stalls. Depending on their tribe, they were often only partly veiled or, rarely, only had a head scarf, leaving the face uncovered, giving the place a more 'normal' atmosphere for us. They were in very brightly coloured, long, loose dresses. Fat-tailed sheep occasionally plodded the streets but had nothing to eat but empty paper cement bags.

After two days of searching the countryside around the town, we identified three possible sites, the best one being about three to four kilometres east out of Muhayil in a small wadi just off a flat plateau of basalt. Dick and I went into town together with Mohammed, who was to act as interpreter, to see the 'Mudira' – director of the local government. He gave permission for us to camp at our preferred site identified as the Dahro Saris area on Wadi Hammarda. We took his engineer out to look at the area and got his nod of approval. After delivering him back to town, we visited the local police station to tell them who we were and show them authorisation letters from the government. There was a police gate on the west side of Muhayil we would need to pass through every so often. We also investigated communication channels back to Jeddah but found no telephones in town. The only way to communicate with Jeddah was through telegrams via the local post office.

Then it was a matter of setting up the camp. The site chosen straddled a three to four-metre-deep, five to seven-metre-wide wadi that flowed west towards Muhayil. It had steep eroded banks of sandy loam and gravel. Three of the steel pole-framed tents were set up for accommodation, and one for an office, on the north side of the wadi; the other four tents, making up two more accommodation tents, the mess and the shower/bathroom, made up the bulk of the camp on the south side where the vehicle track from the main road came in. At the western end of the camp a deep pit toilet was dug, and hessian draped on poles around it to give us privacy. The Somali field assistants were given 'Pakistani' tents, more traditional square hessian type tents with a central wooden pole.

We all had assigned ourselves specific tasks – Dick was to set up the office tent, Chris the shower tent with hot water system – a classic 'donkey engine'– a 44-gallon drum heated by a wood fire. A couple of rainwater tanks were placed uphill of the camp and Musa, on the days he was not driving to and from Jeddah, was charged with keeping them topped up with water from a Muhayil township bore. Our drinking water was from bottles only.

I had been chosen, perhaps unwisely, to install the electrical system. All I had to do was follow minimal written instructions obtained when we purchased the 3.5 kVa generator. Set up involved siting the generator under some shelter and running dozens of wires from a switch board to various locations all over the camp. I had calculated the potential load of the fridge and freezer, the most critical components, so it should have been easy. However, when I first started the generator, threw the main switch and stood back, all we got was a weak 'brown out' light from the light bulbs. I hastily turned everything off, re-read the instructions, swapped what was apparently a critical wire in the switch board (they all looked the same) and tried again. We had light! Quite a relief. No one was going to drive back to Jeddah to get an electrician.

By the fourth day, with camp set-up essentially complete, we started doing reconnaissance trips to our field areas. The program was basically as follows: the area of interest – the Ablah Graben – consists of a package of strongly folded metamorphosed sediments called the Ablah Group, forming part of the slopes and foothills of the Red Sea escarpment. It stretches over 300 km north-south, but we were only to investigate the southern half. To the west lay the sandy coastal plain bordering the Red Sea. To the east, mountains rose to an elevation of over 2,900m (9,500 feet) to form the western edge of the Rub-al Khali or Empty Quarter.

The topography of the area we were to work had a relief of 400m to 1,500m, dominated by bare, sharp, jagged north-south ridges, generally aligned with the geological layers. Shallow dry gullies cut through the scree on the lower hillsides spilling their coarse sediment onto colluvial fans in the wide flat valleys. Blindingly white quartz scree covered older pediments, dotted sparsely with scatterings of flat-topped thorn bushes. Many of the colluvial flats were irregularly divided into small, tilled lots with stone wall boundaries, but rarely with any crops since they were dependant on rare, localised storm downfalls.

We divided the north-south running Ablah Graben into five areas and assigned a geologist to each one.

Work involved very early starts (6 – 6.30 am) in an attempt to beat the heat, with one geologist and one field assistant per vehicle driving out to their assigned field areas and mapping 1-2 kilometre straight line traverses east-west across the bedded sedimentary units using a tape to measure thickness of units. Air photos dating from 1953, which we had a lot of trouble purchasing in Jeddah due to 'security concerns', were used to aid navigation and location. Where we crossed prospective units, we would grab a semi-continuous line of rock chips over 10 metre intervals to make up a sample. On returning to the camp around 1-2 pm we would eat a quick lunch, drink more water and head to the office tent to plot up the day's work.

The heat in this area of Saudi Arabia was blitzing. Our field area was located 60-80 km inland from the coast, well away from any sea breezes, and around 15-20 km from the top edge of the Red Sea escarpment. Every day by mid-afternoon in summer, as the heat built up, the hot humid air drawing up moisture from the coast rose up against the escarpment and formed huge storm clouds. Most of the time we got no rain, maybe a few dusty wind gusts, but occasionally lightning flashed across the darkened sky and thunder echoed between the ridges and it would pour down on our camp, flooding our little wadi and our tents. The rain usually lasted less than 15 minutes. We had a thermometer to track the temperatures. Most summer days oscillated between 40°C (105° F) and 45°C (113° F) with exceptionally high humidity. One day we had 49°C (120°F), which was driven by an unusual easterly desert wind plunging off the escarpment, but it did not feel as bad to me as on other days due to its dryness. Mind you, the two guys out in the field that day returned by 11 am complaining of feeling ill without realising it was the intensity of the heat causing them problems.

A pattern of field work was soon settled into: up at first light to have a quick breakfast, then out to the field which may be a 30 to 60 minute drive away; lay out the tape, start noting the rock types, thickness of each bed, sedimentary structures, alteration and any mineralisation. The field assistant would help with the tape measure, sampling, carrying samples back to the vehicle and explaining to any local what we were doing. By mid-morning it would be very hot. Rocks became too hot to pick up for more than a brief few seconds then we would drop them like the proverbial hot cakes. We carried a minimum of two 1½ litres bottles of water each which we had no problem consuming while out. On one day after coming back from the field for lunch, I drank two 1½ litre bottles of water before starting to eat and then had another ½ litre with the food. This was after drinking the normal amount out in the field. In the office tent after lunch, we would sit at our work desks with several layers of cloth sample bags under our forearms to absorb the sweat that trickled down our arms, threatening to sog up the maps, cross sections and reports we were writing. On the hottest day in mid-June, as the temperature soared to 49-50°C, Chris took a glass cordial bottle from the office tent to refill with cold water from the fridge. It shattered into pieces on contact with the cold water. There were some incidents of heat stroke in the crew when we went on 'fly' camps down south as we put in extra hours in the field to shorten the trip away.

At 5pm, the official workday was finished although we usually completed whatever we were doing, no matter what the time was, then across to the mess tent, grabbed a can of Schlitz non-alcoholic beer from the fridge, and sat and chatted in easy chairs in the shade outside the mess. The ritual of having a non-alcoholic beer became addictive. We felt bereft and upset if we had drunk all the 'beer' and no more supplies were coming in for another day or so! We followed this pattern of work every day, seven days a week for six weeks, before we got a field break. Occasionally the routine was broken by a cold/flu or minor injury, so you got a couple of days in the 'office'.

While we seemed to be ensconced securely enough in our camp

there were some concerns, the principal one being that we were working in Saudi Arabia without an Ngama, the residential permit, which was held up by the failure of Utah, and possibly the Saudi administration, to register the company, Utah Saudi Arabia Inc. We each carried a letter, written in Arabic, which explained we had permission from the Saudi government to be doing what we were doing. However, if we had to pass through Muhayil on our way to the field area, the police at the gates would want to see our Ngama. Luckily, they did not give us any problems as they saw us regularly, but there was always the potential for a new man to be on the gate. Another significant issue was finalising our Employment Contracts, particularly the section dealing with taxation. Genesio and I, being Australian, were taxed both in Australia (whence our pay originated) and in Saudi Arabia. The two English geologists were only taxed by Saudi Arabia (at 5%!). The American manager in San Francisco responsible for progressing the legal papers admitted that his in-tray was filled with three years' worth of telegrams and memos pushing for settlement of the contracts, but he still had not made progress. I submitted my resignation letter at the end of June because of these issues but agreed to work through till the end of the year. The letter never took effect anyway.

Some of the work areas were three to four hour's drive away, so it was more sensible to assemble a larger team, pack the vehicles and spend two to three nights on a 'fly' camp so that the geological traverses could be completed more quickly. One such camp was completed in early April in the Ad Darb area – the southernmost extension of the area we covered and populated by the rather fierce Tihama people – known for wearing garlands of flowers on their heads but beholden to no-one. We drove west to the Jazan-Jeddah road, then south to Ad Darb, in approximately two hours on good sealed roads. Then more than another hour up gravel tracks alongside and in Wadi Rim (Reem) to start our traversing. On these fly camps we worked all day to maximise our time in the field. Fly camps were usually set up on the banks of a wadi, under the stars, near a stagnant pool of water. Consequently, mosquitoes could keep us awake all night. If it wasn't the mosquitoes, it was the half-wild dogs barking all night!

Luckily there was not much else to worry about – except if Chris was with us. He seemed to attract whatever venomous wildlife was about. For example, I was rudely awakened one night by Chris thrashing around while half-wrapped in a mosquito net, trying to levitate himself from his camp stretcher into the adjacent Toyota without touching the ground. My torch light soon revealed that it was a snake under Chris's bed that was the root cause of our disrupted sleep. I used a shovel to lift the snake and carry it off into some nearby scrub. The snake may have been a saw-scale viper as Chris was initially alerted to its presence by the rattling sound it made.

Scorpions seemed to keep a low profile generally, but once again Chris proved that they existed. He decided, one quieter day, that it was a shame to waste the new bright yellow motocross motorbikes that were standing unused in the store tent – so he rummaged around the tent in the dusty boxes for a gallon can of petrol to ready a bike for a hill climb straight up slope from the camp to a nearby ridge crest.

To quote from Chris: "As I took a new can out of a box, my little finger was assaulted by a fury of repeated pinpricks and there, hanging on to my finger, was a tiny transparent pale green scorpion. After I shook it off on to the ground and rearranged it with the help of a nearby geological hammer, I decided to continue my mission. However, this was delayed by the fact that slowly over the next half an hour the glands in my armpit swelled up until my elbow was held up at shoulder height.

The one really disconcerting bit of wildlife was the 'camel spider' (Solifugae). They are buff-coloured hairy creatures, commonly up to the size of your splayed-out hand with ten legs, elongate body and two closely spaced eyes above a set of massive jaws. Eight legs are used for running and the extra two at the front pound any prey (such as a moth) into submission before ramming it back into the fearsome looking mandibles. These desert dwellers sometimes caused consternation at the evening meal as they used our exposed legs as a pathway to rush up onto the table to grab a moth which would then be devoured in front of us. Luckily, they generally ignored us, and we tried to stay well clear of them! Except Chris. He tells the story of one morning heading off for his shower, grabbing his towel off the outside frame of the tent as he went, throwing it over his shoulder, then feeling a nasty scratchy feeling at his neck. To his shock and horror, he realised it was a camel spider.

The 'Solifugae' found itself being launched into near orbit as Chris reacted to its presence.

Out in the field area during the day we saw few people except the occasional shepherd with a camel in tow tending goats or fat-tailed sheep. There also were isolated Bedu encampments with one or two dark brown tents here or there on the flatter pediments of quartz scree, with the owners tending goats and camels on the nearest relatively green pickings. They still had a wandering lifestyle, moving in the wake of summer storms that triggered spurts of green growth. However, these days, moves were often undertaken using Mercedes Benz trucks and four-wheel drives.

On 15 April it was time for me to take my first leave break, having worked continuously in the field since 8 March. Four of us in two vehicles left Muhayil heading south first to Ad Darb, then up the escarpment road to Abha. The road up to the tableland was unsealed and less spectacular than the Jeddah-Taif road. Abha itself was a large 'city' but we were still unable to telephone through to Jeddah. The first section north along the top of the escarpment, at around 2,200m (~7,000 feet) altitude, was very hilly and windy, the road flanked by bare, barren rocks and the odd 19th century mud-brick Ottoman guard tower. Other areas were quite green because of stunted six-foot native pines. The intermontane basins tended to be tilled with tiers of small, terraced fields though we rarely saw much obvious in the way of crops. We also drove through patches of granitic terrane with high bouldery mountains. The final part of the road on the first day of travel was towards Bil Jurisi, and there was near-continuous ribbon development of small towns. We had a very cold night camped under the stars just north of Bil Jurisi in a small valley off the road. The next day, after an early start, we drove down a long valley into an intermontane basin – very broad and still some one to two thousand metres elevation, on to Taif and then down the very spectacular scarp to reach Jeddah.

On our occasional drives to and from Jeddah, we saw some horrendous remains of fatal and near fatal accidents and had a couple of incidents that involved us. The worst case I experienced was as I was driving back to Jeddah in the middle of June for my second field break with Chris and Mohammed crammed in the front seats with me. We were approaching the Mecca bypass, which has a sweeping right-hand bend in a cutting. I noted a vehicle approaching but I had not realised from my perspective that it was on the wrong side of the road. A shout of "Look out! Look out!" from the others coincided with my sudden realisation of a potential head-on so, with a sharp jerk of the steering wheel, I took us onto the wrong side of the road to narrowly avoid the idiot that was coming at speed, straight at us. He shot past – still on the wrong side – and disappeared down the road. I had to hand over the driving, my nerves shot.

The shaky feeling was then compounded by the very aggravating news when we got to Jeddah that we would not be able to fly out (in my case to Greece) as the company still had not finalised the work contracts with the Saudi government, so we couldn't get exit-entry visas. What's more, we were required to turn around and head straight back to camp for another six weeks before our next break. We were "very stroppy" to quote from my diary.

Arabian hospitality

Back at camp, Chris and I had to continue our routine of traversing across various parts of the countryside accompanied by our loyal field assistants. We started doing more fly-camps to reach those places further away from our base camp as the long drives to the start of various traverses were eating into our workdays.

In early June, Chris, Mohammed, Yusuf and I did a fly-camp in the Jabal Sawdah area, which was 40 km south of Muhayil ('as the crow flies'), and accessed the target area through Wadi Rim from Ad Darb. On the 8th, working on traverse JSR08, Mohammed and I got invited to two lunches, two breakfasts and at least three morning teas by locals who were intrigued with what we were doing! It was unusual to get any invitations as the Arabs tended to stay out of our way, but they appeared to be more curious south of Muhayil. It is impolite to refuse invitations, but we could only accept one of the lunches and two of the morning teas. The formality of having a chai or coffee is quite rigorously followed. On arrival at the agreed meeting place, often a nearby Bedu tent with a couple of low platforms slung with rope netting, we greet each other in the appropriate manner; "As-salāmu ʿalaykum" (peace be upon you) then the response; "Wa ʿalaykumu salam" (upon you be peace). Then I'm obliged to let the Somali field assistant carry on conversing as my Arabic was totally inadequate. I'm assuming the conversation covers why we are here and what we are doing but might, for all I know, include questions about why are we so dumb as to be working in the hot sun; why does this Christian wear shorts; what is he writing in that notebook; is he a government spy, etc? Women were nowhere in sight – only a couple of men dressed in rather shabby, dirty thobes, dagger in the cloth belt and wearing close-fitting cloth caps. Cardamom-spiced, watery coffee may be offered, commonly straight out of a large vacuum flask, or very sweet strong chai in small glasses. Only after drinking a minimum of three of these can we excuse ourselves and get back to work.

The one lunch invitation I received came about unconventionally. We were in the Jabal Sawdah area, walking up Wadi Rim, a wide gravelly dry riverbed with well exposed rock outcrops under scattered thorn bushes in cliff banks, when I noticed we were being followed by a local some 100 m back. I asked Mohammed why this guy was doing so. He said that the man was making sure that we would come to the lunch, having accepted the invitation. This came as news to me but then I obviously was not aware of an apparently significant exchange three hours earlier. When we finished the traverse, we walked back to a small cluster of stone houses scattered over the high banks in a bend of the wadi, with some shade trees. We were first invited to sit on a large plastic prayer mat under a tree in the wadi and we discussed our work with some of the men while a cassette tape of Arabic music played in the background. They consisted of an extended family group living in the same valley who normally met on a weekly basis for a shared goat dinner. The goat was supplied by a different family each week. Then after washing of our hands we had prayers – with me sitting quietly in the background. After a long wait, a watery goat soup, tasting a little like a chicken broth, was handed around. Finally, we were called up to a dark stone house with a narrow, arched entrance which we had to duck down under to get inside. It was a single dark room with a wood and mud ceiling and an earth floor covered with a grass fibre mat. Small square windows barely let in enough light to see that there were several men sitting cross-legged in a circle on the mat, with their backs against the walls, and I was invited to sit with them. In the centre of the room was placed a huge enamel plate, almost a metre across, piled high with white rice and, on top of that, large chunks of chopped up boiled goat meat including skull (and somewhere in there an eyeball or two?). Buzzing around the dish were dozens of bluebottle flies. Everyone dug into the meal using their right hand to scoop up finger loads of rice and to select chunks of meat. I, being a guest, was offered select pieces identified by one or two others as delectable and was required to eat so as to not cause ill feeling. I admit eating a minimal amount, digging down in the rice to try and find a non-flyblown morsel. After finishing we washed our hands again with washing powder serving as soap then, because we were guests, we were liberally sprayed with cologne! The villagers were very friendly.

After 2 pm we rejoined Chris, who had completed his traverse and had missed out on lunch, then endured the long drive up Wadi Rim, over the divide into the Jabal Sawdah north area, and back to Muhayil. The road was steep and rough in places but spectacular.

On 2 July, while working about 20km south of Muhayil, Yusuf and I were invited by two attractive young Bedu ladies into the shade, under an awning of a tent for chai. They were in typical loose ankle-length long black dresses with bright splashes of colour and no face covering. And no men around either – most unusual. We did not go into the tent of course which would have been 'haram' (forbidden) – just the alcove where raised stretchers (a cross between a bed and a lounge) provided seating for us as we drank the offered 'tea' and Yusuf made polite conversation. We were treading on dangerous ground according to Yusuf later. If the men had returned while we were there or had found out later, there may have been repercussions.

We could not have done any field work in Saudi Arabia without the Somali field crew. They were an interesting bunch of guys, quite well educated, particularly in languages. All our crew could speak quite good English, Arabic and their own dialect of Somali. A couple of them could speak bits of other languages as well – in particular, Yusuf could speak some French and German! Yusuf had been to university. The unfortunate thing is that this education they received could not be used; here they were working as untrained labour in a foreign country while their own was falling apart under President Siad Barré.

What I did not fully appreciate until doing more reading in later years, is the strength of their clan system and the occasional animosity between different clan groups which, in retrospect, seemed to have been reflected even in our Somali crew where a couple made rather disparaging comments of another.

Somali families back home are often dependant on remittances sent by those in work. The practice of sharing everything they owned with all in their extended family meant that, at times, some of the guys did not want to return home at field breaks knowing that they would be obliged to give any savings to their extended families. This social system can lead to Westerners misinterpreting their requests for more money or material items such as clothing as just being greedy – when in fact they are being generous with what they have.

There was incredibly rapid progress on road infrastructure down the coastal plain and through the Asir mountains. At the beginning of our project in March 1979, it took us 1½ days to drive down to Muhayil with the last few hundred kilometres on gravel roads. By February 1980 the roads were sealed all the way from Jeddah to Muhayil, so it only took 7 or 8 hours. These roads were being built at a cracking pace by Chinese and Korean companies. The Chinese construction camp for upgrading a road running east of Muhayil and up the escarpment was sited on the Wadi Tayyah some 10-15 km east of Muhayil. We were invited one evening to the camp to have a meal and see a film which had been re-named from 'The Prince and The Pauper' to 'Crossed Swords'. I recall meeting a Palestinian guy there and we entered into a lively discussion about the Middle East politics.

The drive down the coastal strip was hot, as the Toyotas did not have air-conditioning and engine heat easily penetrated the fire wall into the unlined footwell adding to the blast furnace feeling. Only a couple of brief rest stops were made, at Al Lith and or Al Qunfudah, to try a bit of spit-roasted chicken or deep-fried fish and chai from roadside stalls. The most common form of animal life seen on our drives seemed to be donkeys which were used as the classic beasts of burden, carrying people or goods around. Donkeys and camels wandered around along roadsides as if ownerless but no doubt were being watched, even if casually. They occasionally became victims of speeding, high-sided Mercedes Benz trucks – particularly on long concrete bridges crossing wadis which did not give the donkeys many escape options. The scene of such tragedies quickly attracted an accumulation of vultures, kites and crows that soon cleared the roads. I saw baboons once, up on some rocky ridge, and snapped a blurry photo of them. The camels looked generally tatty but it's not surprising considering what they had to eat. I was obliged to watch and listen to a couple crunching on the thorn tree outside my tent one afternoon. It sounded like they were chewing up cornflakes. The thorns were in the order of 3-5 cm long.

Safety on the road was our biggest concern. The Saudi's sometimes drove at great speed on the country roads, apparently with a prayer of "Insh Allah" ("if Allah wills it" – or "Allah wills it") but the consequences of the accidents were horrifying. On one of the early drives down to Muhayil, we saw a large American sedan speared right through by a safety barrier. On another we were passed at great speed by a four-wheel drive vehicle and, when it was about a kilometre ahead of us, it suddenly veered for no apparent reason off into the desert sand dunes raising great clouds of dust and sand. As we passed by half a minute later, a couple of uninjured men were stepping out of their vehicle – still upright despite the nature of the terrain the vehicle had just ploughed through. Another time, we came across a fuel tanker which only half or an hour earlier had literally run over a sedan – the car must have rolled several times under the truck as it was mashed metal – and which was covered with some pieces of cloth to hide the unfortunate occupants. The worst case was seen by Genesio while heading to Jeddah for a leave break, driving through a sandstorm on the coast road. He had low visibility due to drifting sand when, suddenly, in the middle of the road appeared the body of a dead youth. The accident apparently had only just happened a short time before. Our strict instructions on witnessing any sort of accident, or remains of one, was to keep driving – which he did.

Generally, the locals were stand-offish, but occasionally welcoming. Only once in Muhayil was I spat at as I walked down the dusty street: we certainly knew our place as 'infidels' and ajnabin (foreigners).

One day Yusuf and I experienced a serious episode, some distance south of Muhayil. We were accosted by a small group of villagers as we were walking to the start of a new traverse. Yusuf said that we had to sit down in the shade of a tree while he explained in Arabic what we were doing. After 15 minutes or so of discussion between the locals and Yusuf, he turned to me and asked for the letter from the Saudi government which we always carried with us everywhere. It was basically a 'to whom it may concern' in Arabic and explained that we had government permission to walk around the place. Unfortunately, the men were illiterate and so we then had to sit there under this tree while one fellow went off and fetched someone who could read. I was wanting to push on while waiting for this guy but was told in clear terms by Yusuf that the villagers, whose numbers had now built up to around 30 or 40, totally surrounded us and would not allow us to go anywhere. The situation did not feel particularly threatening at the time, no weapons were drawn, but my relaxed attitude was probably aided by the fact that I could not understand whatever they may have been saying about us. After about two hours, a fellow finally appeared who, after reading the letter, told the villagers its contents and we were allowed to continue on our way. I made it a quick short traverse in case they changed their minds!

Chris was severely startled on one traverse when a Bedu came out of nowhere and fired his rifle in the air several times. Chris ducked for cover not knowing what was going on, but it was quickly established by the field assistant that the local had just become the father of twin boys and was celebrating!

Traversing continued relentlessly. Occasionally a full day was required in the office tent to catch up on plotting up of field work so this would break up the pattern. Our social life was non-existent. We never went into Muhayil township, for example in the evenings, nor even to the local market, such was our effort to keep a low profile in the district. The weekly supply trip arriving from Jeddah provided our main excitement as it often returned with personal mail and newspapers as well as the fresh supplies of non-alcoholic beer.

By 23 October we had basically completed the final planned traverse in the Ad Darb area on another fly camp and done what we called the 'first pass' operations of the project. For the next few weeks, it was a matter of doing infill traverses over particularly interesting zones and prospects – the 'second pass' work. However, the Hajj period commenced on 26 October so we closed the camp out of respect for our Muslim field assistants (and because there was no way they could work for six hours in the baking hot field while not drinking water). Chris and I worked on reports in Jeddah until returning to the camp on 7 November, while the others went on leave during that time. While in Jeddah we were able to catch up socially with some of the people helping us. For example, Chris and I enjoyed a nice white wine with Helga, our typist, and her partner Hans. We also had alcoholic beer and pork ribs at another friend's place. To quote from my diary "it was like having ice cream in the desert – pork and alcohol in a strong Muslim country." These Western foreigners were able to access these illicit items because contraband products could be brought in through secure channels and they all lived in large expatriate compounds where anything goes, including home brewing and distilling, as the Saudis did not police these isolated havens.

Unknown to us, on 20 November – the start of the Muslim new year 1400, there was an uprising when armed civilians, calling for the overthrow of the House of Saud, took over the Grand Mosque in Mecca. We later heard rumours that there were similar uprisings in Medina and elsewhere. The first I knew of it was while in Jeddah preparing to return to the camp after a three-day working trip there. As coincidence would have it, I had driven up to Jeddah on that day of the revolt and saw nothing unusual. I passed through the two police gates (one at Muhayil and the other near the Mecca by-pass) without a problem. On the way back though, on the 24th, I had to pass through an additional three new police roadblocks. Fortunately, the police were not interested in me as a foreigner – just looking for arms apparently. Six days later Musa, driving back from Jeddah with another load of supplies, was stopped twelve times. A couple of Utah managers cancelled their planned visit to our camp as the fear of repercussions from the turmoil spread. There was no mention by the company of getting us out though!

After another stint in the field, carrying on as if the uprising never happened, Chris and I were due for our next overseas break. We drove to Jeddah on 6 December and I flew out to Greece. I was back in Jeddah on 20 December and worked over the Christmas period including Christmas Day as if it did not exist – which it didn't in the Muslim world. However, in recognition of Christmas, Chris, through some nefarious contact of his in Jeddah, managed to acquire a bottle of 'Sadiqi' ('friend' in Arabic). This was a pure white spirit, illegally distilled in some expatriate compound and sold in second-hand water bottles to disguise its true nature. It certainly had a kick! It was so strong that we resorted to experimenting with adding it to our non-alcoholic wine and beer but that did not improve the taste of either!

Finishing up

I am sorry that Ayaan Hirsi Ali's book, *Infidel*, had not been published before I went to Saudi Arabia. It would have opened my eyes to the cultural background of the Somalis we had as field assistants as well as the rise of more fanatical forms of Islam. Ayaan Hirsi Ali was an eight-year-old girl when she arrived in Saudi Arabia, living in Mecca and Riyadh till sometime in 1979, covering part of the time I was there. Her description of the development and influence of the fanatical Muslim Brotherhood, a movement that spread through Africa in the 70's, is particularly telling. In retrospect this new-found zeal for Islam may explain the observation we made of our field assistants who, during the initial months of employment, were pretty casual about their prayer times. When we observed other Muslims starting to pray on the sides of roads or outside shops, we would ask our accompanying fieldy if he wanted us to stop driving so they could pray. "No, no", they would say – "we do it later". But Mohammed, the tall dark, serious Somali with the dignified tilt of the chin, became the prayer leader at the camp, calling the other field assistants together so that, by the end of my stay in Saudi Arabia, they were much more observant of the prayer ritual. Mohammed was even keen to take up discussions of the Koran with me when I asked him. He brought out a copy of the Koran which had English translation on one page, Arabic on the other – to show me the bits that mention Jesus to show the bond between Islam and Christianity.

On Boxing Day, I was back in the field camp with Chris, Yusuf, Abdullah and Ibrahim for the final three months of planned field work. While it continued, it was broken up by a visit by two officials who worked with the DGRM (Saudi mineral resources department), who flew in to check our work. They flew into Wadi Shari, landing a De Havilland DHC-3 Otter on a gravelly fluvial flat. After giving them a rundown on the project, we visited various field locations including Wadi Baqarah, Jabal Saban,

Wadi Yiba and Jabal Sawdah. They left on 30 January but not before giving us a joy flight over the field area so that we could see from the air the ground we had been traversing for the past twelve months.

We started packing up the camp on 13 March and, after gifting my guitar to Abdullah, I drove out of camp for the last time on the 14th to finalise a few things in Jeddah before I took off for Brisbane on the 19th.

You may recall earlier in this story that our aim in Saudi Arabia was to look for stratiform copper. The results were disappointing. The rocks were definitely deposited in shallow water all those millions of years ago, and they certainly had some copper in them but nowhere near enough to consider mining. We found particular calcareous layers with sometime quite abundant malachite, the green copper carbonate. However, we quickly realised that we were not finding zones which had anoxic conditions during or around the time of sediment deposition. These situations form the best traps for the metals. Anoxic depositional conditions are usually indicated by finding black carbonaceous sediments and/or an abundance of pyrite. The interpretation was that although there was copper present, it had not been 'fixed' or trapped in significant concentrations anywhere. So, the conclusion was that the Saudi project was to be wrapped up, reports written, and we were to disperse to other projects.

My exploration work in Saudi Arabia as an 'infidel' was enlightening, especially as I considered that I was relatively well travelled. It opened a window to part of the vast Muslim world that was to have a large influence on Western society over the next few decades. In turn, the Yemeni people in the areas we worked had very little idea of our Christian world. It would be nice to think that, with the rapid modernisation of infrastructure and communications in Saudi Arabia, mutual understanding of our complex societies will also have improved significantly.

Acknowledgements:

The harshness of the field conditions in Saudi Arabia were ameliorated considerably by a team of wonderful people who maintained good humour throughout while sharing the load, making the program technically a success. In particular I would like to thank Chris and Genesio who very kindly reviewed this story and added some interesting anecdotes.

16

The Cradle of Gold
Saudi Arabia, 2008

Jeff Rayner

My cell phone rang. I was collecting rock chip samples in the Eastern Pontides, Turkey. Good phone coverage in the heights overlooking the Black Sea. "Hi, it's Toni, CFO of a Forbes Rich List family, would you like to come to Saudi Arabia and find gold?" "Yeah, sure, when?"

Saudi had reformed its Mining law in 2006 and was now welcoming foreign gold miners to pillage its mineral riches, to replace the finite resources of black gold, whenever that runs out. In pre-Islamic days, the climate was wetter and cooler, Neolithic man scraped and forged metals from sand-covered quartz veins and gossans from around 3000BC. Gold mining continued in the Abbasid Caliphate from 750 to 1250AD. Over 5,000 historic workings were documented by the US Geological Survey and the French BRGM from the 1970's to the 1990's. Decent shows had established mining camps, instead of tents. The Mahd adh Dhahab ("Cradle of Gold") deposit was mined from biblical times, and in the Caliphate era. Tens of thousands of slaves tunnelled and died in the mountain, the site of the fabled "King Solomon's Mines". The deep and winding cuts in the Cradle of Gold are still visible today and, from a distance, the shadowed workings on the barren mountainside take the form of a human brain. It is estimated that, over the millennia, the mine produced over 1Moz of gold from low sulphidation epithermal veins, coalesced on the perimeter of a Precambrian volcanic caldera. Maaden operates the underground mine there today.

"Sheik wants you to come fishing with him and we can talk about gold mining". Wow, I ring my then wife, she's Italian and in Italy. Offers me some good advice, saying "Jeffe, this is notta a Australian fishing trip you know! You musta notta have to wear flip-flops, a white singlet, your tradey shorts and the zinca-cream, better you getta some good clothes." Where do you buy good clothes in the Pontides of Turkey?

Two weeks later I have a Saudi visa, it takes up a whole passport page and is valid for a stay of one month. Business Class on Turkish Airlines on the Istanbul-Jeddah route is cheap, only $50 more than the seats for the riff-raff at the back end. I sit there wondering what the extra $50 bucks is for, as we share the same cramped seats, no booze and no movies. After customs in Jeddah check me for hidden liquor and funny videos, I am chauffeured to the top floor, the 14th floor of the family tower, a stone's throw away from the Red Sea waves crashing in on the rocks. The private pool outside on the same floor looks warm, but it's still a muggy 27°C and 2 am, got to get some desperate sleep, fishing starts at 6am.

So there we are, on a 26m luxury motor launch, 60kms offshore loitering around coral reefs, the last reefs before the Red Sea rifts cascade and plunge to 2.4km depths, where zinc deposits quietly spew up in the abyss. The Arabian Plate got sick of the Horn of Africa 56Ma years ago and leaned northwards, knocking violently on the door of the Eurasian Plate. Had the Arabian Plate been in a hurry it could have split the Isthmus of Suez in Egypt and flooded the Mediterranean basin 50ma years earlier.

Fishing for me is boring, but this is fishing in grandiose style: we are sitting in lounge chairs on the aft deck, armed with hand line spools 20cm in diameter and hooks with one-inch barbs. The squid bait is so fresh I could eat it raw. Old Sheik is hauling in coral trout, five in total for now. My catch are suicidal 'black fish' that the Saudi's won't eat, but the Filipinos will. They have been unlucky to have got belly hooked as I tugged violently on 'a bite'.

Sheik hauls in another coral trout, a beauty, must be 6lbs. The Indian deckhand delicately unbarbs the hook from the fleshy lips and spears another squid. Sheik leans over towards me and quietly says, "Do you drink, Jeff?" I thought that this could be a trick question, but I'm sure he could see a lie coming, so I reply "Yes". He says, "Do you drink every day?" Wow, even trickier, I say "Mostly on weekends." He seemed happy with that.

After decimating the local coral trout population on that particular reef, we weigh anchor and motor back to the Port of Jeddah. At 10pm, Sheik ascends in his private lift which starts on the 12th floor, no 13th floor. "Jeff, the bar is open, scotch or gin?" Since living in Italy I went off beer and have become a (red) wino, but any port in a storm and, besides, it would be rude to refuse. Sheik continues, "Tomorrow I have the CEO of a major Saudi bank coming, we will talk about gold mining. You will give me all costs, I want to know everything." Shit, I panic: we are a junior explorer, we haven't even got a decent resource anywhere yet, we floated on the London Stock Exchange two years ago, with not very much to show and not very much money to spend. I realise quickly that this has to be a (big) hypothetical mine. Same size as Busang perhaps?

Our little company does a JV with Sheik. We have a small team, just three geologists (two Aussies and a Turk), a Saudi driver and one 4WD Toyota LandCruiser. Armed with GPS, gallons of water, reports and maps we head off into the unknown, looking for gold. Our Saudi friends are baffled how we can just arrive from somewhere else on the globe, navigate their desert lands and return to Riyadh weeks later with bags full of visible gold. We find out firsthand the size of the opportunity in the Arabian Shield. Some old workings are marked by foundation stones of tiny ancient dwellings and grinding stones, all still there, archaeological litter, and any one of them could be a possible world class ore body, untouched for at least 900 years.

As luck would have it, we find 1Moz in our second year of exploration and we have also pegged an outcropping volcanogenic massive sulphide gossan that spans up to 50m width and extends over 5000m in length. The siliceous gossan has never been drilled; the ancients mined the flinty chert for arrow heads, otherwise the gossan has been unharmed since the climate changed and the Arabian desert took its form. This VMS deposit is being drilled right now, over 25Mt of copper-gold resource has been tallied up at modest economic grades.

These rapid exploration successes means that other Directors might turn up. The Prof turns up and hits it off with the Sheik. Great humour and promises of riches to come. We discover that Sheik has a decent man-cave in his Riyadh palace. But only scotch and gin, no tonic water, none to be found in the whole of Riyadh. No Billy Joel either.

There's nothing better than a +50°C day in the desert to clear out the hospitality of the night before. We have an entourage to inspect our new gold find seven hours' drive west of Riyadh. A dozen white thobes, capped in red and white chequered shemaghs, lurk around us, threatening to inspect kilometres of trenches, tentacles of gold rich veins, and what anywhere else would look to be a certain heap leach operation. Pity there is no water though.

Lunch is a dodgy falafel and stewed camel, bought from a Bedouin village along the way. The Prof feels a rumble in his tummy. I say, "bloody hell mate, in London you consume all my gout pills; in the Mid-East and Africa, it's all my Imodium pills! Here take these sample bags, last trench on the horizon, not a blade of grass in shifting sands to hide behind". "Well", the Prof says, "at least it will keep the flies off our water melons!"

It's a successful day, a successful venture and we bond with the Saudis. Drill rigs will soon dot the lands, the local Bedouins will most likely be alarmed, but I'm sure we can work something out. I am not working in Saudi these days. It's a pity because the Kingdom is an easy place to explore, decent sized tenements, no fences, no landowners and, more importantly, no greenies.

Though we are from different cultures, religions, have different number of zeroes on bank accounts, Saudi is probably one of the last great frontiers for explorers and miners.

From Ralph Stagg's story, "I Will not Shoot My Fellow Citizens", p. 270, A party examines drilling of the Tunisian phosphate project.

17

Revolutionary Rumblings
Iran, 1977-1978

Wilson Forte

Wilson Forte graduated with Honours from the University of West Australia in 1972. He began his career in the same year as an exploration geologist working for Asarco in Australia.

In 1975 after being seconded by Asarco to look for base metals in South Africa, he took a sabbatical year travelling overland to London through more than twenty countries in Africa, visiting mines and volcanoes along the way. From London he went to Iran in 1977 to undertake a geological mapping contract.

Back in Australia, he was the principal of Lonarch Geological Consultants from 1978, mainly consulting on gold and base metals exploration to many major Australian and multinational mining companies for 16 years. During this time, he also wrote Independent Geologist and Valuation Reports for listings on the ASX.

He was Managing Director of Hampton Hill Mining NL, which he helped to float on the ASX in 1994, for 18 years. He is currently a director of that company. More recently he pegged the tenements which were floated as AIC Mines in 2017. He still enjoys occasionally going bush to explore for mines.

Shah's survey

Over a period of two years, commencing in 1976, a group of more than 30 Australian and British geologists mapped most of the Makran Mountains in Southeast Iran, mainly using helicopters for access. Paragon-Contech (an Iranian-Australian joint venture) mapped the area under a contract for the Geological and Mineral Survey of Iran. The Makran and its mineral potential had been long neglected by geologists because of its lack of oil and gas potential.

Iran was ruled by the Shah, Mohammad Reza Pahlavi, who came to power in a coup d'état in 1953. This coup was orchestrated by the USA and Britain, when Eisenhower and Churchill were in power, to protect their control over massive oil interests in Iran which were under threat of nationalisation. Interestingly this was the first covert action undertaken by the United States (read CIA) to overthrow a democratically elected government during peacetime.

In October 1977, I joined the mapping group in the second field season during which there was an escalation of bloody protests culminating in the Iranian Revolution and the overthrow of the Shah in February 1979.

Culture shock

After flying into Tehran airport at night from London we felt our first culture shock. A company driver met us at the airport in a large two door, two-tone coloured and two-tone horned 4WD Chevrolet Blazer. In Australia we were still mainly driving Land Rovers with a few Toyota LandCruisers. I had seen these big American 4WDs in Hollywood movies featuring gripping, edge of the seat, car chase scenes, being driven at breakneck speed through busy urban streets by Hollywood heroes; outrunning police cars, with their sirens wailing, red and blue lights flashing, buying time to prove they were setup. The drive into the city had many similarities with these gripping movie scenes! Our driver drove flat out from the airport, gunning through red lights all the way. You can imagine our relief to hear, after recounting our rapid transit from the airport to seasoned hands in the company, that we were lucky. Our driver was one of the good ones!

We regained some of our composure when morning revealed the truly beautiful view of Tehran strikingly outlined against rugged, snow-capped Alborz Mountains.

Identity crisis

Before we travelled down to the job in the Baluchistan Province of southeast Iran, we were issued with photo ID cards to show we were working for the government. I guess this was probably the idea of the Shah's notorious SAVAK secret police. We weren't too sure this was a great idea, because the government had recently raided a camel train of 'import/export' agents, a euphemism for smugglers, in the mapping area and had thrown their leader in goal. The smugglers had retaliated by kidnapping the local police chief, who was still missing.

Even though I always carried my photo ID card in the mountains, I was never asked to show it. Most of the Baluchi in the mountains seemed to know about us, probably via the bush telegraph, and that we were not Americans, who were not their favourites.

The only time I needed to show the photo ID card was when I arrived at Kerman airport to board a plane to go on a mid-contract break. Lining up to walk on to the tarmac, to my horror, I saw an official checking ID cards prior to boarding the plane. I say to my horror because I suddenly remembered that my ID card was back at the bush camp.

A British geologist alongside me told me not to worry, as he would slip me his ID after he had passed the checkpoint. I was concerned, but had no alternative plan. So we moved forward bunched up, milling around the official. This British geologist was now deliberately well in front of me. He passed through the check gate, then nonchalantly peeled off behind the official while I sidled away from the bunch to the security fence, where he passed back his ID card. Armed with his ID, I returned to the middle of the remaining group of geologists, passed the check and boarded the plane with a sigh of relief.

The British geologist was right, any group of foreigners all look the same to locals anywhere in the world. However, this occasion was a bit of a cliffhanger because the British geologist was a pale skinned, redhead versus my tanned Aussie skin, light brown hair and full beard. I would not like to try and get away with it in today's Iran.

Mountain mapping

The best exposure of the rocks to be mapped, using air photos (no GPS) for location, was in mountain streams. At the beginning of the day, we were ferried by Allouette III helicopters to near the top of a previously chosen stream to start the traverse, then dropped off individually. The aim was to be picked up late afternoon, where the stream joined a main river valley.

We used a very effective but simple signalling mirror with a hole in the centre, so you could see where the reflected beam was aimed to alert the helicopter pilot of our whereabouts. He would quickly acknowledge our signal by turning on a rotating light on the top of the helicopter to avoid being blinded by our well-aimed flashing.

Earthquake evidence

Another difference we experienced working in Southeast Iran was that it is a very active earthquake zone.

On one of my traverses, I was surprised to see a break in a young, cemented gravel deposit on the bank of a stream, which had obviously been caused by recent seismic activity. However, what concerned me about this break was that it had been produced by compression, with one block thrust up over the other. As I discovered, most of the rocks we were mapping exhibited evidence of a punctuated continuum from old to more recent earthquakes. These rocks were deposited as a consequence of numerous undersea earthquakes, initiating submarine landslides of vast amounts of sand and clay, over at least the last 50 million years. After accumulating into sequences kilometres thick, they were hardened, then thrust back up onto the land as the Arabian tectonic plate converged on Eurasian plate to form the present day 2400 metre-high mountains we were mapping.

What I had observed in this terrace was just a small-scale demonstration of this mountain building which was still in progress.

Mountain encounters

The tribal Baluchis in the mountains were very hospitable, so, on chance encounters, we had to spend some time with them. There were no roads – just tracks along the contours made by camels and donkeys. Mountain streams, which had narrow channels flanked by steep banks with waterfalls in places, were the focus of life in this desert.

Although I did not speak Baluchi, men I met communicated by pointing and they all knew the word helicopter, particularly when verbal rendition was reinforced by me giving an ocular demonstration with my arm rotating above my head. If it was late in the afternoon, sometimes they would accompany me to the pickup spot to witness firsthand the roar of the helicopter turbine, shattering the silence as it settled in a blistering cloud of dust.

For both cultural and climatic considerations these sensible local people only exposed their face and hands. We came to realise that women in the mountain villages were a bit scandalised by our bare limbs being exposed by only wearing short sleeved shirts and shorts. The men were always hospitable and would offer to share what little food they had. On one occasion, in return I offered a spare orange to a man who dismounted from his camel on meeting me. As he was squatting on the ground with the orange held by the tips of his fingers, his camel peered down over his shoulder to join in the contemplation. I felt privileged to witness the close bond between them.

On some occasions hospitality came with challenges. One of our geologists recounted becoming slightly unsettled, when he looked into a 'special' bowl of goat head soup, to see one of the goat's eyes had parted company with the skull and was staring accusingly up at him.

Sometimes I unwittingly crossed a cultural boundary. I was traversing down a narrow river gully when a young colourfully dressed woman, obviously coming down to fetch water, suddenly appeared on a terrace above. Being a naive friendly Australian, I tried to communicate with a smile and by holding up my hand with my index finger and thumb joined in a circle which we Westerners interpret as indicating everything is OK. Much to my consternation she threw back her head, accompanied by a surprised stare, then turning on her heels, retreated rapidly, uttering a loud high-pitched ululation which sent a shiver down my spine. When I related the incident back at camp, I was informed that this gesture was considered obscene in the Middle East, being similar to a Westerner extending his middle finger with the back of the hand towards the recipient.

I have since read that there are two types of ululations: mournful and, less commonly, joyful. I'd heard multiple voices ululating in the night when I was travelling through Nigeria the year before, while hitch hiking up through Africa visiting mines and climbing volcanoes. From where we'd camped, although we could not see the village we heard a wake in progress, with mournful ululation resonating in the still night. From my river gully encounter, I interpreted the young woman's ululation was a shocked reaction, in response to my unintended inappropriate gesture.

Speaking of obscene gestures in different cultures, I remembered being driven to a bus station on the edge of Lagos in Nigeria through chaotic traffic by a very helpful Australian woman from the Australian High Commission. I noticed a driver coming towards us, who had been abruptly cut off by another car, thrusting up his hand with great vigour toward the windscreen with all five digits spread out. She told us that this was a gross insult as it cast doubt on the legitimacy of the reckless driver's parentage, suggesting he had five fathers. With this in mind I guess I should have been forewarned to tread carefully.

My misadventure in the mountain stream was a good lesson on how it is so easy to miscommunicate across cultural customs – however friendly your intentions!

Irrigation ingenuity

I was impressed by the ingenuity of the villagers who grew food in irrigated terraces they had constructed even where there was only a trickle of water. The terraces were fed by water channelled from natural or constructed dams on the streams. Where the irrigation channels passed over gaps incised by narrow rivulets, they were bridged by ingeniously hollowed out palm trunks.

Great expectations

Just how highly these people regarded us was brought home to me on a traverse which skirted a village. The villagers had obviously seen me coming while I was lost in measuring and photographing textbook sedimentary structures. I was approached respectfully by a young man who beckoned me to the edge of the village to where an old man was squatting, holding his lower stomach, gesturing that he was in pain and appealing for medical assistance.

I wondered whether this expectation of our capabilities was rooted in the past through word-of-mouth stories of the medical expertise of earlier western explorers. I remembered reading that in 1907 (70 years before my encounter), the recently graduated British-trained doctor, Morris Youdelevitz Young, was sent out by a company which became British Petroleum, to Bakhtieri tribal land in southern Iran. His job was to care for expatriate staff on the company oil rigs, who were falling sick in epidemics caused by hot weather. Drilling had started in November 1902 and only struck oil in late May 1908 when both funds and hope were running low. However, this was the discovery of the biggest oil field known at the time, making Australian/British lawyer William Knox D'Arcy a second immense fortune from his interests in the precursor company to BP. His first massive fortune was from dividends and the sale of shares in the initially fabulously rich Mount Morgan gold mine in Queensland.

In the meantime, the brilliant young doctor became a great ambassador for BP for, in addition to ministering to the sick expatriates, he treated thousands of tribal people living in the region surrounding the rig. His reputation was greatly enhanced when, after experimenting on a sheep, he successfully performed cataract surgery on the local tribal chief. Henceforth he was known as the "healer of the blind".

Sadly, I was ill equipped to provide medical assistance to the old man. Initially I tried to indicate he should see a doctor, although I was conscious he had probably never seen one in his life! I then gave him some Panadol tablets and gestured that he should take them with "ap" (water, in Baluchi), to ease the pain.

Road maps

Mapping along roads could only be a small part of the job because there were so few of them. On a road traverse, in addition to an Iranian driver, I had a Baluchi from the nearby village of Gangebad to accompany us. This was apparently because the driver, not being from the area, was nervous of the reception he might receive in chance meetings with semi-nomadic tribal Baluchis in the mountains. His nervousness may have been heightened by the knowledge that our traverse would end on the border with Pakistan. The neighbouring province of Pakistan is shown on some maps as spelled with an "o" instead of a "u", and so as Balochistan, and is inhabited by people of the same tribal group. I was aware the tribal Baluchi moved freely between the two countries along mountain tracks with little regard or, more likely, complete disdain for central government custom controls.

The three of us sat across the wide seat of a large Ford F250 4WD ute, speeding off on the bitumen to the start of the traverse. The journey was reminiscent of my alarming experience of the drive from Tehran airport when we first arrived in Iran. This driver had the similar driving style as our Tehrani chauffeur; two-speed driving, either flat out or stopped. I soon realised this was not the only eccentricity of his driving.

When we came up behind slower moving vehicles on the road, the driver passed them not only on blind bends, but also where the road had been excavated into the mountain with a ravine on the other side. This fatalistic approach to driving was explained as that, on the off chance a vehicle was coming from the opposite direction on the bend, then it was Inshallah, the 'will of Allah', so the driver could do nothing to change that. Fortunately, there weren't many vehicles on the road.

By the will of Allah, or just chance, we continued from bitumen to gravel roads without incident. I knew from experience that this incident-free state of affairs could not last.

Bogged in Baluchistan

The first obstacle to our progress was a dry gravel bedded creek crossing, which I gauged from my own bush driving to be a not much of a challenge for a powerful V8 4WD. Wrong. We started to become bogged, sinking further down into the loose gravel each time as the driver revved back and forth. Although I don't classify myself as a petrol head, I confess the deep burbling roar of the big labouring American V8 engine was very stirring. I was jolted from my brief reverie realising our lack of progress did not add up.

Then it dawned on me, I glanced down at the gear levers to observe the vehicle was only in 2WD. I suggested politely that he engage 4WD. The driver said he couldn't because the front drive shaft was disconnected. Expecting the worst, I calmly as possible enquired as to the whereabouts of the missing drive shaft. To my relief the driver pointed behind the seat, which we pushed forward, and I was happy to see the shaft, with thank God (Allah) the fastening nuts and bolts all attached. Returned to its rightful place, with 4WD engaged, extricating the vehicle from the bog was a breeze. I dared not ask why the shaft had been disconnected. As the price of petrol, in this oil rich nation, was about six cents per litre, saving fuel did not appear to be a reason for the disconnection.

Camping out

In the afternoon the driver asked me if I wanted to return to base camp. We had not completed the traverse, so I said I was happy to camp out, as we had come prepared for this contingency with food and bedding. Sleeping out did not faze me as, earlier in the year, I had just finished an overland trip from Johannesburg to London. In places with no accommodation, we would pitch our small tents after dark in the bush or near cemeteries.

In retrospect it probably wasn't the best idea. We had conveniently forgotten that, earlier in that trip, while travelling on the 1600-kilometre voyage down the Congo River, we were told that the boat was not calling into any ports because there had been an outbreak of green monkey disease that had originated in the Sudan. Later in London we found out it was Marburg virus, which is closely related to Ebola. This near miss with that virus furthered my perception of the Congo River as being enigmatic since Joseph Conrad, who was the captain of a steamboat on the river in 1890, wrote his prescient novella on the ills of colonisation in *Heart of Darkness* after witnessing the appalling conditions of the Congolese while under Belgian colonial control.

Having decided not to return to base camp at Gangebad, we stopped mapping to look for a camping spot, as the wind was picking up and it looked like it might rain in the night – a rare event in the desert. I suggested we camp out of the wind near the base of a high, very steep-sided peak. Our Baluchi became agitated so, without querying him, I deferred to his choice of sheltering from the wind behind a small hill nearby.

It started to rain shortly after our evening meal. The next issue was where each of us would sleep. The driver gallantly offered me the seat in the ute out of the rain. I said I was happy sleeping out of the rain under the high clearance F250 on my trusty blow-up mattress which I'd used through Africa. So, the driver slept on the seat and the Baluchi drew the short straw to sleep on the open back of the ute. He didn't appear too perturbed with the rain beading on his greasy wool blankets.

Early in the morning I was wakened with a start by desultory volleys of loud cracking, like rifle reports echoing in the hills. This shattering of the desert silence was followed by the clatter of what I realised were slabs of rocks tumbling to the base of the nearby high, steep peak I had suggested as a camp site.

I had not experienced this phenomenon of exfoliation of rocks before but remembered the explanation from my first-year physical geology text. High up the peak, some of the rain that had fallen early in the night pooled in cracks in the rocks. After the rain and the insulating cloud cover had cleared, a rapid drop in the air temperature to the more normal early morning level of below zero followed. The pooled rainwater in the cracks then expanded as it froze, prising the slabs of rock from the hill.

Another cross-cultural lesson: trust your local guide's recommendation on bushcraft.

Extraction and mateship

Next morning, we were just completing the traverse at the border, with Pakistan in sight. An ornate medieval mud brick fort, topped by a parapet with classic rectangular gaps through which to fire weapons, marked the border. It was more of a statement than a defence. While the fort may have given some refuge during an attack with small arms, it would certainly have been demolished even under a small-bore cannon barrage.

Our timing in completing the traverse was perfect, as above, we heard the approaching roar of one of our helicopters coming to check on our progress.

With mission accomplished, I boarded the helicopter to fly back to base camp. As the contrasting images of medieval fort and modern Ford F250 rapidly receded in this vast, ruggedly beautiful, Middle Eastern desert, I was reminded that this part of the world was so different to where I came from in Western Australia. The spur to that came while reading a letter from my niece in high school back in Perth, and which Tony Gates, another Western Australian geologist (refer to Tony's "Letters Home from Iran" in *Rocks In Our Heads*, Ed.), had kindly brought for me in the helicopter from the main camp.

Back there for a few days, later the driver/interpreter took some of us to visit the local village. The village comprised a group of small mud houses with minor ornamentation and wooden shutters on the windows but no glass. It was a colourful scene with scattered rubbish, amongst which donkeys, dogs, goats and chickens rummaged happily. The driver pointed out two likely looking characters who apparently belonged to the fabled band of import/export agents, aka smugglers, of whom we had heard about when we arrived in Tehran. It was interesting to meet them and view their wares for sale including brass trinkets, printed cotton and red and black decorated donkey carry sacks. I think this may have been all that they were willing to display in public, because I'm sure these innocuous goods were not the target of the government raid on the band I previously mentioned.

I fell behind the other geologists while changing the film in my camera. My Baluchi mate from the road traverse gestured for me to follow him into his backyard, where an old lady was squatting smoking a hookah. His wife was baking flat bread by dexterously flinging the flattened dough against the wall of a wood fired beehive-shaped oven in the ground. Both sides were cooked in a very short time, producing delicious bread.

Using an underarm shovelling gesture with his arm, my Baluchi mate, sporting a rare grin from ear to ear, was letting me know that he was telling his family that I was the crazy Westerner who had slept under the vehicle on the road traverse. This time I had successfully bridged the cultural gap and had won a friend.

After finishing the mapping, we all moved to the city of Kerman to complete the compilation of the data collected in the company's office compound. We were billeted in the compound and told it was dangerous to venture out because of civil unrest, which we later knew culminated in the Shah being deposed.

Touring

At the end of the contract most expatriates flew out of Iran. Tony Gates and I reasoned the disturbances were probably mainly directed against the Shah rather than Westerners and we wanted to see a bit more of this very interesting country, so we caught a bus from Kerman to Isfahan.

Any misgivings we harboured about our safety were dispelled in bazaars in Isfahan where, although our height marked us out, friendly people gathered around wanting to practise their English. The most fascinating cultural revelation to me was when visiting a mosque: the guide took us downstairs where they had excavated remnants of an ancient Zoroastrian fire temple. It reminded me of how in many parts of the world incoming religions assimilate by constructing new places of worship on top of old holy sites and by the renaming of pagan festivals.

Not only was Shiraz the most westernised-looking city we had visited, apart from Tehran, it was also the most colourful. Shiraz appeared to be very multicultural with people from different ethnic backgrounds, that I guessed, from their appearance, to include Kurds and Arabians in addition to the Iranians. Many of the women wore layers of multi-coloured dresses and skirts decorated with gold and silver glitter. The dresses reminded me of Berber women in the Rif in Morocco. People were again very friendly and, in one restaurant, they wouldn't let us pay for the meal.

The most interesting conversation we had in Shiraz was with an Iranian who had returned from Sydney and was planting shiraz grapes to make wine. I have wondered how he fared in his venture after the Revolution.

The main reason for visiting Shiraz was its proximity to the magnificent ruins of the ancient city of Persepolis, dating back to 515BC. Interestingly, some of the tent city, that had been erected six years before by the Shah to celebrate the 2500th anniversary of the founding of the Persian Empire, was still standing. The party lasted for just three days in October 1971. I have since read estimates of the cost of this massive undertaking, with royalty and high-ranking officials from around the world attending, to have been about 170 million British pounds: with the Shah declaring it "the biggest party on earth", to prove that the Pahlavi era was "a period of renaissance for Iranian civilization".

It was probably the most expensive party on earth, with most of the food, wine and chefs flown in from Paris, but it caused a great smouldering resentment among Iranians. I think this resentment grew, not only because of its cost, but that it was showcasing European culture, rather than the Iranian one it was supposed to be celebrating. Some commentators suggested this party demonstrated how out of touch the Shah was and that it marked the beginning of the end of his reign. They believe it crystallized the opposition, led by the Ayatollah Khomeini, leading to the deposing of the Shah and his exile in 1979.

Leaving Tehran

Back in Tehran, where we had heard more rumours of demonstrations against the Shah, people were still very friendly towards us. This friendliness again even extended into the labyrinth of lanes in the Grand Bazaar. When Tony Gates was buying a small gold coin for a mate back in Australia, we were reminded of the great wealth of some of the Iranians.

We found a lane in which shops specialised in selling gold and chose one that was crowded, working on the principle that the shop was busy because he was a fair trader. While Tony was concerned about the cost of the purchase of one small gold coin, a woman in front of us bought 10 large gold ones. The gold merchant told us the coins were gifts for her servants to celebrate the Iranian New Year on 21 March in 1978, coinciding with the spring equinox and the day we flew out of Iran.

It was a great privilege to have worked as a geologist in that country. Despite it being a period of bloody protests in some cities, the experience confirmed my observations that in most countries ordinary people you meet in everyday life are open and friendly.

18

Tourist! Tourist!

Turkey, 1987

John Nethery

John Nethery was born in 1946 and raised in country New South Wales, before attending UNSW in Sydney to graduate in Science (Geology) and with a Diploma in Education in 1968. He joined AOG Corporation, rising to General Manager – Minerals of AOG Minerals Ltd, exploring throughout Australia, Oceania and West Africa for gold, base metals, uranium and diamonds. Since 1987, as a consultant, he has explored, mainly for gold, throughout Australia and elsewhere in Oceania, South-East Asia, the Mediterranean region and briefly in western USA. In 1993, after 16 years living in Sydney, he moved to Chillagoe, North Queensland, and is still exploring for gold and base metals at this time.

In mid-1987 Geoff Loudon, Gavin Thomas, Ian Plimer and I had arranged a joint venture between Niugini Mining Ltd, Plimer and me to apply our experience of discovery of epithermal systems in Papua New Guinea and Australia to elsewhere on the planet. The idea of finding massive deposits of 'invisible' microscopic gold was rather appealing particularly in areas where mining dated back thousands of years. The Mediterranean volcanic arcs were very tempting. I was appointed operator.

The discovery in October 1987 of a substantial swarm of epithermal gold-bearing veins on Milos in the Cyclades and indications elsewhere in Greece encouraged us to expand our horizons east and west, while we waited for Athens to make up its mind about whether or not it would allow foreigners to explore and exploit their precious mineral deposits. It was all very hush-hush. Turkey was particularly beckoning and, after we established that we could operate there, I got to work on published metallogenic literature and geological maps to draw up a priority list of reconnaissance targets and areas of interest.

In 1989 I arranged for Mike Barr and Charlie Georgees, two very experienced and observant geologists, who had assisted with blanket 'eye-balling' reconnaissance in Greece, to join me in a similar first-pass exercise in Turkey. The concept was to initially define districts with favourable geological parameters and indications of serious hydrothermal activity. We had decided to avoid the far east of the country, for political reasons, and to split the remainder into two; with Mike and Charlie working together in the west, and with me exploring in the central Black Sea Coast, Pontides, Central Anatolia and the southern Taurus Mountains.

Having drawn up a list of areas of interest, we still faced the formidable task of navigating, because the government, for security reasons, did not allow access to adequate topographic maps. I knew that they existed and had naively thought that I could purchase them in Ankara. Of course, we had no Turkish language whatsoever. This could be considered serious 'seat of the pants' reconnaissance. Trying to navigate a commonly highly populated country using quarter-million-scale geological and topographic maps would have been an impossible task.

Enquiries led us to a series of district road maps produced by local governments that showed most villages, and a rough pattern of linking roads: this in a country with villages at about five kilometre intervals. The maps were reasonably accurate in terms of village location as black dots, but the roads were drawn as straight lines between those villages, which of course they were not. It was rather like the old board game of "Snakes and Ladders". If you landed in the wrong square you basically slid back several levels, or even had to start again. Imagine yours truly in a small Suzuki Samurai 4WD way out in the back-blocks, not being able to speak the language and with only a vague idea of where I was supposed to be heading, trying to work my way through villages whose streets were built for equine quadrupeds of the donkey variety. It was quite daunting, but fun all the same. I travelled about two thousand kilometres in three weeks and barely spoke a word of English. Tourist! My English to Turkish phrase book often saved the day.

One day I was up the side of a hill in the Pontides peering at some good-looking rocks. It was fairly steep, but the Suzuki Samurai is a useful little 4WD, being very light and manoeuvrable and the vegetation was shin-high and clumpy, so I had managed to work my way up the trackless slope. Why walk when you can drive? I'm not sure Eurocar would have agreed. The open country was obviously used for grazing. Several flocks of sheep and goats were dotted here and there with their attendant shepherds.

I figure I was about half a kilometre off the road. It was a lovely peaceful sunny day. Quiet except for the tinkle of sheep bells in the distance. Then "... there was movement at the station for the word had passed around ....". I could see them coming for a kilometre or so at a fast canter. Two wild-looking hirsute horsemen arrived, with rifles slung over their backs. Horses heaving and snorting. They were yelling at me and, when I didn't respond, they tried yelling to each other. I am not sure if they were angry or just surprised, but they were making a lot of noise. I offered them the universal Mediterranean gesture. Arms spread with palms upwards. A shrug of the shoulders. "Tourist. Merhaba. Güzel bir gün". "Hello. It is a nice day". They stared. Shocked looks on faces. Then at each other. Shrugged their shoulders, wheeled the horses and headed off whence they had come.

Two days later I was in the back-blocks out from Sivas, about as central as one can get in Turkey. I could see the prominent silica-rich outcrop way up on a ridge above a village and aligned along what had to be a fault-controlled valley. An unsealed road obviously terminated at the village. Will I, or won't I? It looked too good to miss. So, I drifted up to the village as quietly as possible, drove to the end of the street then continued up the steepening low scrubby hill. It was pretty good going, but 4WD nonetheless. A bunch of kids spotted me and started chasing and yelling. Obviously trying to get me to stop but I motored on and soon left them behind.

Having reached my target and confirmed that it was indeed silica-flooded, and hence prospective, I then had to decide what to do next. Do I return through that village to face a grilling or do I continue on from my vantage point to the next village I could see down the opposing fault valley? I chose the latter. As I was crawling down the slope, I could see two shepherds with a flock of socialist sheep all in a bunch and several stray anarchist goats floating around doing what they wanted to do individually. That's the thing about goats. They have a very different mentality to sheep. I prefer goats for their free spirits.

As I approached, I could see that the shepherds were confused. They could hear a vehicle and were straining to see down the valley towards the village, wondering where on Earth it was. I was within a hundred metres of them before they realised the vehicle was coming down the hill – not up. They were beside themselves, shouting at each other and shouting at me. Obviously, nobody in their experience had ever driven anything over that saddle. Obviously from my clothing, I was not a local. Not one of them. I grinned back and said: "Tourist". They kept repeating "Tourist? Tourist?" to each other and both leaped in the car forgetting about their sheep and goats. This was more important. They signalled for me to continue on to the village laughing and joking to themselves "Tourist? Tourist?" I spent the rest of the afternoon in the tea-house drinking copious quantities of that ultra-sweet Turkish brew 'Çay' with the men of all ages, conversing with the help of the school teacher who had spent time in Germany and had a few words of English. I certainly overdosed on tea that afternoon. "Çay?" "Çay?" It seems I was indeed the only person to have ever taken a vehicle of any sort over that saddle. Tourist! Tourist! Mad!

Two days later I was in the foothills of the Taurus Mountains east of Nigde. It had started sleeting and was bitterly cold. I had spent the afternoon convincing myself that I was looking at an ancient surface gold mine which had shed material on the gentle slope at the base of a steep ridge. I could clearly see the remnants of a water race along the break of slope and the furrows leading like the shape of branches of a pine tree in a regular geometric pattern towards the central trench. At the base of the hill there were piles of boulders clearly stacked beside the central trench. The ancients had been channelling water from a race at the top through the series of trenches to concentrate gold in a sluice box at the base. The box had probably been lined with a Golden Fleece! I confirmed later that there was some serious gold there.

Heading down a track towards the village, with about three kilometres to go on my way back to Nigde, I came across a young woman, stumbling along with a heavy bundle on her back. A young boy about ten years old was carrying a digging tool. It was almost dark, and they were clearly somewhat distressed by the sleet. I stopped and beckoned for them to get in. The woman pushed the boy into the front seat beside me and quietly climbed in the back with her bundle. The boy started talking to me and was rather shocked when I responded with: "Tourist! Tourist!" and patted my chest. He turned to his mother with a wide grin, but she hid her face. So we carried on quietly down the track with the boy grinning all the way. Near the edge of the village she motioned to me to stop. No worries. I understood immediately. They disembarked. She bowed in thanks. The boy waved. I drove on through the village past the teahouse, with the men all standing yapping. I assume her husband was there among them.

Almost the end of my journey and back in the big smoke. Nonchalantly easing my way along a gridlocked street in Ankara and munching on a Muska Boregi, a spinach and cheese filo triangle. I'd bought it from a street vendor weaving his way among the traffic. The reason for the gridlock was a busy roundabout intersection with a traffic light in my direction. A policeman or soldier (I couldn't tell the difference in Turkey), with a sub-machine gun slung over his shoulder, was madly gesticulating at the traffic, which of course was ignoring him. He was obviously getting very frustrated. The traffic light would turn red, but drivers were ignoring it and continuing on. How dare they ignore him? It came to my turn. Light turned red. I propped but the cars behind were blasting their horns so, in a fit of indecision, I started again. That was it. The young soldier had seen enough. He stepped in front of me hand raised, came around and started yelling at me, pulling a book from his bag. I grinned. Patted my chest. "Tourist! Tourist". The lad went bright red, ripped his gun from his shoulder, threw it on the grass verge, jumped on it then stomped off. The cars behind kept beeping so I drove off.

19

Working for Gaddafi
Libya, 1992-1997

Greg Ambrose

Greg Ambrose is a retired petroleum geologist of 45 years' standing, employed by exploration and production companies for 30 years and by government agencies for 15 years in Australia and Libya.

From 1980 Greg worked for Santos Ltd, Central Petroleum and Lasmo Plc in Australia, and with the Arabian Gulf Oil Company in Libya. He has been involved in numerous oil and gas discoveries both in Australia and Libya and has authored over 65 publications relevant to these areas. In the period 1998-2007 Greg worked in the NT Geological Survey as Assistant Director Petroleum Prospectivity. In 2007, he joined Central Petroleum where he held the position of Manager Geology until 2014.

I had been by employed by Santos Ltd in Adelaide and Brisbane over the period 1985-1992, but towards the end of that time I had become disenchanted with the company as it then was.

A colleague and very good friend of mine, Roger Grund, was similarly disappointed with the situation at SANTOS and we discussed our options over lunch on most Fridays. He had worked in Libya during the 1980's for 7-8 years and had made enough money to set himself up financially. He suggested I write to his previous manager in Libya to check out the job situation and that I should use him as a referee. The strange thing was that Libya had always been in the back of my mind ever since my days in the SA Government where I worked in the 1970's. I had become friendly with my boss Terry Watts, who had also worked in Libya and who had shared many of his adventures with me. Now it seemed my future could include a stint in Libya which could provide an escape route from SANTOS, which was not a great employer at that time.

As it turned out Roger's reference was pivotal and, at the start of 1992, I was soon offered a position as Senior Staff Geologist with Arabian Gulf Oil Company (AGOCO), based in Benghazi. Three months elapsed before I flew into that city, and I used this time to make preparations, consolidate the family situation, and generally psyche myself up for what would be the biggest adventure of my life. I was very excited and I could tell Jesus was close by, guiding me through my life. My wife Penny, who had no such faith, was of course apprehensive, but wasn't planning to come out to Libya until after New Year, 1993. She wanted to do a year-long course at Flinders Medical Centre specialising in intensive care studies. Thus from March until after Christmas I would be on my own in Libya – a very tough call really.

I spent a lot of time on the phone to my Libyan contact in London who was organising my transfer. He was an Englishman who was fully aware of the possible pitfalls I could face, and he was a great help. Just before I was about to leave for Benghazi, the Libyan supremo, Colonel Muammar Gaddafi, was again in the international headlines as the Lockerbie bombing was still lingering fresh in the minds of western countries, especially the US. As it turned out, the UN had ordered sanctions against Libya, including suspension of all international flights into the country. This was very disconcerting for me as can be imagined, and my contact warned me this situation would probably drag on for some time. He booked me a flight from Adelaide to Benghazi via Singapore and Athens. As it turned out we cut it fairly close; my flight from Athens into Benghazi was almost the last one into Libya and I was lucky to get into the country without having to go overland from Egypt or by boat from Malta. My departure from Adelaide is lost in my mind but I remember the tears, although I don't think the kids really realised what was happening. I would not see them or Penny again for 10 months … I didn't know it then but it was the beginning of the end for my marriage. Even to this day it is very painful to think back to these times.

I had never informed anyone from AGOCO about an illness I had, but I had to get 10 months' worth of medication into the country. Penny and I decided the best thing was to sew packets of pills into the lining of my coat. Roger had told me I would also need some US dollars, as there was a flourishing black market in Libya and the dollar was very strong against the Libyan dinar; these were also sewn into my coat lining. I took about USD5000 into the country which would last me the 10 months.

My intermediate flights to Singapore and then on to Athens were uneventful, although when I arrived in Athens I was missing home already and wondered what I was getting myself into. I remember as we flew into Benghazi I looked down and noted a conglomeration of pale cream/grey buildings abutting the blue waters of the Mediterranean. The surrounding foliage on the appropriately named Green Mountains reflected the end of winter, while I had left a brown Adelaide at the end of summer. On landing, I was met by a burly Libyan dressed in green AGOCO overalls. He seemed friendly enough and spoke English quite well. I recall looking skyward as I watched what was possibly the last international flight out of Benghazi disappear out over the horizon. Never again in my six-year stay would it be this easy to get into or out of Libya: the UN sanctions would see to that.

We travelled through the suburbs to the AGOCO guest house which was recently built. Everything was new and unspoilt, which was great. The rooms were decked out with all mod-cons, including air conditioning. The dining room was spotless, and I was pleasantly surprised by the food which was fine. On my first day in the office, I met all the Libyan bosses including Ahmed Asbali (Manager, Exploration Department), Abdullah Mansouri (Geological Manager) and Yunis Fatouri (Geophysical Manager). These people were friendly enough as long as you played within the rules and, generally, I found the management at AGOCO compared quite favourably with that which I had encountered in Australia.

The first thing I had to do on my arrival in Benghazi was to ring Penny. This turned out to be a frustrating exercise and it took me several hours to get through to Adelaide and speak to her and the kids. The weekly trip to the telephone exchange was indeed a stressful exercise but eventually I organised a system whereby Penny could ring me in the office through the AGOCO switch board but, as with all things to do with AGOCO, there were frequently delays and stuff-ups. My first three months were spent in the guest house, which was great because I was fully kept in most respects, and it did give me time to adjust to living in that city. Gaining basic knowledge of the road system, shopping areas, markets, hospitals and traffic rules was essential, the last being very rudimentary. The saying in Benghazi was "the definition of a nanosecond is the time span between the lights changing to green and the Libyan driver behind you bumping you from the rear". I soon learned that Libyan hospitals were places where you went to die rather than anything else, and that one should look after one's health, especially on the roads. I also quickly learned that driving at night was dangerous. I was surprised to see so many apparent motor bikes on the road at night, when they weren't there during the day. It soon dawned on me that what I was seeing at night were cars with only one headlight. Well, this of course was just great – not only were camels and sheep competing with drivers for road space, but most drivers were totally reckless and the attention to safety detail was minimal.

Travelling on the roads was indeed very dangerous. I can't recall how many times we did the 800 km road trip from Benghazi to Tripoli, but the road was one long traffic hazard. Dead camels and wrecked lorries and cars littered the road and were a constant reminder of what we were doing – risking our lives. In the later years I was there, particularly 1997, there was another major risk – namely carjacking. Gaddafi's enemies lurked in areas where expats could be easily encountered on the roads – the rebels would simply pull out a gun, hijack the vehicle and leave the hapless expats in the desert to find their own way home. The rebels were always causing trouble in the Benghazi region; they were basically local tribes from the east of the country who had been warring for decades against tribes from the west, who were loyal to Gaddafi. He finally met his fate in the summer of 2011 when he was murdered by the rebels, but more of that later.

Now Libya of course is a Muslim country where consumption of alcohol was outlawed. However, I remember in my second week in Benghazi a bunch of my expat friends took me to a local beach we called Los Palmos and which was about 15 km out of the city. Here on this beautiful, palm-festooned beach we munched on barbequed pork chops while downing copious amounts of potent home brew and also homemade wine. If you really wanted to get hammered, there was an illicit spirit, distilled from sugar by the Polish guys and known as "flash", which would certainly do the trick.

It was from this beach that a hapless expat, who owed the Libyans a lot of money and hence couldn't get an exit visa, beat a hasty retreat across the southern Mediterranean to Crete. He started wind surfing every weekend at Los Palmos, so as not to arouse any suspicions amongst the locals, until he was fully proficient. Then one weekend, he donned a small backpack and life jacket and wind surfed north from there. Sometime later he turned up in Crete, located some 120kms to the north. This disappearing trick probably saved his life and is the stuff of legend to this day.

Everyone went to the beach on weekends (there were four or five different ones) during the summer, with our numbers varying from 20 to 50. Generally, you could transport beer quite easily to the beach without any problems as the police tended to turn a blind eye. However, if you were caught with flash in your possession, you were in serious trouble. Flash had to be cut with water and mixed with soft drink. The Libyans had a taste for it but tended to drink it straight and very strong, and cases of blindness were not unknown. Thus the police, whilst frowning on beer consumption, became extremely upset if significant flash was found in your possession. Sometimes you could just be unlucky when carrying alcohol and be in the wrong place at the wrong time. This happened to a friend of mine whose Hash House Harriers' ("Hash") name was "Virgin". The police pulled him over at about 2-3 am on his way home from a party – they recognised his vehicle and, sure enough, it contained a large amount of beer so they threw him in the cells for the night. There the police on duty proceeded to drink the beer in front of him and had a great night. The fact that Virgin was driving while well under the influence was beside the point to them. There were no stereotypes in Libya, and certainly Virgin didn't qualify, but he was far from being on his own.

During my first year in Benghazi, I made many friends amongst the expat community and what could have become a very difficult year for me turned out to be a lot of fun with plenty of laughs, although there were times I missed the family terribly. There were plenty of expats in the AGOCO office and I got to know most of them over the year. Canadian and European geoscientists included good friends Doug and Kasha, Mercer, Jim, Bill, Randy, Alex, Guy and his family, David, Chris,

Roy, Rutger and Hermann. Guy's family comprised his wife Shirley, daughters Samantha and Kym, and son Andrew. They would become very close friends to Penny and my kids when they eventually arrived in 1994. I would end up sharing an office with Dave Huffman for five years and he and I became close over that time and I still keep in contact with him. Over that time he offered warm friendship and wise counsel. With those new expat friends, it was a bit like being in a war together and one formed a closeness with mates far more quickly than would normally be the case. I also had many expat friends amongst the corporate fraternity in Benghazi, namely Patrick, John and Clive at Brown and Root; John, Andy, Sarah, Paul, Tony and Katrina at the International school; Oyenn, Biddy and their family and also Frank and Oliveira, at the Irish International Corporation.

Despite missing Penny and the kids, my first year in Libya was a hoot and Jesus played a role in paving the way forward for me. My work role with AGOCO saw me jointly responsible for exploration by a small team of professionals in key acreage in the vicinity of the giant Sarir and Messla Oilfields. We drilled two exploration wells in 1992 near these giant fields, and both were discoveries which together delineated a new 300 million barrel oil trend. This was a great start to my time with AGOCO and stood me in good stead for the six years I was to be with the company. The credit for these two discoveries was split between myself (the geologist) and Rutger Gras (the geophysicist). As a reward Rutger, who worked as a consultant for Schlumberger, was transferred to Paris on an expat package – what a plum. I, on the other hand, received a small pay rise but I was happy with that as these discoveries would set me up for the rest of my stay in Libya.

Our discovery was just the news my Libyan managers wanted to hear.

So with work under control, I was able to concentrate on my social life in Benghazi which revolved around the weekend beach outings, the Hash fun runs, the darts competition and various barbeques and parties. I was having a great time, and this was enhanced when I left the guest house and moved into an apartment which was only a three-minute walk from the office. The apartment was great; it was located on the fourth floor of a block and comprised two spacious bedrooms, kitchen, bathroom-laundry, bar area and a living-dining room. One of my neighbours, a Pakistani geophysicist named Mercer, sold me his bar for $200. It was great and included a dart board, score board and bar fridge.

Its main function was to host "darts practice" on Tuesday nights, 'knock-off drinks" on Friday nights, and various spontaneous events which could occur on any night of the week or on the weekend. My bar was certainly a focal point for social activities and people were always dropping in for a drink or a chat, or both. Tuesday nights were probably the biggest night of the week when everyone would turn up for darts practice – I would have up to 20 people crammed into the bar playing darts games like "Mickey Mouse" and "500". The nights were lubricated with beer and flash and really were a lot of fun. The darts competition venues were usually large villas with their own entertaining areas, the competition was quite intense and the subject of much speculation and gossip from week to week. I was not a great player but enjoyed it immensely and indeed it was quite an honour to be selected to play on a team.

In summer on weekends everyone went to one of the beaches surrounding Benghazi for a barbeque and water sports. There was surfing, swimming, snorkeling, petanque and wind surfing to occupy the time and the social craic was great and, as always, well lubricated. Most of the Europeans were from former Eastern Bloc countries like Poland, the former Yugoslavia, the Czech Republic, Bulgaria, Romania and Russia. There were also ample representatives from western countries like England, Scotland, Ireland, Canada, South Africa, and a few from Australia. There were several Americans as well, but they were few and far between after the Lockerbie bombing incident for which Libya had been blamed, quite rightly as it turned out. Ironically the Libyans liked employing Americans, but few of the latter were keen to take the risk as the US authorities were very harsh on anyone caught working in Libya. This included anybody, not just Americans. An English friend of mine, who worked for a Korean conglomerate, flew to the US to buy hydraulic pumps for a project in Libya. As soon as Immigration saw his passport had been stamped in Libya he was arrested, thrown into prison for three months, and was only released when his company paid a $3m fine to the US Government. Another Australian friend of mine held a US and an

Australian passport and he was also nailed by US authorities who immediately sentenced him to work on a chain gang for three months. The Libyans tried to alleviate these types of problems by not stamping the passports of US citizens travelling into or out of their country, but few were willing to take the risk.

After I had moved into my apartment, I decided I should support social harmony and begin brewing beer to satiate the thirsty masses. A friend of mine in the apartment block was leaving the country so I bought all his beer-making equipment including 150 glass bottles with screw caps, and of one litre capacity. There were many Canadians rotating into Benghazi on a fly-in fly-out basis. Usually the cycle was two months on, one month off back in Canada. The point was that these guys were critical to beer production as they brought back to Libya large quantities of malt and hops, both essential ingredients. These "mules", as we called them, were more than happy to do this as they attended many of the functions, like darts practice and the beach events, where vast quantities of beer were consumed. The only ingredient for brewing we could purchase in Benghazi was sugar. You could buy 20 kilo bags at certain souks (markets), and access to local sugar was essential as you could hardly bring in by air the vast quantities needed, whereas the volumes of hops and malt were relatively small. Sugar was also the key ingredient and feedstock for the flash distilleries set up by the Poles, Russians and Filipinos.

The brewing process wasn't complicated, but it was effective, and we produced some fine beers of various types. The process started with boiling up sugar, hops and malt in a large saucepan filled with water, for a few hours. This brew was then poured into a very clean plastic container which had a volume of about 50 litres. Cold water was then added, and then yeast to trigger the process of fermentation. The timing of bottling was critical: usually the fermentation process formed a head in the top of the barrel as the brew bubbled away but one was risking trouble if it were bottled before the fermentation was complete. Unfortunately, the latter resulted in my first brew being a complete disaster. I enlisted my good friend Rutger to help. We did everything correctly; except we didn't think a tiny stream of bubbles emanating from the top of the barrel was significant. This turned out to be very foolish, but we went ahead and bottled and stacked our precious brew in crates in my second bedroom. However, several days later, just as I was nodding off to sleep, there was a shattering boom which emanated from there. I thought I knew what it was there and then, but I was paralysed with fear and just lay there for several minutes. Sure enough, there was a second blast shattering the peace followed by a third and fourth, etc. Yes, you guessed it, we had bottled prior to the completion of fermentation and the inevitable result was that most bottles had blown up. When I raised enough courage to enter the room the beer crates looked like they had been hit by a mortar bomb. There was glass everywhere and several plastic bottles had shredded vertically under the tremendous pressure. There were even glass shards embedded in the ceiling, which was particularly unnerving. Rutger and I wondered how these mini "bombs" could be disarmed. The first idea was to fill the bath with water, wrap the bottles one by one in a towel and then try to loosen some of the screw caps. We hoped we might even salvage some beer from this disaster, but this proved impossible as the tops would not budge. We decided to write off the entire brew and we placed the remaining bottles on the flat roof of Rutger's villa which was located a few kilometres from my apartment. It was the middle of summer and our plan was that the glass bottles would explode harmlessly in the heat of the day and this part of the plan worked fine; we had to get something right. Rutger did tell me the whole villa shook when the beer bottles exploded one by one. We, of course, became a laughingstock amongst the expat community and, even many successful brews later, the farce that was my first brew still drew plenty of laughs.

One event which was also good for a few laughs was the Hash weekly fun run followed by drinks and food. Like all social events in Benghazi, it was very well lubricated and a lot of fun: it was the highlight of the week. The run was rugby club humour at its best. It usually comprised a 3-4 km cross-country run followed by a ceremony highlighted by a series of beer "down-downs". Every "hasher" had a hash name – my name was "Rice Pudding", after the one produced by Ambrosia. The hashers were dominated by Irish/English expats and also by a bunch of eastern Europeans, including males and females. The trail was set by hashers using water-based paint to lay it out in the countryside, and it could vary from beachside courses to ones in the Green Mountains. The Libyans didn't seem to mind these strange Westerners daubing their farms and countryside with paint and they seemed to have become used to it over the decades. However, we always familiarised ourselves with the local police checkpoints near the trail as they often pulled us over to examine our documents. It was a good idea to have a spare pizza in the car as the police were very partial to them: they always happily accepted the donation and waved us on without even noticing the driver had imbibed a skinful of beer. That's just the way it was in Libya.

The Hash was held at one or another of about a dozen various local venues we named, including "Los Palmos", "Container City", "Red Balls", "Green Mountain", "the Wadi" etc. The last was interesting as it was a linear, shallow depression about 10 km long which wandered through Benghazi suburbia. Unfortunately, there was a hospital nearby and its personnel had the habit of dumping used medical equipment in it, presumably because there was nowhere else to put it. A lot of the plastic tubes, douches, bandages, etc were blood-stained and totally disgusting. Luckily, I never came across any bodies, but it wouldn't have surprised me had I done so. Some of the sick bastards who set the Hash trails thought it was a bit of a lark to run the trails through this discarded paraphernalia. Very funny I don't think: this was just the way it was in Libya.

I can remember driving through a checkpoint on the way to the Hash one afternoon when an armed soldier pulled me over and jumped in beside me: as he cradled his AK-47 I noticed it was pointing straight at me which was a little disconcerting. With the few words of Arabic I had learned, I tried to gauge what he wanted, remembering I had 20 litres of beer in the boot – perhaps he would like a piece of that action? The soldier, realizing I was just one of those crazy expats, then pointed his gun in the direction he wanted to go and I meekly headed down the road back in the direction from where I had come. I drove for about 20 minutes before we finally pulled up at a small group of buildings – he jumped out and a few minutes later he came back with a box full of ice-creams with which I presumed he was going to treat his mates back at the checkpoint. This indeed was the case. I arrived back at the Hash a bit late just in time for the down-downs. The hashers all had a good laugh about my escapade, especially the Virgin who had occasionally spent the night in the police cells. For myself, I consumed several "down-downs" to calm my nerves after my little adventure. Again, this was just the way it was in Libya.

English was the chosen language of the Hash but the eastern Europeans, especially the girls, had a very limited vocabulary. Still with so many expats in Benghazi, including single and married males, the girls were always a likely target and many liaisons survived the language barriers. Some expats even married their ladies under Libyan law unbeknown to their real wives back in the UK or wherever. This was quite common but, unfortunately all too often, the marriage back home would collapse under the pressure of these bigamous liaisons. I for one did not partake, except for minor flirtations which were pretty harmless.

One funny incident occurred in Tripoli airport while I was waiting for a flight to Benghazi in a beat-up Libair 727. The flight was delayed, as usual, and it was eventually announced that family groups should board first. I was just considering my situation when a young lady I knew from the Hash in Benghazi came up to me and said: "Come on Greg, let's get on board as man and wife". We got away with this, no worries, and this was one way of expediting the trip. We shared a down-down at the next run and had a laugh.

Many Benghazi expats worked for international companies who had constructed large camps to house their employees. These included companies like Brown and Root, Motherwell, Dong Ah, and the Irish International Corporation. I spent a lot of time at the Brown and Root camp where I had quite a few friends. I was always popping over for barbeques or to play squash with a friend of mine named Patrick. Similarly, their guys would come over to my place for darts practice and various barbeques and, of course, every week we met up at the Hash for the run and a few down-downs. What a great bunch they were! Aside from squash and basketball, there were also tennis courts and even a golf course we named Royal Benghazi Palms. The latter was just a patch of desert with scattered palm trees and rows of white pegs marking the fairways. The greens comprised black oil sand which needed to be raked for each put. If you raked the sand the right way the ball often had no choice but to be channelled into the hole. The Koreans had built the golf course some years before and it needed some TLC as camels and a few sheep were usually the only inhabitants: green keepers were hard to find in Benghazi.

Brown and Root was an interesting company, having its head office in the US. The Libyan operation was run by a fully-owned UK affiliate (Brown and Root UK) which circumvented the ban on US companies operating in Libya. This I found very ironic considering how the sanctions were impacting on US citizens and others who were caught travelling to Libya as I have already mentioned. Admittedly, as far as I knew, there were no US citizens working for Brown and Root in Benghazi, but everyone knew where the profits from their operation were ending up – New York. Never mind that, why let a few sanctions upset a lucrative money trail back to the US?

One of the main social activities for expats and Libyans alike were charcoal-fuelled barbeques. There was plenty of scope for these, whether the get-together was at the beach, the Green Mountains or in someone's apartment. There was plenty of fresh meat supplies in Libya as food was subsidized by the government. The local lamb was delicious and cheap in the butcher shops. The beef was imported from Ireland and was also very cheap. The weird thing about it was the Libyan butchers didn't seem to have a clue on how to butcher a whole beef carcass. If you went to buy, say, 5kg of beef, the Libyan butchers would hack it off from just about anywhere on the carcass. This presented an opportunity as the butcher charged the same price for beef no matter which cut was on offer – they didn't know any better. I knew my way around a beef carcass pretty well because of my years working on Anna Creek Station in South Australia and eventually I schooled the butchers in cutting out the fillet, rump, and sirloin cuts for us. No matter what the cut, the price was the same. I've never looked a gift horse in the mouth so, yes, we did eat a lot of great fillet steak bought for a couple of dollars a kilogram. Fresh fruit and vegetables were available from a number of local markets and the quality was fine and the prices were cheap. There was also a thriving fish market near the port area where you could buy all types of bream and snapper, as well as squid and Mediterranean crayfish. I spent my 40th birthday in Libya and my Pakastani friend Mercer put on a bit of a feast for me in his apartment including fillet on the barbeque. There were plenty of liquid refreshments, great music and bonhomie which lifted my spirits, as it was now August, and I hadn't seen the family for six months.

Overall. I found the Libyans to be "laissez faire" Muslims who were not attached at the hip to their faith. However, during the Ramadan fasting period, one had to be careful. During that time, Muslims fast all day and start eating at sundown so expats should not openly eat, drink or smoke in front of them during the day as some could become a little bad-tempered. So what does my Pakastani friend Mercer (theoretically a Muslim) do during Ramadan? He wanders down the hallway at work with a cup of coffee in one hand and a cigar in the other and comes into my office to have a chat. The big boss, Asbali, happened to walk in at the wrong moment and caught him cold. Mercer said he was just holding the coffee for someone else and promptly put the cigar out. Asbali, who was a good guy, didn't say anything and just walked out.

The months rolled on and I started to think about booking my first trip home. The travel sanctions were being applied very strictly, but there were several ways to exit Libya and pick up an international flight to Singapore. One could travel by bus to Cairo where you could get a connecting flight to Singapore; or you could bus/taxi it to Tripoli, then catch a boat to Malta and connect onto Singapore via Frankfurt, Amsterdam or Zurich. A third way to exit Libya was via Tunisia – one could take a taxi to Tunis and pick up a connection from there. Alternatively, there was a great little tourist town at Jerba, which had its own airport, and flights to Tunis could connect with flights to Europe and on to Singapore. None of these exit routes were ideal and, although some of them may seem exotic, the novelty soon wore off I can tell you. Crossing Libya's eastern border (with Egypt) or western one (with Tunisia) were extremely frustrating to say the least. The boat trip to Malta was the same, although Malta was quite a nice destination as were Cairo and Jerba. The taxi trip from Benghazi to Tripoli was fraught with danger, as I have already mentioned. The trade-off was to fly to Tripoli on the local airline, Libair, but their beat-up old 727's didn't inspire much confidence either. The sanctions were on, and hence the spares for Libair's antiquated planes were a little less than satisfactory, plus there were other political risks on which I will elaborate below.

Finally, my exit date came around and I rose early to catch a bus from the bus station, where the crowd was more like an Arab rabble at a strip tease than anything else. After much pushing and shoving, and payment of a 30 dinar bribe, I finally got my suitcase loaded onto the bus, and took my seat without further hassle. I started to think about the four mates I worked with who were flying out to Tripoli on a Libair flight that same morning. I thought I might try that the next time I went on leave. The bus trip to Cairo was uneventful enough except for a prolonged hassle at the Egyptian border which lasted four or five hours. It was obvious the Egyptians and Libyans weren't soulmates. The whole trip took about 12 hours, but I found a decent hotel and was resting on a bed with a cold beer when I heard a news bulletin. Apparently a Libair 727 had crashed just outside Tripoli and there had been no survivors. I figured it was the previous morning that it had happened which meant it was the flight to Tripoli my friends were on. I went into shock for a while and spent some time praying. My prayers were answered when I bumped into a colleague in the hotel lobby who assured me no expats were on that flight and that they had caught another later one that morning.

To elucidate what exactly had happened to the doomed flight I had to wait nearly two decades. However, firstly I will refer to the gossip that was around in Benghazi when I returned from Australia in mid-February 1993. The news about the crash was that Gaddafi was responsible and that it was supposed to be some kind of crazy gesture to satiate the West's desire for revenge for the Lockerbie bombing, even though no one had been charged at that point. The Libair flight number was the same as for that doomed Pan Am flight and there were no western expats on board – only Libyans. Supposedly a bomb had been placed on the Libair flight, but it had failed to detonate, so air force jets were scrambled and it was shot down. Gaddafi immediately had the site bulldozed and the government treated the incident as an accident but made no further comment. So what was all this insanity about? Some expats suggested the murder of Libyans in this "accident" was to illustrate that the crash was due to a spare parts crisis engendered by the sanctions against Libya and that, indeed, Libyans were suffering greatly because of them. I knew Gaddafi was insane, but I didn't think he was this unhinged and hence I dismissed this version of events. It wasn't until 2013 that the true events became apparent to me via a TV documentary put to air in Australia which included consideration of the assassination of Gaddaffi in 2011. The program was directed at changes in Libya since the onset of the "Arab Spring". One point of interest during the program was the assertion that all the unrest and violence was a telescoped result of the Global Financial Crisis – one portion of the program showed a very senior Libyan army officer talking about the effect on the armed forces as he walked through a military hardware dump. There were various remains of MiG fighters but, significantly, the officer explained that a large piece of fuselage from a 727 which "crashed" in late 1992 was indeed the remains of a domestic jet shot down under Gaddafi's orders at that time. I nearly fell out of my chair when I heard this, and I was sad to think that this was the only evidence of what had really happened so many years before. Gaddafi was indeed definitely unhinged during his tenure as leader and, in the end, deserved to be taken out by his own people.

My stopover in Cairo, in late 1992 on my way home to Adelaide for Christmas, was fun as I teamed up with an Egyptian friend of mine and did a few "touristy" things. I rode a horse around the pyramids and visited the famous Medina Hotel where Sadat and Jimmy Carter held a seminal meeting in the late seventies. We visited some great markets where I bought a few presents for the kids and Penny, and we also had a good look around the Cairo Museum which includes fantastic exhibits from Egypt's ancient culture. My connection was a direct Singapore Airlines flight from Cairo to Singapore – this was a far cry from Libair and I was very impressed with the service, food and entertainment. I'll always remember bopping away to the musical comedy "Four Weddings And A Funeral" which had just been released at that time. It was December 23rd when we touched down at Changi Airport and I had a day to kill. This would mean I would get home the day after Christmas, which was a big disappointment for me and the family. Never mind, I rang Penny, Georgie and Sam from the Raffles Hotel on Christmas Day and then settled into a few beers as I soaked up the flavours of a sophisticated city as compared to those of Benghazi: no comparison really, but I had had a good year healthwise, and I was very pleased to be home for a good six-week break.

My QANTAS flight touched down in Adelaide on Boxing Day and my reunion with the family was very emotional – I can hardly remember it now as my mind was so fuzzy. Sam and Georgie hadn't seen me for 10 months and I could see the changes in them since I had last seen them in March. However, blood is thicker than water and the old bonds were quickly re-established as we made the most of our time together.

The family had a great holiday on the west coast around Port Lincoln where we soaked up some of the best seaside scenery in Australia. We fished for King George whiting off the beach and enjoyed a variety of seafood which was always the case when we stayed there with our friends Scott and Jenny. Penny and I also had a getaway together at an old manor house located in the hamlet of Padthaway in the southeast of SA. My mood was buoyant and everything seemed to be going well as far as the family's reunion was concerned. My six-week holiday in Adelaide was mainly spent catching up with family and friends and everyone noticed I seemed well and "quite chipper". I just hoped it would last as I never wanted to go back to the lows of my SANTOS years. It seemed Libya was my saviour and I hoped things would continue to improve for the rest of my life. Unfortunately, there was fat chance of that happening as things evolved through 1994 and later decades.

Penny wanted to continue her nursing studies at Flinders University so she and the kids stayed in Adelaide through 1993 while I "kept on truckin" in Libya. I flew back to Cairo at the beginning of 1993 and then caught the bus to Benghazi. I had USD5000 again sown into the lining of my coat along with my medication. Those times are a little vague to me now, but my health was still quite good and work was going satisfactorily. My social agenda was split between the Hash, darts, brewing and various parties at villas owned by major contractors. Brown and Root had a great camp not far from my apartment, replete with squash court, tennis courts, and a swimming pool. I played squash there once a week with Patrick, who worked with that company. In the summer months I went to the beach most Saturdays which was great – swimming, barbeques and lots of beer were always the going fare. I also played golf occasionally at Royal Benghazi Golf Club when it was free of camels and sheep.

The main task in Libya for Brown and Root was the "Great Manmade River Project" which aimed to bring fresh water from deeply buried aquifers in the Sahara Desert to irrigated farms stretching from Benghazi to Tripoli along the coast. I can remember driving between those cities and marvelling at the sight of fleets of massive semi-trailers stretching many miles from horizon to horizon. Each of the trucks carried one massive piece of pipe with a diameter of about 20-25 feet and which were lined with copper for protection from corrosion. Each of these sections of pipe, one per truck, weighed many tonnes and were valued at $50,000 each (1993 dollars). Deep under the desert there were sandstone aquifers which held fresh groundwaters entrained from Permian-aged melt waters dating back to the thawing of that much earlier ice age. The general idea was to pipe this fresh water to where it was needed most near the coast for farming and water supply. The only problem was the Libyans weren't really beholden to a long history of farming per se; they were historically nomads and herdsmen from desert areas. So, what did they do? They brought in Egyptian farmers more attuned to cultivating crops, etc. The other thing about this project was that the freshwater resource was ultimately limited and obviously there was no potential today for recharge by meltwater from a thawing landscape. How this volume restriction affected things in later days I do not know and, unfortunately, it wouldn't be safe to go there today to find out how the project has progressed over time.

Libya is famous for its ancient ruins and the nearest of these occurs at Cyrene which is a couple of hours drive east of Benghazi. The port city to Cyrene was named Apollonia, which these days is almost completely inundated by the Mediterranean Sea. Whenever we visited these ruins, we usually took a barbeque lunch and either had it on the beach at Apollonia or on top of the giant ridge where Cyrene was perched, looking down on its sister port city. It was a very powerful and beautiful vista which I would recommend to anyone. The ruins themselves were fascinating and in pretty good condition. The only disappointment was that the miniature Italian railway which ran around them was no longer in working order. The Libyans were not known for their maintenance or engineering skills sets and, as in too many Middle Eastern countries, many expat-driven projects quickly became dilapidated once under local control.

My good friend, Rutger, was a frequent visitor to Appollonia and Cyrene. On one trip he was approached by two young teenagers who had a few coins in their possession. They enquired whether Rutger would be interested in buying these Roman coins to which he replied in the affirmative – a price was negotiated, and Rutger thought no more about it. A few hours later however, who should turn up but two of the local police? They invited him back to the local police station. Here they threw him in the cells and charged him with procurement of antiquities which, at the end of the day, were the property of the Libyan Government. Being quite bemused by this, Rutger had to cool his heels in the cells for a couple of nights. Eventually he paid a small fine with US dollars and everyone walked away happily – except Rutger who was quite miffed he had fallen for this little scam, but he had learnt his lesson.

In mid-1993 I decided to bring out the family in early 1994. I was to trip back to Adelaide via Malta and Singapore, pick up the family and then return to Libya via Malta. The Maltese people were very friendly and the whole place had a warm feel to it. I must admit it was always better to pass through Malta on the way out to a holiday rather than coming back to Benghazi from one. One inviting thing about Malta was that it was reasonably inexpensive with good quality restaurants and hotels. The main reason to travel via Malta was that, because of the sanctions, the best way to get to Tripoli was on water. I wouldn't refer to the passenger craft as ships as they were more like large boats. There were other overland routes via Tunisia and Egypt, but the Malta connection was the most popular.

However, getting on and off the boats in Tripoli was an ordeal. First of all, you had to physically battle your way off the boat with all your gear which wasn't easy – then you had to go through customs and get all your luggage searched. The final hurdle was to find a taxi capable of getting me, and the family, to Benghazi – a trip of some 800 kms. Getting on board the boat in Tripoli was also a nightmare. Once you were on the boat, there were other hurdles to contend with. I remember that, on a later trip, one of the boats had a list so severe my daughter Georgie was petrified of falling into the Mediterranean. If there was any food available on board at all, it was pretty ordinary and usually the sanitary conditions left a bit to be desired. If you were too late boarding to get a bed in a cabin, you were relegated to a deck chair on the main deck out in the open air. There was no air conditioning on the boat, which was acceptable in summer when the cabins were stifling – that's assuming you can sleep in a deck chair.

I remember a time when the whole family were trying to board a boat in Tripoli Harbour. First we had to get a boarding pass by getting in a queue – no, I am wrong, it wasn't a queue, it was a melee which went on nearly all day. After about six hours of trying, we finally succeeded, and Penny and the kids at last got off the stinking hot wharf and onto the stinking hot boat. Whilst boarding I was stopped by a Libyan security official who indicated I had to go back to the wharf to get my passport restamped. I knew from experience this was bullshit, so I just walked back into the crowd for a moment, waited a minute or two, and then walked straight past this guy again without stopping. He let me do it without a word – he was just playing a little game with an expat. Meantime we had been so late boarding we had missed out on a cabin and were relegated to the deck chairs. This was just great – we were all exhausted but, as luck would have it, an official had spoken to the captain who welcomed us and allowed us to have the last family cabin on the boat – Jesus had smiled on us again. We actually got some food and drinks on this trip and rested easy that night even though we knew we would have to go through similar nonsense when we disembarked in Malta the next day. Getting in and out of Libya during the sanctions was a "pig fight" all right; no matter if you went via boat to Malta, taxi to Tunisia, or bus to Cairo.

Back in Benghazi in early 1994, everything was fine and I quickly settled back into my routine of work and social activities. It was great to have the family with me and they were enjoying Libya and all the ancillary activities like the Hash, the darts competition, trips to the beach and also exploring the Roman ruins. I was playing squash and tennis regularly which, along with some jogging and the odd game of golf, were together keeping me reasonably fit. They were obviously having a great time and Georgie, in particular, although she was only seven, seemed to catch the travel bug, which stayed with her all of her life. The flight to Cairo was uneventful but we didn't have time to do any "touristy" things before we flew to Malta. We stayed in a great hotel there right on the water and boarded the boat the next day. We were to land in Tripoli the day after that.

The boat trip entailed the usual hassles, but we did get a family cabin, and successfully disembarked in Tripoli. AGOCO had a family guest house there where we could overnight before catching a taxi to Benghazi the next day. The guest house was pretty basic, but Penny took it in her stride, as she did most things, which was exactly the right approach if one were to survive in Libya. We hired a taxi the next day, a Peugeot 404 station wagon, and made the eight-hour trip to Benghazi without incident, although Penny and Georgie didn't think much of the Muslim toilets. As we entered the city, Penny reminisced about her father who was in the Australian Navy during the war when his ship had shelled the town during the days of Rommel's sorties into Cyrenaica in the early 1940's. It now seemed a strange coincidence that here we were now, 50 years later, returning to the place of all this wartime activity. There were indeed a number of war time cemeteries in the town, which included a lot of Australian graves, and every year the expats met for a Remembrance Day ceremony to honour the fallen.

On arrival in Benghazi, I showed Penny around my apartment and she was quite impressed as it was quite roomy. We threw a "Welcome to Penny and the Kids Party" where 50 guests feasted on barbeque fillet steak, chicken, and lamb, with accompanying salads and, naturally, there was plenty of beer, wine and flash. This was the largest party we ever threw in the apartment and it was a great night; of course there were many others to come before we parted company with Benghazi. Penny and the kids would eventually leave Benghazi in 1996 while I departed at the end of 1997. There were many adventures before I got back to Australia but these await a second story.

I can look back at the period I was employed by AGOCO with a considerable degree of professional satisfaction. I was responsible for the geological exploration in key acreage of the giant Sarir and Messla Oilfields. As mentioned earlier, I provided the geological model for two successful wildcat wells which delineated 300 million barrels of oil – worth almost USD20 billion in the ground today. I also recognized a previously by-passed oil zone in the giant Sarir Field, which contained another 90 million barrels. A complete revamp of the "Nubian" geology of South East Libya was authored by me and published internationally: it focussed on the stratigraphy and hydrocarbon habitat of the Sarir

Sandstone, the main petroleum reservoir in that field, and the paper has been cited in 64 publications to this time.

Over the years I have kept an eye on Libyan affairs and watched the ebb and flow of its political standing in the world scene from the terrorist scandal of the Lockerbie bombing in 1987 right through to its present status as a failed state. Events in Libya were particularly tumultuous in the period following the murder of longtime leader Muammar Gaddafi in October 2011 during the height of the Arab Spring. For some time, Libya remained a country without any semblance of reliable, stable government. Armed militias roamed unchallenged and ISIS and Al Quaeda became entrenched in the country. I have read that, in a private telephone call between Tony Blair and Gaddafi, the dictator informed Blair that, if he were deposed or assassinated, it would trigger a mass exodus of Muslims from North Africa into Europe via the Mediterranean. This is indeed what happened, which is reminiscent of the aftermath of the disintegration of Yugoslavia after Tito's death in 1980, with wars and an exodus. A more amusing topic for consideration was Boris Johnson's (then Britain's Foreign Secretary) 2017 visit to Benghazi to meet one of Libya's strongmen, namely Field Marshall Khalifa Haftar. His arrival at Benghazi Airport (which I know well), was a bit of a farce. Here is this Field Marshall in a rag tag uniform supported by an equally shabby brass band trying to play "God save the Queen". The five band members looked like something out of "Sergeant Pepper's Lonely Hearts Club Band" but their version of "God Save the Queen" was hysterical – what a noise. The news services in the West know a good laugh when they see one and this incident got repeated coverage on all the news channels there.

However, deep in my heart, I wish Libya and its people well and hope they can formulate a unifying and stable government soon. This saga is still unfolding and a recent meeting of leaders from Europe and Africa (in Brussels in early 2020) made some progress to this end. The faction in Tripoli seemed to have most support but negotiations continue between it and that of Khalifa Haftar in Benghazi. In 2020 Turkey offered troops and some stability as a Tripoli-based strongman was installed as the leader of this troubled land and the Benghazi lead rebels took a back seat in this power struggle.

20

A Helicopter Gunship in the Karatau Range
Kazakhstan, 1997-1999

John Hammond

John Hammond graduated as a geologist from the University of Western Australia in 1969. His career took him to more than 25 countries, primarily as a grassroots explorer, but he also had significant involvement with feasibility studies and corporate acquisitions. He began working with AMAX, exploring for nickel in the Eastern Goldfields of Western Australia and then spent nearly eight years on the Namosi porphyry copper project in Fiji. In 1988 he transferred with Australian Consolidated Minerals to Perth and from that base he managed exploration in Western Australia, the Northern Territory and the Mediterranean region. In the latter area he led further exploration success in Turkey and Greece. Before retirement in 2013, John spent the previous 20 years working as an internal geological consultant, firstly for Normandy Mining and, later, in a similar role with Newmont.

At the request of our employer, in early February 1997 I travelled to the Republic of Kazakhstan with the late Bob Gunthorpe, who was then chief geologist for Normandy Mining. We obtained visas for a one-off entry with no problem, provided conveniently through the mail from the Kazakh embassy in Sydney. Our mission was to investigate an exploration opportunity offered by Dabney Industries Corporation in south west Kazakhstan. Now it was the middle of winter in the northern hemisphere, and very cold on the Central Asian steppes. When we pointed this out to our masters and said we may not be able to do much effective field work due to snow cover, we were told that we should go anyway, assess data and make contacts in preparation for a later field inspection, if warranted.

I flew from Perth to Frankfurt in Germany, then on to Almaty which at that time was the capital of Kazakhstan: it is now Nur-Sultan. The last hour of the Lufthansa flight into Almaty was spectacular, as it was early morning with clear skies and the plane flew parallel to the snow-covered peaks of the southern Tien Shan Mountains. This was in stark contrast to my arrival into dimly lit Tashkent in the middle of the night a few years earlier. I cleared immigration and customs without much delay and found a driver with a sign and a car waiting for me. After a twenty-minute journey through the light early morning traffic, I was dropped at quite a nice hotel and met up over breakfast with Bob, who was already there.

Next day we met the team: Mark Wilson, Dabney's in-country representative; and from their Kazakh partners, Shaukut, interpreter extraordinaire; Yulia their in-house petrologist; and an Almaty base office manager. As I recall, we did not see much data relating to the project in the next two days except for the location of the tenements (which covered some 64000 Km²) and the various prospects within the area. We did spend a lot of time, rather prematurely, looking at potential new office accommodation!

Anyway, on the third day, as the weather was clear, it was arranged that we would travel to the tenement area around Kentau (which is situated about 670 Km west of Almaty) with Mark and Shaukat to meet the field exploration team and examine data pertaining to some of the main prospects. We took off in a Russian Mi-8 from a helipad close to the city. The helicopter had a crew of four and capacity for up to 24 passengers. Needless to say, we had plenty of room for luggage! In the clear and sunny skies, the scenery was awesome as we flew along the northern side of the Tien Shan. We arrived at Kentau in the early afternoon, after a four-hour chopper trip, and were taken to the hotel. The first thing we were told after checking in was that the power came on only twice a day; from 7.00 to 8.00 am, and from 7.30 to 8.30 pm. We learned later that this facility had opened just for us, and only for the period of our visit.

Kentau is located at the foot of the Karatau Range, 30 km northeast of the city of Turkestan. It was developed as a major lead and zinc production centre in 1955, around the village of Myrgalimsay, and was considered a model mining town in Soviet times. Initially the population of the city was mainly from Russia, the descendants of the repressed: Greeks, Russians, Germans, Koreans, Jews, and Chechens, as well as Uzbeks and ethnic Kazakhs. By the time of our visit, the mines were all closed and less than ten percent of the population of 80,000 was employed. The town had intermittent electricity and water supply, only when pumps were running for care and maintenance of the mines; hence the limited power in the hotel.

Despite the fine day, it was snowing at night when we walked from the hotel with Mark and Shaukat for dinner. The restaurant, a couple of blocks away, was run by two enterprising Korean ladies who opened it up the day we came into town and closed it again on the day we left. There, over dinner, we met the Kentau geologists, Viktor, Kassim and Babajan. The meal was very good, accompanied by a few shots of vodka to warm us up.

After dinner, it was still snowing as we made our way back to the hotel. There was no street lighting as the power was off. I should note here, that for the duration of our visit, daytime temperatures were commonly just below 0^0C and at nighttime they dropped to about -10^0 C. I was very glad I had packed a torch as the hotel was dark by the time we returned, and some light was needed to get back to the room on the second floor and to get into bed. The latter sounds a trivial matter, but it was not. The beds were not made up as we would be used to; to keep warm, one had to negotiate the body into a sort of sleeping bag made of woolly animal skins. It was convoluted and complicated, not simply a sack with an opening at one end. Anyway, when I finally got inside it was extremely warm, and that was essential. Next morning, after I woke up, I tried to have a shower. No luck there though, as the pipes were frozen. I went down to the front desk and managed to get a bottle of water to clean my teeth.

Every breakfast, lunch and dinner were at that Korean restaurant as it was the only one in town. It was snowing and we spent the first day at the cold geology office, with Viktor the chief geologist giving us an exploration overview of the large tenement holding. Shaukut was of course translating for us. In subsequent days Viktor, Babajan and Kassim, all in turn, described their 20 or so priority prospect areas. Now in Soviet times, it seems that an exploration geologist would be assigned to an area for a very long time, maybe even a lifetime. He or she would map it and carry out exploration activities such as geochemical sampling, trenching, drilling et cetera. All the Kentau geologists had worked on their areas for substantial periods, so they knew them in great detail. There was, as we had predicted, no chance to go to the field to examine prospects as they were all covered in deep snow.

At the end of day four, Mark advised we should return to Almaty while there was a break in the weather. We had seen enough data, so next morning we took off for Almaty but, this time, the helicopter was packed to the gunnels. As a goodwill gesture, Dabney allowed some locals a free ride to the capital. The chopper was full of people, bags, produce (from greenhouses?) and even a few chickens. With no more to be done in Almaty we headed for home, after about a week in the country. At that stage we were not allowed to copy any maps or documents – which were all notated in Cyrillic script anyway.

Bob and I recommended that this opportunity did justify further follow-up and so, about a month later, I returned to Kentau with Barry Davis (another experienced Normandy geologist) and several geologists from the French company La Source, (a Normandy/BRGM Joint Venture). The routine was much the same as on my initial visit; meetings in Almaty, a helicopter ferry accompanied by Mark and Shaukat, the hotel and restaurant opening especially for our stay, and exploration briefings in Kentau. I had hoped that the snow would have thawed enough to visit some key prospects on the ground, and we did fly by helicopter out to a few areas. However, they were still mostly snow covered, so only limited outcrop could be inspected and that was often not over the main targets. On one of these sorties we were caught in a snowstorm on our way back to base. With the ground covered in snow, it was almost total white-out and this was a very dangerous situation for an airborne helicopter. The pilot maintained his horizon by flying low next to a telegraph line that was still visible. The whole episode lasted for nine minutes but it seemed very much longer at the time.

The second trip to Kentau concluded without any further excitement, and Barry and I headed home to Perth, still without having significantly advanced the case for a joint venture with Dabney. When we got back and discussed things further with Bob Gunthorpe we felt that, as Kentau had been a centre for lead and zinc production, and there had probably been only fairly cursory attention to gold exploration in the wider area, we should make another trip to carry out due diligence when the snow thawed.

So, in April 1997 Barry and I returned to Kentau and managed to do enough field reconnaissance (that included visiting and sampling two known gold prospects) to convince us that a deal was justified. We recommended accordingly, but it took some time before our recommendations were accepted. Consequently, it was not until July in the following northern summer that I returned to Kazakhstan to kick off the exploration programme with him.

The planned field campaign was two-pronged. We would conduct stream sediment sampling over the whole tenement block, using a technique suited to low level gold detection; and detailed mapping and sampling would be carried out to prepare the two most advanced prospects for drilling. To stream sample the entire 64000 Km^2 we needed helicopter support. The only machine available was a big Russian Mi-8, very different to the smaller, nimbler, Bell JetRangers that we generally used for such work elsewhere. This particular helicopter still had the odd bullet hole in the fuselage from days when it had served as a gunship in some far-away foreign conflict. Our field team consisted of the three Kentau geologists and casual labour who they hired, plus Barry and me. Mark and Shaukat provided all the logistic and translation support that we needed.

It was now summertime and Mark had organized a rental house in Kentau that served as our base. It appeared to be of Greek-style architecture and had an outdoor area where all the crew could get together for a beer and a meal after a long day in the field. (It is interesting to note that there were still enough Greek people living in Kentau to warrant a bus service once a week to Thessalonica in northern Greece; a road journey of over 5000 kilometres.) Mark engaged, as cooks, the two Korean ladies who had run the restaurant during my previous visits. The house also had a banya (sauna) which was great after a strenuous day's fieldwork. We had no communications with the outside world and no budget to cover such a necessity. Mark made a unilateral decision, completely justified on safety grounds, and purchased two expensive satellite telephones. In 1997 these were bulky and looked like large laptop computers; one was to stay at the base house and the other in the helicopter when we were out working. Now these 'sat-phones' could communicate well with each other locally, but they needed a fixed, tuned, dish (aerial) if we were to be able to make international phone calls. Shaukat, who had many skills, undertook the task of setting up the dish.

Here I must digress; Shaukat (nicknamed Shortcut, referencing his translation skills), was a very talented and funny guy who often made us laugh. Sadly, he passed away several years ago. He spoke about 10 languages which he had taught himself by listening to shortwave radio in the Soviet era. If he had been caught, he most certainly would have ended up in prison. During his compulsory military service, he had been an engineer on Mi-8 helicopters and often, on our arrival back at base, he would evict the engineer from the chopper and then conduct the five minute shutdown routine himself. He also took great pride in the fact that he had, during Soviet times, been sent to West Germany and received training by Siemens to install, operate and service a CT ('Cat Scan') medical imaging machine that the Kazakh government had purchased for the main hospital in Almaty. This was the only one outside Russia in the Soviet Union and he kept it going for over 20 years, without ever having to go back to Germany for any backup training. Shaukat was also a little eccentric, he owned a black Mercedes Benz car which he kept in his garage in Almaty. This was his pride and joy, but he never drove it as he had no driver's licence.

Anyway, back to the satellite dish installation: Shaukat had spent the whole afternoon finding the optimum position and fine tuning the dish on the roof of the house. Now Barry had taken some Aussie rules footballs with us from Perth as a novelty for the locals to try out. (The local game is soccer.) We were playing kick to kick in the yard at the house and unfortunately one kick went astray and hit the dish. This of course knocked it out of alignment. Shaukat's painstaking tuning effort was all down the drain. His only reaction was "shit happens". He was a very patient man and, by the next evening, he had re-tuned the dish and the 'sat-phone' system was working perfectly. After that no footballs were kicked near the house.

The geochemical sampling went better than I expected using the big helicopter, which had the disadvantage of not being able to get into tight spots. However, it did have the advantage of being able to lift heavy loads so we could carry all our team for a day's sampling in one trip, then 'leapfrog' samplers working around three sample sites simultaneously. Once Barry and I had instructed the locals how to carry out the sampling, we spent much of our time examining geology both on outcrop and in the streams near the sample sites. The summer weather was perfect with warm, clear days but cold nights. Wild tulips were flowering, and some ridge lines were a sea of red. One afternoon I was up on a ridge examining an outcrop and Shaukat had come with me and was about fifty metres away. Suddenly he was yelling and waving his arms about. I thought he must have broken his leg or been bitten by a snake. I ran towards him and, as I got closer, I could see he was holding his little short-wave radio. When I reached him, he calmly announced that he had Radio Australia and I could listen to the news. Almost every day on our way back from the field we would put down near a stream and a couple of the guys would get out and collect bundles of grass. When we asked what they were getting, Shaukat told us it was just "some weeds" but he didn't explain any further. Later we found out that the grass was for the cooks and was the key ingredient for a delicious soup, served nearly every night before our main meal. From then on, we called it "weed soup".

A few other things are memorable from that July 98 field trip. After we had been in Kentau for a couple of weeks, Mark and Shaukat wanted to go back to Almaty for the weekend. Barry decided that he would go too. I didn't particularly fancy two more, four- or five-hour chopper ferries as we were already flying plenty on the job. I opted to stay in Kentau without any minders and assured Mark I would be all right with the locals. Just before he left, he gave me an envelope without saying what was in it. He simply told me if I had to leave in a hurry, not to catch the train as all the 'low life' travelled by train in that part of the world. Otherwise he was sure I could figure out what to do. After they departed for Almaty, I checked the contents of the envelope and was astonished to find ten thousand American dollars in cash.

On another weekend, Kassim invited us to his house for a small party to celebrate his daughter's second birthday. The apartment was decorated with many beautiful, large, woven wall hangings. It was an enjoyable afternoon experiencing the generous local hospitality. His wife had put in a great deal of effort and served the guests with lots of fantastic food and Kumis (fermented mare's milk). The latter, a national drink in Kazakhstan, was something I never really took to. As I recall we ate sitting on mats on the floor.

All up, I spent about a month at Kentau on that trip and then left Barry to manage the project. On the return chopper trip to Almaty, I had an experience I could never have imagined. The Mi-8 has dual controls and the chief pilot got me into the co-pilot's seat. Under his very watchful eye and instruction and with his hands close to the dual controls, I flew the helicopter for about half an hour in perfect weather. That was a real buzz!

The last bit of this story for me played out in early January 1999 when the company brought the Kentau geologists and Shaukat to Perth to show them some Australian mining operations. Aside from the mine tours, there were many things that we take for granted that they found amazing. I remember talking with them about visiting Bunnings and they just could not believe the size of the shop and the selection of tools and hardware. As they were here on Australia Day, they saw the Sky Show from the banks of the Swan River and enjoyed the fireworks display. Afterwards however, Kassim commented that he couldn't understand how so much money (around half a million Aussie dollars) could be blown up in less than half an hour. One evening they came to our home for a barbeque and, among the pre-dinner snacks with beer, my wife had cooked papadams. Babajan commented that these were the largest potato chips he had ever seen. She explained to him how the papadams were made and that they couldn't be potato chips as no potatoes are so large. His retort was, "anything is possible in Australia Mrs. John."

21

"She's Apples in Central Asia, Mate"
Tajikistan, 1997

Mike Kitney

Mike Kitney MSc (Mineral Economics) graduated as a metallurgist in 1971 from the WA Institute of Technology. In almost 50 years in the mining industry, he has worked as a process engineer, study manager, project manager, project consultant and has held several non-executive director positions, He has worked in Australia, and in Tajikistan, Sub-Saharan Africa, Morocco, Malaysia and Indonesia. He is currently trying to achieve true semi-retirement.

April 1997 seems a long while ago when your scribe packed his bride (of many years) and a few belongings and headed into the bosom of Nelson Gold Corp, listed in Toronto, headquartered in London and operating in Tajikistan as a partner with the Republic of Tajikistan in the Zeravshan Gold Company. ZGC was in the throes of expanding a once-defunct Soviet tungsten and gold mining operation and needed a little help, which kept me occupied for about five years thereafter.

Whilst staying at the Taror Project in Tajikistan was like living in the pages of *National Geographic*, being based in England offered some relief from the rigours of FIFO to that exotic Central Asian destination.

Borders

She:

FIFO London – Taror involved flying Uzbek Air to Tashkent, driving 600 km south through Uzbekistan, one's first glimpse of a real-life police state, to the Uzbek – Tajik border post at Jartepa and then 50 kms, via Penjikent, to Taror. Crossing a civilised border is a painless experience. Not so there. Paperwork is key to the existence of border police and military in those places. This includes an identification form and a currency declaration. Complicated enough, but entirely in Russian and in triplicate tends to try the patience of a neophyte traveller.

My wife was the first to get caught out, on her first trip in. The prospect of dealing with strange people at a strange border in strange lands filled her with trepidation. So to be prepared for her encounter, she filled out her forms before leaving home. Upon inspection at Jartepa, the Uzbek border guard demanded to check how much of which currencies she was carrying. USD – OK; AUD – OK; GBP – 40 short!! She may as well have been caught carrying drugs. Threats of imprisonment and deportation flew through the air. She was aghast, and faced with few options, she took the best one and burst into floods of tears and sobs of helplessness. The hapless border guy gave in immediately, gave back her papers and money, put her back in the car and sent her into Tajikistan. Problem solved!

He:

My turn didn't take long to come round. It was common practice for returning personnel to bring a Company mailbox with them to site. Mail protocol clearly stated NO CASH! I got the duty on one particular trip and duly appeared before the Tajik border guy with said box. Now I have never been able to smell money, nor do I have X-ray vision, but I'm sure this chap was supremely talented on both counts. He bade me open the box and, with the precision of a smart bomb, reached into it and, with a look of sly triumph, extracted a single envelope. The prize? One thousand US dollars in nice new bills. Having waited in a bare room for several hours as ordered, my man reappeared with a metre-long form in Tajik language and ordered me to sign it. As my driver explained, this was my confession to smuggling money, which would see me do three months in the Penjikent pokey. Not a thrilling idea. Naturally I refused to sign. He left. It was a long afternoon in that bare room. At 5 pm, which was coincidentally his knock-off time, my captor indicated I should reboard the mini-bus, with mail box intact. Shortly after, my driver motioned to give him $100, which I took from the incriminating envelope. A few minutes later the border guy appeared and politely asked if I would give him a lift home to Penjikent, 20 km distant. Gaol time averted and money smuggling employee $100 poorer!

Locals

We hadn't been in Sogdiana long when we managed to borrow a vehicle and set out to explore its environs one Saturday afternoon. We found and followed a rough two-wheel track that crossed some fields on its apparent way into the nearby hills. It passed through a hamlet comprising a collection of adobe and stone dwellings and sheds that looked as if it had been lifted from the pages of the National Geographic. As we drove into the village, a group of children and a few adults and a couple of attendant donkeys gathered, the humans waving and beckoning to us to stop.

What do you do when invited to join an unfamiliar group of people in an unfamiliar place? You stop. We were immediately surrounded by locals and, while they inspected us, we inspected them amid much arm waving and chattering. We were made so welcome, so much so that someone appeared carrying a bowl of thick white liquid with tiny bits of green grass mixed through. The invitation to drink this concoction was obvious. My mind went immediately to visions of acute gastric upset followed by crippling regurgitation. So, we drank. It turned out to be a delicious preparation of natural yoghurt flavoured with wild dill. Upon signalling our approval, we were promptly invited into the residence of a couple of local elders.

Bashorat (the lady) and Nemat had both grown up in this village of Hungaron and had worked in the commune fields for most of their lives, finding time to marry and produce nine offspring who ranged in age from late teens to mid-forties. Although in their sixties, the years of bucolic toil had made their mark. They looked more like 90!

Their living room was adorned with rugs on the walls and floor and housed a low table – the sort that challenges the knees of most 50-something Westerners. A scene from the Arabian Nights perhaps. But the remarkable scent of apples immediately caught our attention. Nailed into the cornices all around the ceiling was a line of Golden Delicious apples. Room fresheners that are replaced during every harvest.

We struck up an easy friendship with our new friends that lead to many afternoons of conversation – with the aid of my local translator, good food and bloody awful vodka. Tajikistan is a poor country with little in the way of arable land, so growing potatoes doesn't happen there. Instead they make their vodka from corn. As best I could determine, there isn't much difference between this and paint stripper. It could be that paint stripper would taste better!

We wanted to repay their kindness and hospitality, but there wasn't much our friends seemed to need. However, on occasional return trips to the mine we would drop in some clothes for the grandkids and we presented the old folks with the first portrait they had ever had taken, which impressed them mightily. Such easy pleasure to give.

And the boys – they got Absolut vodka!

Central Planning

ZGC had a licence to export its bullion product on condition it was refined in the Tajikistan refinery located in the northern city of Khujand. Once exported, the refined bullion had to go through a western refinery to meet LME specifications. Naturally ZGC had to pay the Tajiks for this unnecessary privilege.

The refinery was housed within a larger uranium treatment complex and had been designed by someone far, far away to treat 10 tonnes per year of gold and 40 tonnes per year of silver by electrolytic means. The gold refinery worked to a plan, or loosely translated from Russian, a 'project'. Operating procedures and conditions were laid down and no-one dared deviate from the plan.

ZGC bullion was characterised by significant silver content. Apart from providing the refinery the opportunity to filch the less valuable metal on occasion, it interfered with the gold refining process, causing delays to turnaround and providing them with an additional smoke-screen to cover minor losses of gold.... .

Any fire assayer will tell you that silver-rich doré can be readily refined by first adding more silver to the receiving melt, and then dissolving all of the silver from cast-augmented doré anodes in the silver electro-refining cells. "Try it" I said to the chief refiner. "Impossible; it is not in the project", he said. We played verbal ping pong for a while and I finally cajoled him into trying it "for just one time". "If it works, you can say it was your idea: if it fails, blame me". He agreed. It worked. He was a hero.... .

22

"I will Not Shoot My Fellow Citizens"
Tunisia, 2005-2013

Ralph Stagg

Ralph Stagg graduated as a geologist from the University of the Witwatersrand in 1970 and subsequently completed a Masters degree in Mineral Exploration at the Royal School of Mines, University of London, in 1977. He began his career as a geologist with Falconbridge exploring for nickel in Western Australia, returning to Southern Africa and working mainly in base metals for Cominco. In the 1980s he became involved in mining geology and then engineering consulting for a number of different companies in Australia, South East Asia and the Pacific. In the mid-1990s he spent a number of years in China working for a private investment group.

In the early 2000's he became involved in the Middle East and subsequently founded Citadel Resource Group, a copper and gold company in the Kingdom of Saudi Arabia; and later Celamin Exploration, a phosphate and base metals company in Tunisia and Algeria. He is currently a director of listed companies exploring in Chile and the Murray Basin of Australia.

I spent a number of years travelling in and out of Tunisia through the Global Financial Crisis; the Arab Spring Revolution and the violent overthrow of self-styled 'Colonel' Gaddafi in neighbouring Libya and the Civil War that followed; as well as through the Icelandic eruption of the Eyjafjallajokull volcano. Whilst only the Arab Spring revolution, which started in Tunisia anyway, had a major direct effect on my activities, the others caused moments of thought and concern and changes of plan.

Why was Tunisia chosen? Well, it occurred in part and as part of a strong personal involvement in many Arab World countries such as that one, Oman, Saudi Arabia, Algeria, and Syria. All of these countries more or less cluster around, or are close to, the Mediterranean and are in North Africa and the Middle East. To a greater or lesser extent, they had not at the time been subjected to what we in Australia call 'Modern' mineral exploration. They are all primarily Arabic speaking and many of them speak either French (Algeria and Tunisia) or English (Oman and Saudi). Some, but not all, of them have relatively modern mining codes enabling operations to proceed as part of a strong rule of law.

This previous paragraph does not really answer the question of "why Tunisia" directly, so perhaps a small bit of history will. Around 2002, I was asked by a good friend to help him in the Sultanate of Oman. He had discovered, through excellent geological, prospecting and mineral exploration work, two copper deposits and he needed help to bring them forward, hopefully ultimately into production. My background as a geologist had been for many years working on the feasibility and development of these types of projects. As part of this, I contracted some engineers and later formed a closer business relationship with them. The deposits ultimately were mined successfully and profitably using the plan we designed.

This led in 2005 to my first trip to Saudi Arabia and I spent five days in Jeddah at the Ministry of Mines. Subsequently we formed a company that was very successful in that it found some gold projects and a large copper project. Its board decided in about 2007 to see if we could expand the assets of this company by looking elsewhere for similar situations to those we had found in Saudi Arabia when we first went there, namely that only early mineral exploration work had been undertaken, and it had to have a robust economy and a reasonably modern mining code enabling security of tenure. After discussion by the board, it was initially decided to look at the mineral prospectively of Algeria.

One of our directors had experience in the oil and gas business in that country but we needed some advice on how to progress there. A former associate of his, an Algerian-born Frenchman, was asked to act as an advisor; his business was in introducing and guiding commercial enterprises in Algeria. His base was in Paris so we spent some time there in successful negotiations. I cannot recall exactly, but I think it was his suggestion that, when we undertook a preliminary visit to Algeria, we should have a look at Tunisia as well.

This we did on our first trip. Later, over several years, we did many trips that usually included both countries. Tunisia, for those who are not aware of its location, is sandwiched between Algeria to the West and Libya to the East. Anyway, Tunisia is easily accessible from Europe and, although we never used them, there were at the time major ferry services between it and several European countries on the Mediterranean (France and Italy notably). Tunisia's economy was very much geared to European tourism and more than 10 million tourists per annum visited the country (compared with its own population of less than that number) principally visiting for the sunshine and beaches, but also for adventure tourism in the northern Sahara desert, which covers the southern part of the nation.

Tunisia has a long history and was part of the Phoenician Carthaginian Empire that, during the 7th-3rd centuries BC, extended over much of the coast of North West Africa as well as of substantial parts of coastal Iberia and the islands of the Western Mediterranean sea. The Carthaginians clashed with the expanding Roman Empire and this was the cause of the three Punic wars: the third (149-146 BC) resulted in the defeat and destruction of Carthage and the growth of the Roman Africa Province. Lands surrounding Carthage were annexed and re-organised and the city of Carthage rebuilt. After the decline of the Roman Empire, the history becomes more obscure.

The geological attraction of Algeria and Tunisia became apparent after our first trip, when I was introduced to the broad and similar economic geology of both countries. In summary, older rocks in the north hold many base metal deposits. Further south there are extensive phosphate deposits in younger rocks. Even further south into northern Sahara, there is gold potential but access is both difficult and expensive and so we did not explore there at all. Both countries have large deposits of sedimentary phosphate. Phosphate mines in North Africa account for 75% of the World's reserves and many of these deposits are very large. Although our interests initially were in base metals, the potential of phosphate came to the fore, particularly as one of the shareholders in our company in Saudi Arabia was an expert in that commodity and encouraged us in this direction.

For Australians, countries such as Tunisia make us suffer somewhat from the 'tyranny of distance'. It takes almost 24 hours' flying time, not counting layovers, to get there from Perth. Because of the resultant jet lag, I developed a bit of a reputation for falling asleep at all sorts of unsociable moments. In Algeria, the custom is to take coffee at the beginning of a meeting and to discuss anything except business in French (with which I am not particularly conversant). I used to take advantage of that by having a short nap and our Algerian associate respected that: I was told he lowered his voice during that time. He was also most impressed that as soon as they mentioned the project I woke up. In Tunisia, the custom was to head for dinner at 9pm and start it at 10.30pm. I regularly napped between courses, probably not the best thing for my health, but I had little choice in the matter until my body clock adjusted later on the trip.

All airports are different, but the Tunis-Carthage airport is relatively easy to navigate compared with those in countries such as Saudi Arabia and Algeria. This has been designed primarily to encourage foreign tourists which provided at the time a substantial part of the country's foreign exchange. On our first trip, we acquired a 'visa on arrival' as required for travellers on an Australian passport. This involved providing payment in Tunisian currency that I did not have. No problems: I was allowed escorted entry through immigration and customs to one of the numerous foreign exchange dealers (who do NOT recognise Australian currency) and was then escorted back to the beginning to pay (in Tunisian Dinars) then was able to again front Immigration and Customs – but with proper papers! We soon learnt that entry using a British or European passport was visa-less and we did not have to partake of that rigmarole when using one.

Tunis is a very welcoming and safe city; the tourist hotels are excellent and there are many local-style hotels that are less expensive and quite adequate with one caveat – these ones are alcohol-free. Given that at the time I was spending a lot of time in Saudi Arabia, I was used to what we called our 'liver-cleansing diet'! In Saudi illegal alcohol (hooch) was readily available but you consumed it at your peril. In Tunisia, wine, beer and spirits could be purchased at some 'speciality' shops, which are discreetly located.

The first flight I took into Tunis flew over the Lake of Tunis, which is a shallow natural lagoon that was the original attraction to mariners seeking a safe harbour. The modern harbour is called La Goulette and is on the other side of the lagoon on the Gulf of Tunis in the Mediterranean Sea. We spent a number of days at a tourist hotel about an hour's drive north of the City of Tunis during our first visit. We met a local Tunisian who ran a mining services business but who also mined and sold some mineral commodities as well as being the only game in town for drilling services. He had a number of drilling rigs as well as access to chemical laboratories. We discussed joint operations, as there were significant advantages to both parties. From our viewpoint, it would give us access to a local expert who knew the geology and projects and who was politically connected, but not with the ruling 'Ben Ali elite'. We were concerned about our ability to navigate these tricky waters in a country that we didn't know and in which we had few skills. The other advantage was his knowledge and expertise in the local mineral industry and his ability to speak English. We both agreed that it was up to both parties to learn more English/French in order to be able to operate more easily.

If I had no prior business in Paris, I used to fly from Dubai to Tunis. This flight originally stopped in Tripoli, but we were unable to disembark there. The airline used Tripoli as its base because of the then political situation in Tunisia. But, after the Arab Spring revolt, Libya became very unstable and all landings at Tripoli in Libya were cancelled, particularly after the death of Colonel Gaddafi. The flight route was altered so we flew from Dubai north of the North African coast over the Mediterranean directly into Tunis, which became the new base for the airline. I was interested in visiting the country as, not only did it have considerable mineral potential, but also my father was there with the British army during the Second World War and I was interested in the possibility of visiting some of the historic sites. The Libyan sector of this part of North Africa was where many of the famous battles of the North African campaign were fought. Sadly, that never came to fruition for me.

In the early part of the 20th century, Tunisia was a Province of France and considerable lead mining was carried out to the extent that it was claimed that, at one point, it had the largest production in the World of that commodity. Output had subsequently dwindled and by the time we arrived there were no significant operating mines for base metals in the country. The potential was good and, in recent times, some significant deposits such as Bougrine (10 million tonnes at 10% combined lead and zinc) had been discovered and mined. Later, one of our senior geologists undertook an evaluation of 64 of the old mine sites (out of more than two hundred and thirty) and we managed to acquire rights to four of them which had remaining resources and showed promise for more, subject to successful drill testing.

Our other interest, in phosphate rock, did not develop on the first visit but we knew there was potential there. So, we reported to the board that there was potential to target reasonable sized projects in both countries with a focus on base metals and further investigations into phosphate. Whilst the older rocks in Algeria also had an excellent potential for gold discoveries, because of the remote Saharan location, we considered the target too ambitious for a small, developing company such as ours.

Restaurants in Tunisia are excellent: a favourite developed early in the piece located on the cliff tops north of the city centre with views over the ocean and near to the dictator's compound. Your choice is (of course) fish brought on display to your table for your selection. All the fish I ever ate in Tunisia and, indeed any of the food, was very fresh and well-cooked and generously served. The ubiquitous baguette plus harissa always starts any meal. The Tunisians like their harissa very spicy and everybody has a unique recipe that is of course 'the best'! The trick for survival for wimps like me who like their chilli milder is to add olive oil. You can then partake with a blend that doesn't blow you away. Tunisians, I was told, eat a tonne of chilli a day. On another trip, we spent half an hour stuck behind a one tonne utility struggling with an overload of chilli on one of the lesser roads in the hinterland. Another day – another load.

Wine is plentiful and, whilst imported French is available, domestic is very adequate and plentiful and well priced. One of our mining engineers liked it so much she tried to bring some back to Australia via Dubai where they confiscated it. A nasty incident nearly ensued: luckily, I persuaded the immigration official that she was insane, and I was trying to get her back to her husband – although she was at the time in fact not married – but nonetheless a perfectly credible story in that part of the World. It doesn't create any advantage to abuse immigration officials, and a reasonable excuse generally gets better results!

Tunisia is a quite small country and one can drive from Tunis to the town of Le Kef (East to West) on a reasonable quality two-lane, sealed highway in less than a day. Closer to Tunis this road broadens and becomes a tollway with some enticing coffee stops (Espresso of course!) along the way. Le Kef was not too far from our main phosphate project and became a regional base for some time.

The first drive out of Tunis only occurred on the second trip and it was a revelation. Northern Tunisia was one of the Roman granaries and even today, or at least at the time I was there, enough grain was grown to supply all domestic demand. So you drive up a highway through kilometres of wheat fields reminding me very much of many broad acre farming scenes in Australia. What you don't see in Australia though, and is very spectacular when you first drive past it in the early morning mist, is the Roman aqueduct that was built to supply Roman Carthage with water. I was told the remains are still 70 kilometres long and that it worked until the 17th century; a huge tribute to Roman engineering.

All along the road to Le Kef there are small and large ruins of Roman settlements, triumphal arches, buildings, remains of villas etc. This is a result of the Pax Romana that enabled isolated settlements to develop confidently in the political stability of the region at the time.

We started to form a good working relationship with our Tunisian associate, and he introduced us to the largest undeveloped phosphate resource in the country, and which lies near Le Kef. The scale of the project, at more than five billion tonnes, was very interesting but too big a bite for our relatively small company at the time. Soon after that, events turned against us, the Global Financial Crisis started to bite, and the board took a decision to cease all 'foreign adventures' and to focus solely on our Saudi Arabian assets. That made things difficult for me and my fellow director who had been working on our foreign exploration initiative with me. After discussing the matter, we asked the board if they had any objections or conflicts with us talking to our contacts in North Africa and asking them if they would be interested in continuing to work with us initially individually, and ultimately with any company we formed, based on those assets. There were no objections from anyone, so we ended up with a series of contacts and the ability to put together some Algerian and Tunisian projects and form a new exploration and mining company.

That started an adventure that led to acquiring two phosphate rock projects and several promising base metal projects in Tunisia and one base metal project in Algeria. All the former ones were in joint venture with the original Tunisian we met on our first trip, whilst the Algerian one was acquired via a joint venture with the company that owned it. This last project took three years and many trips to Algeria to negotiate, whereas we were able to progress much more rapidly in Tunisia owing both to the more favourable system there as well as the contacts of our local associate. As an example he went to school with the Director of Mines.

We self-funded the projects for a while until we had some certainty with respect to the tenement situation and then we managed to raise funds through a listed shell company that we introduced to our portfolio and by effectively merging with it. As we grew, we were able to employ staff and start some real fieldwork. The first focus was on our two phosphate projects that were much smaller than the giant first project we had had a look at but sufficient to become large enough projects for us. Both needed drilling to develop resources and an engineering feasibility study to determine the development parameters and marketing of product. The locations were good with an existing railway leading to the port of La Goulette where deepwater berths were available and there was the ability to negotiate access to them.

Phosphate rock is not a commodity that can in general terms be simply dug out of the ground and sold. Like many commodities, it is unusual to find it initially has sufficient grade for that. Considerable and expensive beneficiation needs to be done. In order to design the appropriate equipment and infrastructure the market needs to be accessed, and a product with saleable characteristics needs to be produced. Phosphate is a necessary part of agricultural fertilisers. It strengthens the stalks of crops and allows them to stand up straight: in short, the 'Viagra' of the plant world. We therefore approached the largest European fertiliser company and asked them if they wanted work with us on this project. Considerable negotiations ensued; ultimately, they agreed.

Drilling advanced and we decided to kick off a feasibility study in Tunis and invited a team of engineers to an initial meeting in Tunis so that, amongst other things, they could undertake a site visit and meet the Tunisian members of the team. I was flying from Australia to Dubai when I got the news that Tunisia was in revolt and I was strongly advised not to come. It was a bit difficult at that point and I was concerned with stopping another five or six Australians who were shortly also about to arrive, and I was very concerned about our Tunisian friends. This was the start of the 'Arab Spring' revolt.

Tunisia at that time had a very young population; more than 30% were under 25 years old. The education system was good, both boys and girls were encouraged to gain a tertiary education, and there were plenty of colleges provided by the dictator to allow this opportunity. However, once they graduated from college there were very few job opportunities. This, not unsurprisingly, led to unrest. A graduate who could not find a job and was eking out some sort of a living in the markets of a regional town self-immolated in protest. This led to a rapid and huge wave of unrest that spread right through the Arab world. In Tunisia, the people decided it was time to get rid of Ben Ali the dictator and his family and cohorts. He initially declared martial law to try to control the situation and tried to put troops on the streets. He instructed the general in charge to fire on the people if necessary. The General responded: "I will not shoot my fellow Tunisians". The end for Ben Ali then came quite swiftly and he and about 25 of his closest family and allies fled with an aeroplane full of treasure to Egypt which had agreed to give them sanctuary.

Much of this became public knowledge after the event, but I had decided, after I managed to stop my colleagues from leaving Australia, to head for Paris and hole up there and await events. As it turned out this evolved quite rapidly. I was talking to my Tunisian associate two days after Ben Ali left and he advised me five days later that it was safe to come to Tunisia. So, I was in-country less than 10 days after the revolt. It was very calm and, although there were many military people in the street, business was largely back to normal, except for the tourist trade. The other striking aspect was to see flowers in the muzzles of the guns on the tanks, the people's chosen way of thanking the military for their restraint during this difficult period.

A week later, we were able to mobilize our team and start our feasibility study.

A new government was formed with promises to hold democratic elections within a period once the country had stabilised. No changes were made to the Mining Act and the Director of Mines was unchanged. So, for us, it was business as usual. After a period, it was found that Ben Ali and his immediate entourage had been robbing the country blind and the new government decided to confiscate all his 'illegally' acquired assets and place them in a Sovereign Wealth Fund for the benefit of the Tunisian citizens.

There were a few subsequent political bumps. At one stage we looked to introduce Minemakers Ltd, a larger Australian phosphate explorer, into our projects. Its chairman and chief geologist came to the country and viewed them. Quite naturally, they were keen to get a take on the political situation. Everyone we met at the time assured us that the problems were all over and that the political situation was now quite settled. They left to report to their board and, three weeks later, the Leader of the Opposition was assassinated: the concept of stability is relative, of course!

The company we formed in Tunisia was successful for a time and still holds a major phosphate resource. I am proud of the fact that we were, and still are to the best of my knowledge, the first private company to acquire rights to a phosphate project in all of North Africa and the Middle East. This shows the forward thinking of the Tunisians. Many other countries will not allow private enterprise into this business.

Sadly, all of this took its toll and I decided to retire from the board although remaining a shareholder. Subsequently, the Tunisian mining associate who was a partner in the business illegally acquired all the rights to the phosphate project and a long, but successful, battle ensued in the Court of Arbitration in The Hague. The lesson is probably that if one has acquired a good project, it needs to be well guarded by management, as someone will covet it.

SUB-SAHARAN AFRICA

From the following, "A Horror Story ..."

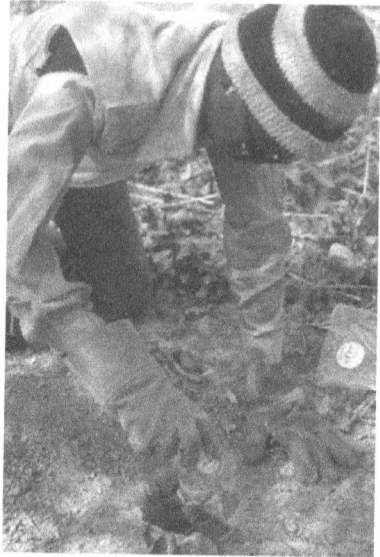

Planting the Track-etch cup and marking the spot. Unfortunately both cups and aluminium tags were often taken by the locals. We had to make up a story that it gave cancer to men and caused infertility in women to stop the thefts.

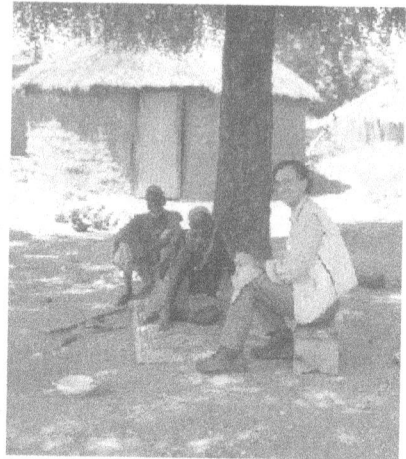

Left: Sebastian and Bakanga riding ahead marking a grid with paint and flags for diggers & track-etch cup 'planters'; Right: Discussing access and compensation with a village chief-of-fields (as with us, there are many chiefs in a community). Here too, I was warning that cancer and infertility would result if cups and aluminium markers were interfered with. Until that story was widely circulated, we found many of our cups for sale at the local Farafin-Dougou open market. These of course were discarded from our survey!

23

A Horror Story of People You Don't Want to Meet
West Africa, early 1980s

Max de Vietri

Now retired, Max de Vietri has had some 45 years' international experience as a geologist in the petroleum and mining industries, the last fifteen years assisting African governments attracting investors from the mining, petroleum or agricultural industries and helping foreign companies adapting to operating in Africa.

He holds a 2014 Doctorate in International Affairs and a 2009 Master of International Relations from Curtin University for Technology. He was also awarded a Graduate Diploma of Business (1988) from WAIT and a Bachelor of Science in Applied Geology (1975) from the University of New South Wales.

Max was awarded an Officer of the National Order of Merit for the Islamic Republic of Mauritania, granted in 2006 in recognition of his efforts in starting petroleum and mining exploration in that country. He has also served as the Honorary Consul for Mali in Perth between 2005 and 2016.

This is not an easy personal story to write. It is not a pretty story but one that must be told as a warning that life is not always rosy for a young field geologist and bad turns can unexpectedly come out of left field at anytime in one's career.

It is about my first trip to Africa. It could be classed as a horror story, but at least it has a happy ending!

For a few years after graduating from uni in hardrock mineral exploration geology, I "went to the dark side" working in the petroleum industry. I travelled all over South East Asia, Australia and several countries in Africa, first as an oilfield mudlogger on numerous land and offshore rigs for a geological and downhole engineering service company.

This job at first principally entailed identifying the rock chips that came back with drilling circulation fluids or 'mud'. After some time as a mud-logger, I began working as a pressure engineer, a more demanding job, as monitoring drilling parameters and their interdependence on changes in formation pressure is a most important monitoring job in drilling into overpressured hydrocarbon basins.

This serendipitous, very well paid wandering into 'soft rock' in many interesting regions of the world went on until, on 21 November 1979, I received a telephone call out of the blue from an international mining company. A much better paid and interesting expat position was opening for a uranium field geologist in Mossiland,[1] and my name had been presented to them.

Wow! An opportunity to get back into mineral exploration with a great pay packet, as well as "a rare opportunity to see Russia!" I told myself, and agreed to meet up at the end of my next hitch (or tour) offshore that would end on 7 December.

I returned home after that hitch and stayed home on Saturday 8 December going to bed early to be fresh to fly that Sunday for an interview on the following Monday. Laurel, my now ex-wife, went to a quiz night with some friends and came back late that evening all excited. She woke me up with a declaration that "Mossiland is a French-speaking country in Africa and not a region of eastern Europe or Russia, as you thought. It is because French is your first language that you have been selected for an interview!!".

On Monday 10 December 1979, having read up and being the full bottle on Mossiland, or at least roughly knowing where it was located, I had a successful interview with the appropriate executives of the mining company. At the end of the interview, I was immediately offered a very well paid "exciting position" as Senior Exploration Geologist based out of Farafin-Dougou, second largest city of Mossiland after its capital. My only companion would be my boss, Martin Pamplemousse, a "nice, hard-working and level-headed French geologist who had worked for several years for the company" as the African Projects Manager put it… This sounded fantastic and most adventurous!

---

[1] Pseudonyms are used for most places and people for obvious reasons, as you'll understand as you read on.

Prior to leaving for this first trip to Africa, I was sent to the Jabiluka Project in the Northern Territory to spend a week absorbing all I could on uranium mineralisation of the Middle Proterozoic Unconformity style. This type of mineralisation was expected somewhere along the 200 km long Foraban Cliffs in Mossiland, an expression of the Middle Proterozoic Unconformity marking the edge of the large landlocked ancient Upper Proterozoic to the much younger Quaternary Taoudeni Basin. As an aside, three decades later several gold exploration companies found economic gold deposits in that same region, but unrelated to the unconformity, within ground we had traipsed over many times without any idea of its hidden treasures.

At Jabiluka, I met a very welcoming Geopeko geologist, Ian Ruddock (real name), and in Darwin the very pleasant BMR geologist, Clive Pritchard (ditto) as well as Elliott Dwyer (ditto), a geologist at the NT Geological Survey. All three discussed at length their ideas on NT uranium mineralisation which was beginning to receive world attention. I could go on at length about the very interesting conversations we had on uranium, a very mobile element, and the techniques used in its exploration at the time, but I must get back to the main story....

Laurel and I flew off to land in the capital of Mossiland, early morning 21 January 1980. We were to be the only Australians in Mossiland and among a handful of English-speakers in the whole country until the gold rush of the late 1990s. This was fine for me as I am a native-French speaker, but difficult for Laurel who had never done French at school.

We were met by a big-statured, sleepy-eyed, hippie-looking guy in sandals with long Henna-dyed hair, beads around his neck and wearing a beautifully clean, sky-blue, loose traditional Japanese robe, a kimono, tied with a blue sash. He had many bangles on his hairy arms protruding from wide flowing sleeves. I recorded this in my ever-present diary and emphasised the 'clean' because in that region of Africa, as I quickly learnt, dust and crap all over your clothes, your car and everything else is the norm.

Yes, it was indeed Martin and he was not dressed in his pyjamas!

A Martin who had obviously undergone a 'psychological change' after a bad divorce the year before, given the description I had been given of a "nice, level-headed hard-working geo...".

While we were waiting in a plush hotel for two field assistants arriving from Paris the next day, Martin was friendly enough. I began to feel ok about him in spite of his strange garb. He certainly quickly made it clear to me that, in spite of me being Australian, he was "The Boss" but I respected that as I knew little of Proterozoic uranium exploration and nothing about Africa. The two who were to be our field assistants, Sebastian and Jean-Pierre (again pseudonyms), were old friends of Martin's.

On the morning of 23 January 1980, two days after our arrival and a day after meeting Sebastian and Jean-Pierre, the 'horror story' began!

At breakfast, Sebastian, greying beard and hair to his waist, was very quiet; but Jean-Pierre, a loud and intense speaker, together with Martin, were very jovial and quite quickly turned the conversation to "the sexual inhibition of people" and especially of Australian women!

Jean-Pierre declared that he was an expert on that subject as "he was a film producer". Both Martin and Jean-Pierre laughed and talked around me to Laurel and questioned her about 'sexual freedom' and 'liberation', etc. At first I thought that this was to avoid excluding her from the conversation… which in any event, never turned to why we were all there at all, sitting by the pool of the plush hotel. We went out discoing that evening, but I very much hung onto Laurel, as she hung onto me.

The next day, on our way to Farafin-Dougou where Martin lived and the project HQ was located, a direct invitation was made to Laurel (and me by implication?) by Martin and Jean-Pierre to go skinny-dipping with them at a pool at the base of the Foraban Cliffs just out of town, since they now understood that "we were indeed liberated Australians"! I felt extremely uncomfortable, but Laurel and I sort of looked at each other and we went along with it. The pressure on Laurel that day from Jean-Pierre and Martin was relentless, with continual innuendos about sex, liberation, freedom, wife-swapping. I was focused on not messing up my new job with my new work colleagues, but recorded all this in my diary, leaving Laurel to spar with, what could only be called their advances.

The days passed as I settled into the exploration work directed by Martin who seemed to do little field work himself, always dressed in a showy and always clean kimono, beads and all. I worked closely with

Sebastian as my fieldie, but felt and recorded an 'angry aura' directed towards me whenever I was with Martin and Jean-Pierre. Whenever Laurel and I met with those two, sexual innuendos were constantly dropped at Laurel. She – and I – became frightened, but I must admit that I was too weak to directly confront the two, afraid of losing my 'dream' job.

After a while, it was revealed that Jean-Pierre's main activity back in France had been filming and producing pornographic movies and, for some reason or other, he had had to suspend this activity and had been forced to hang around outside his country for a while… Sebastian on the other hand, was a hippy continually smoking 'Ghanaean Gold', a strong West African-grown cannabis and, to make cash, had been regularly driving stolen French cars through Spain to Morocco or Algeria and on down through the Sahara to a ready and lucrative market in Abidjan, capital of Ivory Coast. He also 'ran' small drug-hauls back to Europe during these trips, but admitted that this was not as lucrative as taking down "Au revoir France" high-value vehicles to Abidjan.

On the work front, our principal activities consisted of mapping the prominent unconformity along some 200km of the 350 km long Foraban Cliffs, liaising with and seeking permission from the locals in various locations and 'planting' track-etch cups on a pre-set grid to collect them some three weeks later. Radon-222 is a radioactive gaseous by-product of radium, itself part of a decay series beginning with uranium-238, the radioactive uranium isotope we were looking for, present in natural uranium deposits. Radon-222 has a half-life of 3.8 days, and with the emission of alpha particles decays into polonium-218, then bismuth, eventually to become 'stable' non-radioactive lead. Track-etch cups are plastic cups with a strip of material sensitive to alpha particle damage. They are placed upside-down in a 75cm hole on a grid pattern and left for some 17 to 21 days. Recovered after that period, they are collected in batches and, at that time, sent to a specialist lab in Australia for track-etch counting, the data possibly giving a proximity reading to any uranium mineralisation that may be present in subsoil primary hardrock or as secondary deposits in the nearer surface formations.

The fieldwork was done riding 175cc Yamaha trail bikes and by mid-April we had already negotiated permission with locals, gridded, planted, collected and sent for analysis some 1,500 track-etch cups; as many as had been achieved by Martin the whole of the previous year. My back, already weak from an injury in my teens, was often excruciatingly painful, but Martin accused me of slacking off if I dared to mention I needed any rest.

I know Sebastian and I worked hard, but a reciprocal animosity between Martin and I had quickly grown and, by end-March, he began pushing me to work longer and longer hours in the field, while he swanned around town in his kimono, lounging around 'doing business' in the many bars of Farafin-Dougou. Jean-Pierre began to also wear a kimono and the pair looked like a right pair of dandy twits as they paraded around the town in their oriental garbs.

This came to a head on 19 April when, during a field visit to Mossi-land by the company's African Projects Manager, he was confronted by our arguing over our individual work commitments. That evening, with Laurel in tears, I let it rip with the Projects Manager and read aloud through my diary entries beginning with the first fears we had at Jean-Pierre's arrival, and the turn the relationship took with the collusive pair of Jean-Pierre and Martin from that very day. The Projects Manager seemed to be understanding, and said he would look into the matter.

He did the very next day, confronting Jean-Pierre and Martin directly... which they of course denied!

On the Projects Manager's departure a few days later, Jean-Pierre and Martin made it hell for me with Martin specifically saying that "he would make it hot for me!" I made it plain that the only relationship I wanted with him was a work one, and that I thought his choice of fieldies was very questionable. For his part, Jean-Pierre accused me of wanting to undermine Martin and that he would write "a letter this long" to the company about my slacking off. Laurel supported me strongly at the time, urging me to "hang on and stick it out until next year and not show weakness, give in or run away".

The Farafin-Dougou office was located in the front rooms of Martin's large house where Jean-Pierre was also staying. As I have always been an early riser, I usually turned up at the office by 7:30am in order to prepare for the hot day's work. Sebastian, who was staying in a nearby flat, would join me at the earliest. We regularly saw various 'girlfriends' who would pass through the office part of the house on their way out. Both Martin and Jean-François would emerge bleary-eyed much, much later, if at all. Sebastian and I did the fieldwork.

While I was concentrating on work, Laurel was beginning to get very bored with life in Farafin-Dougou. For reasons of employment laws for nationals, we had to have a cook, a gardener-houseboy and a 24hr house-guard, leaving almost nothing to do for this 27-year old, young, intelligent and active woman. Yes, she did try to get engaged in community/charity enterprises, but not knowing French at all on our arrival in January, she found little with which to occupy herself. I did try to take her out in the field as often as I could, but… how satisfying could that have been for her?

Monday 5 May, discussing a coming trenching program with Martin that would involve the employment of some 100 casual workers, Martin admitted that he was not happy with Jean-Pierre's level of work activity and that "Jean-Pierre would not be coming back next year"!

Breakthrough!

A friend of Martin's, Alain, arrived from Paris a few days later and he was introduced to Sebastian and myself as a 'trainee replacement' for Jean-Pierre. A few days later, unexpectedly returning with Sebastian from the field around midday in order to pick up yet another spare tyre, we found Alain repainting Martin's drained swimming pool: Martin was casually having breakfast in yet another kimono I had not seen before and with two local girls I had seen in the bars in Farafin-Dougou, finishing what had been a large breakfast for four. Jean-Pierre's trail bike was parked in the corner of the garage; and his field gear, in the corner of the lounge, had stayed untouched for the last several days, as confirmed by Martin's houseboy. This behaviour continued and made my and Sebastian's work all the more difficult as we had to perform for both teams to keep our schedule of gridding and track-etch cup planting, of churning Track-etch cups for analysis back in Brisbane to plan further for a major trenching programme that had already begun and for a follow-up drilling programme that was to begin just after the wet season, in September.

Sebastian was turning out to be a hard-working assistant in the field and and a good companion back in Farafin-Dougou. He and I were overburdened with the endless problems we were having with our hundred or so casual workers spread as a couple of dozen teams digging trenches on radon anomalies throughout a region stretching over some 200km. Together with our total lack of understanding of tribal, ethnic and land divisions and the fault lines that existed in the Farafin-Dougou community, it meant we were becoming increasingly inefficient. We were firing and hiring people without understanding why some groups worked well and achieved objectives, while some were absolutely unable to even dig even one foot of dirt without major fights, thefts, walkouts and landowner disputes. A most distressing moment was when I was hauled up before a panel of public officials from a government department that dealt with expat employment of local people, to justify my firing of a group of inefficient workers from a certain village. Martin, who should have known better after his several years in the country, gave us no guidance to the social aspects of working in Mossiland.

I am skipping many, sometimes very funny or sometimes sad and harrowing, events that occurred during that time that made my first experience in Africa most memorable and would in themselves make interesting, amusing or sad tales.

My focus is to get this horror story off my chest and I will continue.

28 May, to my surprise, Martin announced to me that he would be meeting Jean-Pierre that evening as he was not happy with his work performance, but that the guy "may do something nasty" if he fired him. For a few days away in the South near Foraban with Sebastian, we had no contact with Martin or Jean-Pierre but, on our return late afternoon to the Farafin-Dougou office for supplies, Jean-Pierre's gear and bike had not moved. As we were discussing funds and liquidity for some compensation payments I had to make to some village chiefs the next day, Martin mumbled that he was trying to kick out Jean-Pierre!

A few days later, after I had strong words with Jean-Pierre – in spotless kimino – on his refusal to assist me in unloading my bike from the back of my four-wheel drive on my return from a difficult few days doing radiometric traverses, he blurted out he would make trouble for both Martin and me if he had to go before end of year 'as per his contract'.

Tuesday 3 June, a high-level corporate guy from the company head-office arrived in Farafin-Dougou. It was Martin's duty to take him around and he did but, in view of the problem I had highlighted to the Projects Manager a few months before, I was surprised at this corporate guy's aloofness towards me and seemingly total lack of concern with the predicament that I had, as I thought, formally reported a few months earlier. Nevertheless, something must have been said as, on the day of his departure on 7 June and while he was packing his bags in his hotel room, Martin called Sebastian and me to act as witnesses to the firing of Jean-Pierre, offering him his economy-class ticket back to Paris and pay to that week – in spite of not having worked much at all since his arrival in late January! Jean-Pierre's reply was that he wanted a cash payment for his employment until the end of year and only cash in lieu of a business-class ticket back to Paris.

Situation unresolved, Martin went to pick up and take the corporate guy to the capital and his flight back to 'civilisation'.

I am not sure what transpired between the corporate guy and Martin on the trip to the capital, nor between Martin and Jean-Pierre once Martin got back to Farafin-Dougou, but Martin was certainly very upset and showed it over the next few days.

On 9 June, my strong feeling that Martin was being blackmailed by Jean-Pierre was vindicated when Martin showed me a letter from the French fieldie demanding one million CFA (or some AUD150,000 in 2021 cash), as well as an equivalent of AUD8,000 cash in French francs in lieu of his ticket back to Paris, and stating that otherwise "it was going to turn very ugly for Martin!!!". Throughout June, Martin remained sullen and withdrawn, rarely speaking to me except to receive reports on the progress I was making in the field.

On Monday 30 June, Jean-Pierre the porno-film producer finally departed Farafin-Dougou for a flight from the capital back to Paris. I was not privy to whether he got away with the cash payment he was trying to extort from Martin, but Martin was very sullen and totally uncommunicative for several days afterwards.

Over June, and with the excuse of the occasional rains, and apparently in poor health, Martin rarely went to the field, preferring to 'carry out business' in the bars of Farafin-Dougou. I had had to employ various assistants, including Don Monteverde, a very helpful American geologist who was taking a break from his normal work in the USA and trail-biking through Africa. I was very lucky to make friends with him at a chance meeting at the single petrol station in Farafin-Dougou one morning. He helped greatly during his brief time with Sebastian and myself, both with the trenching program and preparations for the coming drilling to begin after the July-August monsoonal rains.

By the beginning of the 1980 rainy season, we had achieved well over 4,000 track-etch cups surveyed in, planted and collected; more than twice the number Martin and his team had achieved the previous year. We had completed some forty 5m-deep trenches, dug in extremely hard indurated laterite, over the strongest radon anomalies. The trenches were scattered throughout a region from Foraban to well north of Farafin-Dougou, spanning a length of some 200kms of prospective Middle Proterozoic Unconformity along the Foraban Cliffs. It was not an easy job!

For the drilling program after the rains, we used a reverse circulation rig brought from Abidjan, Ivory Coast, but the many drillholes completed over some two months, and following up the radon anomalies, returned lateritic sections down to depths of over 100m and absolutely zilch uranium.

Either our exploration efforts were somehow flawed or indeed no uranium was to be found!

Anyhow, I could go on trying to explain how the exploration program ended up and hypothesise on why we discovered no uranium. One thing is for certain... we totally overlooked and missed out on the potential of the incredibly high gold results we were occasionally receiving from both surface rock chip, trench and drill samples. We were analysing for several elements but only rarely including gold assaying, an expensive analysis in those early days of the 1980s gold hike.

With my attention totally focused on work and especially problems around it, I was insufficiently aware of my wife's lack of anything meaningful to do and her rapidly increasing boredom. A group of some twenty English-speaking Canadian helicopter pilots, engineers, mechanics and admin guys with their families – or single – in Farafin-Dougou, to carry out a River Blindness eradication program for the World Health Organisation (WHO), 'rescued' Laurel and me from our isolation.

This rescue did give us much needed social interaction, but it was both a blessing and a curse in disguise, in that ... but that's another story!

We left Martin and a fast closing-down uranium exploration project in Mossiland, finally arriving back in beautiful Australia on 26 February 1981.

As the access door of the plane opened to let us step down onto the tarmac at Sydney's Kingsford Smith Airport, we were hit by a wall of eucalyptus scent that brought tears to our eyes ... It was not the cleaner's detergent, but the smell of summer drifting down over Sydney from the Blue Mountains!

One good thing that did come out of Mossiland was that our first son was born on 14 July 1981, a last hurrah by Laurel to celebrate surviving our first African experience!

I returned several years later, and fell in love with Africa ... But that too is another story!

24

Oh No! Manono
Zimbabwe and Zaire, 1981

Roger Thomson

Roger Thomson graduated with Honours in Mining Geology at the Royal School of Mines, Imperial College, London in 1967. He migrated to Australia as a "£10 Pom", on the basis that it would take a couple of years to have a look around, and is still based here after more than 50 years. He has worked in all States of Australia with stints in Brazil, Ecuador, Argentina, West and Southern Africa, Indonesia, the Philippines and Europe. Roger has held the positions of General Manager Exploration with Delta Gold Ltd and Sons of Gwalia Ltd, and has been closely associated with making economic discoveries of gold, tantalum and lithium in Australia. He successfully managed the programme that led to the discovery of the multi-million-ounce Sunrise Dam gold deposit in Western Australia. He was a founding director of AIM-listed Mariana Resources and ASX-listed Image Resources, Meteoric Resources, Magnetic Resources and Emu Nickel. Since 2011 he has operated a Perth-based geological consultancy.

My association with Greenbushes Ltd started in 1970 when I was engaged by an innovative mining engineer, John McSweeney, to provide geological services to the Greenbushes tin mine south of Perth. At that time Greenbushes Tin NL, as it was then, was dredging alluvial tin and tantalite using an old bucket wheel dredge called the Jim Crow that had been dis-assembled in Victoria, transported and re-assembled at Greenbushes. I was fortunate to be involved in the exploration of the source of the alluvial tin mineral, cassiterite, and tantalite at Greenbushes, a very large and deeply weathered series of granite pegmatites some 3.5km in length which had not been subject to any systematic modern exploration. It was an exciting time to see the mineral potential of the pegmatites unfolding. The Greenbushes pegmatites were eventually proved to be one of the world's significant sources of tantalum and are now one of the world's highest-grade sources of lithium in the form of spodumene, and hence a world-class supplier of the lithium now so in demand for rechargeable batteries.

A few years later, after an 18-month stint as site geologist at the Telfer gold exploration project (then owned by Newmont) and two and a half years in Brazil looking for copper for St Joe Minerals Corp, I was retained by Greenbushes to assess and acquire gold and rare metal prospects in Australia and overseas. An additional brief from the Greenbushes board was to investigate and assess competitor tantalum deposits around the world, which could affect the company's strategies in producing and marketing its tantalite concentrate. This brief took me to China, Canada, Zimbabwe, Namibia, Brazil and Zaire (now known as the Democratic Republic of the Congo or DRC). It is the journey to Zimbabwe and Zaire that is the subject of this little story.

In 1981, having been tasked with assessing tantalum mines at Bikita in Zimbabwe and at Manono in Zaire, I duly contacted Bikita Minerals in Harare and arranged to meet the managing director in Harare. Dealing with Belgium-based Zairetain, the operators of the Manono mine, was a bit more complicated and at first there was some reluctance to arrange a visit but eventually the director of the Manono operation said I could accompany him on a charter flight to the mine if I could meet him at his hotel in Lubumbashi, DRC, on a certain date.

So off I flew to Harare via Johannesburg and met the MD who kindly drove me (at very high speed) to the Bikita mine, some 370km south of Harare, showed me around and drove me back to Harare. It was only on the way back the next day that I asked him why he was driving so fast, to which he replied: "it's hard to hit a fast-moving target". He also explained that if we did hit something, or someone, we would not stop but report it to the police in Harare, because it was simply too dangerous for two white men to stop at a traffic accident in rural areas. It was then that I realized that the security situation in Zimbabwe at that time was not that good. The next challenge was to get myself to Lubumbashi to meet the Zairetain director in time to catch the charter flight.

From Harare I flew to Lusaka in Zambia and then on to Kitwe, a mining town on the Zambian Copperbelt close to the Zaire border and about 150km south east of Lubumbashi, where I thought I could catch a bus to my destination. I remember the impact on landing the small twin jet at Kitwe was the hardest I have experienced, but the undercarriage survived. It reminded me of a small charter flight I had taken from Perth to Jerramungup in Western Australia to look at a gold prospect, where the undercarriage collapsed on touch down and we skidded down the length of the gravel airstrip on the belly of the plane with a great deal of noise, but we survived. To my chagrin I found out that there was no public transport from Kitwe that could get me to Lubumbashi in time to catch the charter flight in a couple of days' time, and that taxis were not allowed across the border. That evening in the bar of the motel where I was staying, I was asking around about how I could get across the border the next day when somebody pointed to a fellow in the corner who owned a trucking company. I got talking to him and he said he had a fleet of trucks carrying fuel from Zambia to Zaire and that his African foreman was driving to Lubumbashi the next morning and that I could travel with him providing my papers were in order, which they were.

So 6am the next day the foreman collected me from the motel in a ute and off we went. He had been driving the route for some years and explained that the security situation in Zaire was not good and that one of the problems was that the border guards and gendarmerie who policed the highways were often not paid for months at a time and had resorted to collecting revenue from travellers in various ways; but not to worry as he was a regular user of the highway while checking on the company's trucks. However, he did advise me to put any cash I was carrying into my boots!

It's about 200km by road from Kitwe to Lubumbashi and at that time it was around a five or six-hour drive with the border crossing about halfway. The Zambian section was no problem nor was the Zambian border post but, on the Zaire side, the border policeman waved me into a back room and said that my vaccination certificate was not in order and that a yellow fever injection was necessary, pointing to a large metal syringe lying in a dirty enamel dish on the desk. I was horrified and was about to protest when the foreman intervened and quietly gave the policeman US$10 whereupon he waved us through. Relieved to be across the border and on our way, it was some time later that we were stopped at a roadblock manned by a wild-looking bunch of police in ragged uniforms and armed with AK47's. I was starting to feel very uncomfortable when they kept pointing at me and quizzing the foreman in French. It was clear that the senior officer knew the foreman and they were talking and joking. After a while they wandered off behind a hut and the foreman came back giving me the thumbs up and off we drove. I then asked the foreman about what had transpired at the check point and whether he had to pay any money on my behalf. He laughed and said that the one thing that the senior officer in charge valued more than cash was black hair dye! It was because he was going grey and he felt his seniority and position would be challenged by the younger men if he let his age show. The foreman was a regular supplier of the required dye and thus a favoured visitor.

It was with some relief that I arrived at the hotel in Lubumbashi where I was to meet the Ziairetain director who turned out to be a rather large gentleman sunning himself by the swimming pool in a tiny thong of a bathing suit. Not quite what I expected. However next morning we took off for Manono, situated about 600km to the north apparently in the middle of nowhere. After an hour or so's flight I recall seeing what appeared to be a giant white pyramid looming out of the tropical haze. It turned out to be a very large heap of oversize material derived from the processing of partly weathered pegmatite. Manono turned out to be an extremely large pegmatite system spanning some 10km in length and mined by open cut in several locations. Manono at that time seemed to be on its last legs with very old mining and processing equipment struggling to produce any amount of cassiterite and tantalite concentrate. The mine village, which once comprised neat bungalows with European-type shutters, was overgrown and being reclaimed by the jungle. How times have changed: it is now a major lithium resource in the form of disseminated spodumene and is currently under evaluation.

I returned safely to Lubumbashi and was able to catch a commercial flight to Johannesburg where I was planning a charter flight to meet a prospector who held claims over mineralized pegmatites in a place intriguingly named Tantalite Valley in southern Namibia. But that's another story.

25

Finishing School
Zambia, 1969-1970

Dan Greig

Dan Greig graduated in geology from St Andrew's University, Scotland, in 1969, and now resides in Perth. Initially he was involved with exploration in Southern Africa (Zambia, Botswana and South Africa) for nine years, before arriving on Australian shores in the late 70s where he worked with US Steel, Esso Minerals, Selection Trust/BP and Sons of Gwalia in Western Australia. Latterly he moved into the consulting world with Minproc (and successor companies) between 1995 and 2012, and then part-time for Behre Dolbear Australia. Work has now largely been abandoned in favour of bowls, bridge and orienteering.

Foreword

When beginning to put the modern equivalent of pen to paper to contribute to this book, a lot of possibilities sprang to mind. As with most retired geologists, I have worked for numerous companies on many projects spread around the world, and I have been exposed to a variety of cultures and incidents of more than passing interest, at least to myself. On reflection, it seemed to me that none of these were truly unique or life-changing in themselves, but that all of them contributed small pieces to who I became and where I finally ended up as a geologist and person. However, nothing was more fundamental to this maturing process than my first year following graduation, spent in Zambia. Memory foreshortens and blurs the actual timing, but it was certainly a kaleidoscope of people and events set against the exotic sights, colours and smells of Africa, all of which remain with me to this day.

Arrival

Midday in early September 1969, finds me walking down metal steps onto the tarmac at Ndola airport, somewhat discombobulated following a sleepless overnight trip from London via Lusaka, into the oven-like heat of a Zambian spring day. Actually, as I am to find out over the next 50 years, it isn't really that hot at all, but to a Scot who hasn't previously set foot below 40 degrees N, it is definitely not your average day temperature-wise. Things look different, too, on the drive over to what is to be my base for the next few years in the town of Kalulushi on the western edge of the Zambian Copperbelt: mainly wooded country, strangely khaki-grey and smoky this late in the dry season; with scattered, corrugated iron- or grass-roofed mudbrick dwellings, wandering people, dogs, goats and chickens; and intermittent large termite mounds and still-smoking charcoal middens. We skirt the industrial town of Kitwe with its many small, colourful markets, pass a headframe, slag heap, waste dumps and the giant, yellow-smoke-belching smelter stack of the Nkana mine, and cross a railway line that is heading, improbably it seems, towards the strife-torn Republic of Congo. Another 15 minutes and we are entering the self-contained little world of Kalulushi, all neatly laid out with whitewashed stones, palms, acacia trees and lawns. We pass a golf course, the underground mine, hospital, primary school, mine mess, a small cluster of shops, and finally pull up at the Exploration Office. I've arrived at the start of my working life.

Why Zambia? Well, while the job market isn't totally dead (that will follow within three or four years), opportunities in 1969 are limited, and none of them are in the UK. For me it comes down to a choice between the very contrasting options of three years with the British Antarctic Survey, or joining Roan Selection Trust (RST) on the Zambian Copperbelt, also on a three-year contract. Africa sounds more exciting, with its black people (almost unknown in our part of Scotland, though West Indians are certainly appearing down south in England), exotic animals and sunshine. Perhaps exploring for copper appears from afar to be more interesting than mapping coal sequences in Antarctica, but I'm not sure that I gave much thought to the technical side of things at this time.

Being newly graduated, I am designated a Learner Geologist for the first six months. I am issued with a thick 'Bible' which, on inspection, carries detailed instructions on how to do just about everything a Zambian exploration geologist can expect to encounter, from collecting, recording and submitting soil, pit, trench and rock samples, to marking and cutting core. It includes brief sections on the geology of the Copperbelt, and acceptable (to the company) terms for lithology, colour and texture to be used when mapping and logging. Puzzlingly, large parts of it are devoted to man-management and logistical matters such as how to organise road cutting, soil sampling, pitting and auger drilling crews, make out pay sheets, issue rations, conduct radio schedules and the phonetic alphabet. It doesn't make a lot of sense sitting in Kalulushi, but, within a few days I will be shipped out to bush camps in the northwest of the country where I quickly discover that a) rocks are a rarity, as pretty much everything is covered by thick soil or laterite, so most of our painstakingly acquired university geological knowledge will not be required, and b) that much of our time will indeed be spent organising and paying our workers, constructing access tracks and bridges to the more remote parts of our areas, and arranging fuel and food supplies (a nightmare in the wet season).

The Kalulushi Exploration office of 1969 covers both on- and off-Copperbelt exploration, the former being effectively reserved for married geologists, whilst we single people are dispatched further to the northwest and west. Technical staff includes perhaps five senior people (i.e. over the age of 30) led by Curly Ellis, a middle-aged American geologist who has been here for many years, while the other 25 or so of us, all male, are only one to five years out of university and all on our first job. Hence the 'Bible', with its detailed prescriptions on how everything should be done, serves to pass on hard-earned company knowledge of how to operate in this strange environment. Our number includes one or two geologists from Holland, Italy and Germany, an Indian, and a couple from Rhodesia and South Africa, but in the main we are drawn from the UK. Strangely, in hindsight, there are no Australians or Canadians.

There are important support services in the Kalulushi office, too, not least of which is the drafting office manned by five qualified draftsmen who seem to have been around a little longer than us and give us quite a hard time for the first year or two. Every request is met with several reasons why the job cannot be done for a week or two, longer if something more important comes in, and don't dare ask for any changes to be made. In their defence, this is before the days of computerisation, so everything is hand-drawn. Similarly, the typing pool operates in the old-fashioned way with manual typewriters, and white-out for minor corrections (otherwise a complete re-type), although the ladies prove to be much more helpful, being drawn from wives of employees, often of geologists that we work alongside and with whom we socialise. In addition to this, there is the research and development section, although most of their work is in support of the RST mines and, particularly, the processing plants across the group. Finally there is the 'Top Yard' a few hundred metres away, where there is a mineralogy lab, a wet chemical laboratory, some sheds containing all the historic exploration drill core and, finally, the Stores, run by a large, peppery, ex-army type who likes things just so and who reluctantly issues boots and geo-picks, sampling equipment and road-building tools, and the large quantities of rations (mealie meal, beans, small dried fish known as Kapenta, bully beef, tobacco etc.) that are sent out monthly to the various camps to feed the several hundred African workers and crew bosses.

My town accommodation consists of a quarter of a breeze-block flat, equipped with an old iron-framed bed (bedding provided), a small wooden table and two chairs, a cooker, minimal crockery, cutlery, pots and pans. No air-conditioning, light shades or even curtains, but at least it has running water, electricity and a fridge. None of this is remotely comfortable or welcoming, so we spend as little time as possible here, preferring the mine mess and Rec Club for eating and drinking. This very basic set-up has the added benefit of making the bush camps an attractive long-term option, this being well before the days of FIFO in any form.

On the move

So, after perhaps five slow days of settling in, suddenly there is action! I am sent for a week to a fly camp, a couple of hours of dusty driving west of Kalulushi, to learn some of the ropes from Tony, a slightly older geologist (all of 23) who is putting in a small camp and airstrip to support follow-up of a soil anomaly and copper flower occurrence (look up Bible, there it is, but easily overlooked unless in flower), previously identified by reconnaissance exploration. I learn to get up really early, as it is now moving towards October and getting HOT and sticky, trudge half an hour to the airstrip site and watch a crew that knows pretty much what it is doing anyway as it clears away the scrub and some not so small trees from a slight plateau. As far as I know, nothing becomes of the prospect, and the airstrip is not used in the next five years.

But already I've been moved further northwest and on to better things; a posting of indeterminate length at Katandana camp, a short distance out of the small regional centre of Solwezi, from which RST is operating a couple of diamond drills, some pitting crews, and a small and underpowered Bell G2 helicopter for reconnaissance stream and soil sampling. The camp has been there for a few years already, and has a nice permanent feel to it. My accommodation is a decent army-style A-frame tent, which comes complete with wash-stand and basin topped up with hot water at seven every morning by a servant (along with a cup of tea)! The mess and office are fine, large, whitewashed, hessian-walled pole buildings with thick grass roofs that overlook a small river with a waterfall and a pool. There are four geologists here (five if you count me) and the place is a hive of activity during the day, with the helicopter buzzing in and out picking up and dropping off sampling crews. Evenings are very fine, too: a swim, a hot bath and a beer in a galvanised iron hip-tub, good dinners produced by Joseph (rumoured previously to have been a chef at a hotel on the Copperbelt), another beer and then bed by 9pm, only to do it all again the next day. Weekends are even better: Saturday morning is part of the working week but, after midday, we are free to go to Solwezi where there is outside company including several young, female expatriate teachers and nurses. Also, a small resort/bar on the edge of town, with plenty of beer, music and late hours. I don't recall being assigned any particular geological tasks in this time, so I guess it was just another part of the overall familiarisation process but, in any case, it is an enjoyable but very brief interlude, perhaps 10 days, before it is decided that I have seen enough and am ready to take over my own project.

Out on a limb

So, I fly back to Kalulushi, collect my own trusty vehicle (an aging and very basic Series II Land Rover) complete with driver, and five really uncomfortable hours later we pull into a very small cluster of tin rondavels, tents and whitewashed hessian huts that constitute the Kipushi camp, close to the Congo border. It's a one-geologist camp, and the departing Italian geologist is in a hurry to leave as his contract is up and he is clearing out. However, we do spend a couple of days touring the area, during which time I am shown a handful of minor soil anomalies (OK, numbered pegs in lateritic loam), a line of pits that had returned nothing of geochemical interest, and several deteriorating bridges that need to be fixed so that as yet undetermined activities can resume in the next dry season). I also discover that what little is known of the geology up here is based on Belgian work immediately to the north, and it is all in French! And what's more, with one exception, the crew speaks only the local, incomprehensible language plus a smattering of French. The exception, Andre (the crew boss), has quite good English and, moreover, he has worked across the Congo border as a geological assistant, so he knows a whole lot more than I do about the regional geology and my job in general. But I get along one way or another, barely supported by my smattering of schoolboy French.

There really isn't a lot of interest shown in my project area by Head Office; no-one comes out to inspect progress or offer advice, budgets are small and exploration is low-key and desultory. The only copper I encounter comes in the form of malachite specimens from the Congo being offered to me by local dealers.

I come on the radio twice a day to confirm my continued existence and to listen to the other camps – there are eight in all – reporting exciting things like diamond drilling developments, complete with codes to describe any mineralisation encountered, and just to listen to the general chatter. However, there are some benefits out here, not least of which is the very good mine mess in the Congolese town of Kipushi, just 15km away, which is more than happy to serve me. I become such a regular visitor over the next few months that the Zambian customs post issues me with a local African pass, recognised also by their Congolese counterparts, so that they don't have to keep on stamping my passport. There is also a very decent golf course a little further on at Lubumbashi, complete with a hospitable French golf professional and his Scottish wife, so I get some opportunity to speak English. Despite taking regular prophylactics, I also catch a solid dose of malaria which confines me to my tent for several feverish days. By this time the rainy season is upon us; average annual rainfall up here is about a metre, pretty much all of it between January and March, and so it buckets down day after day until we are all but cut off on the Zambian side – except for the airstrip where our supply plane appears once a week, as the weather allows, with my food order, mail and occasional passengers travelling through to other camps.

By late March it is becoming a bit depressing. To pass the time after work, I have the crew construct a large hessian net into which I can blast golf balls, but after a very short while the fabric rots, the balls continue on their merry way, my supply runs out and I admit defeat.

I start to understand why some others break their contracts and go home, but then the rain stops, a few of us take a short holiday down south to recharge at Victoria Falls and, on return, things start to look up a little as we can move around again.

Moving on, moving up

Then in May, out of the blue, suddenly things get a whole lot better. The project geologist running the Lumwana project, 200km to the west, is packing it in and heading back to the UK to do an MSc. I'm to get my ass across there for a quick briefing before he goes, and then take over the position. So now, finally, after this slow and haphazard start, I leave behind my L-plates and become a Geologist.

This is not our average exploration project as copper mineralisation has been located and diamond drilling undertaken at a number of locations around the Mombezhi Dome. Some of this drilling dates back to the early 60s, and there is even a PhD thesis on the Malundwe deposit by one of the early exploration geologists (McGregor, 1965). No formal resources have been published, and the project has been on the backburner for several years, RST having more than a sufficiency of reserves on the Copperbelt at this time. Possibly the Government has given the company a prod, but now, for whatever reason, there is a push to drill up the two main deposits. There are also two or three other geologists based at Lumwana, looking after exploration further afield, supported periodically by a helicopter, so there is no shortage of company, and in addition we have a string of visitors from Head Office, passing on knowledge, ideas and support. My days are suddenly extremely busy and challenging as there are diamond drill sites to be designed and pegged, two or three rigs to monitor, downhole surveys taken by the then Tropari technique, which has subsequently and thankfully been superseded, core collected, logged, marked up for cutting (at least the latter is done by one of the crew bosses), sampling, hand-written core logs prepared, and cross-sections hand-drafted and sent to Kalulushi. Fortunately, the rigs are not set up for wireline drilling, and are small, old, underpowered and prone to breakdown, while the bits struggle to cut the harder gneissic units, so production typically averages 30 feet per rig per 12-hour shift (we are not yet metric) and it is possible to keep up. Drill results typically don't come back out to me for four weeks from sample collection but, after a short while, I become pretty good at visual grade estimation of the relatively coarse mineralisation to the level where I can confidently request repeat assays where results don't match expectations.

The weather for the next several months is spectacular, with cool occasionally frosty nights and warm days of endless sunshine. The camp is eminently liveable, too, with the mess/office set over a pool on the perennial Malundwe River, meaning that water supply is no issue. We have an excellent cook and an experienced work crew due to the long-term nature of the project. For entertainment we put in a concrete pad – notionally for the helicopter – and take up badminton and table tennis in the early evenings. After dinner we have cards or music, but everyone is in bed by 9pm unless it's the weekend. We can still visit Solwezi, but it is a bit of a flog over a lousy road, so we don't often bother. In any case, there is a fair amount of wildlife around and lots of interesting spots to visit in the district, or we can explore downstream on the Malundwe using the bathtubs for boats, so we really don't want for things to keep us amused.

The end of the beginning

Suddenly, the first year is over, and it is time to go back to Scotland – got to use up at least part of my annual 55 days of leave. I have no memory of what I get up to in my month off, but I do remember wondering what was going on back in Zambia in my absence, and keenly anticipating getting back to work.

Postscript

Looking back, that first year provided a necessary, albeit often chaotic, transition between university and the real world, which left me confident that I could handle the work and the lifestyle of a geologist from there on.

I called Lumwana home for the next four years, during which time I learnt not only the subtleties of designing, budgeting and managing regional and detailed exploration programs, but also became involved to varying degrees in resource estimation (manual), copper metallurgy, mine engineering and other aspects of project evaluation, all of which were to provide a solid basis for the remainder of my working life.

For the more technically-minded, the geological world has of course changed immensely since 1970, including radical changes in both exploration (and related) technologies and the understanding of geological processes. For one thing, the genesis of the broadly stratiform Lower Roan Copperbelt deposits, held by RST in 1970 to be unarguably syn/diagenetic, is now known to involve multiple phases of mineralisation progressively developed during basin maturation and orogenesis. To fit with the accepted RST wisdom of the time, the Lumwana deposits were interpreted by us as high-grade metamorphic equivalents of the shale-hosted Copperbelt deposits, whereas post-2000 research and subsequent mining activities at Lumwana suggest that the host rocks may well be part of a basement sequence, while also confirming that structural deformation and hydrothermal activity played major roles in ore formation. Nonetheless, regardless of the now outdated RST genetic model, our earlier exploration methods successfully located, delineated and quantified the two deposits that were subsequently mined.

26

Eating My Words
Namibia, 1997 and 2010-2012

Andrew Drummond

My two experiences in Namibia were separated by more than a dozen years, and each one ended in failure to achieve my aims. It is an African country which is enjoyable and very much worth visiting. I found it safe and friendly, perhaps surprisingly so given its convoluted and often violent history, and someone referred it to me as "Africa-Lite".

A brief history seems appropriate as an introduction to the story.

Much of the country, particularly the western sector facing the Atlantic coast, is comprised of the Namib Desert. Further inland is grazing country and, on its northern border, the Okavango River supports a higher population as farming is possible along its length. With few rivers, little other land obviously worth ready cultivation, and historically no known mineralisation of consequence, this desert area was one of the last chunks of Africa to be seized by a European power during that great colonisation drive of the 19th century. The Germans did so around 1850, and named it South West Africa. Its capital was established inland in farming and grazing country and was called Windhoek. Walvis Bay port and the town established nearby, Swapokmund, provided the country's connection to the outside world. That latter town still has some lovely German style architecture, particularly its old railway station. Communications were enhanced by the eventual laying of a submarine cable to there and, initially, a series of heliograph towers enabled messages to be relayed from the coast to the capital. Later, there was a rail connection.

Diamonds were discovered along the Orange River, which is at the southern boundary of the country and, later, rich deposits of them were found in the sand dunes along the Atlantic coast to the north of there and to the south of Swapokmund. This area became known as the *Forbidden Coast*.

With increasing numbers of German settlers in the late 19th century, competition for resources began between them and the Herero tribe which rather sparsely occupied the bulk of the country. The Herero rose against the Germans at the start of the 20th century and were ruthlessly suppressed by the latter in what was later termed as the first genocide of that century. Interestingly, the governor at the time was one Heinrich Goering, one of whose children was named Herman and who ultimately became second to Hitler in Nazi Germany, and commanded its Luftwaffe. Perhaps it was a sad trial run for the horrific Jewish Holocaust 30 to 40 years later.

This explains why one of the main streets in Windhoek was named Heinrich Goering Strasse, although probably does not explain why that street name still exists. And another is named after Robert Mugabe, which again makes original sense as he also had to lead a revolutionary group to wrest his country from its colonial master: but surely his standing amongst black Africans has been irrevocably ruined by his subsequent greed – and that street has not been renamed either. I guess it says something about the forgiving nature of the Namibian people – or that matters evolve very slowly there.

At the outbreak of World War I, the British navy seized control of Walvis Bay and Swapokmund, and South African troops defeated the Germans inland and occupied the country. The territory was mandated to South Africa by the League of Nations in 1920. South Africa requested that South West Africa be incorporated into it after World War II, but this was denied by the UN.

In 1966, the UN revoked the South African mandate, but the latter refused to accept that, and extended its apartheid laws into the country. The Southwest African Peoples Organisation, or SWAPO, had been founded in 1958 and, from 1968, began an armed struggle for independence. The fighters were based in Angola, the country's neighbour to the north, and were supported by Cuba. A ceasefire was shakily implemented from 1988. After a transition process, independence was gained in 1990 and peace broke out.

So Namibia is a young country defined by a mix of races, tribes, religions and an episodic bloody history. But, to me anyway, it is doing rather well all things considered, on the surface at least. It has a good economy supported by tourism, commercial fishing and mining. And the generated wealth does not have to be spread too thinly as Namibia is one of the few African countries which is not over-populated. But apparently still waters can conceal underlying turbulence, and matters can prove to be more difficult than might initially thought would be the case.

I first went to Namibia in 1997, one of several African countries which I visited when looking to pick up uranium projects. I had fairly recently been managing director of Uranium Australia NL and my enthusiasm for that commodity had been reawakened after Malcolm Fraser had killed it off in Australia in the 70s.

I had made contact with the multinational company, Gencor, as I had been advised that one of their major uranium projects, Langer Heinrich, may have been available for a deal. After some corporate mergers or whatever, Gencor had decided to divest itself of its uranium interests and, as it had been a very depressed scene for that commodity was some years, they were not expecting to get much for it. I visited the country and was escorted to the project by one of their senior geologists, Joe Hartleeb. He had had a long history with the project during its exploration and evaluation phases and was a fount of knowledge about it and Namibia in general: I found he was ready to share and his tour and company were both educational and very pleasant over our few days together. It resulted in my making a lowball offer to Gencor on behalf of my syndicate but, as luck would have it, with no other company having shown interest in a deal on the deposit for years, another ASX-listed company turned up a couple of weeks later and succeeded in acquiring it for a fairly insignificant amount of cash while our offer lay on the table. That's the exploration and acquisition game!

But an anecdote to indicate those subsurface swirling waters I mentioned previously.

Langer Heinrich is only an hour or so's driving inland from Swakopmund and we overnighted at a hotel at the latter, with the plane for Johannesburg due to fly out late that afternoon. With time to kill, Joe showed me the town and he took me to a spot which still had the concrete footings for the guy cables which had held the old heliograph tower vertically in place. He explained that, upon the outbreak of World War I, a British destroyer arrived offshore from the town and, if I recall his story correctly, it lobbed a couple of shells over it. The mayor then put a white flag on a rowing boat and rode out to the destroyer. He advised that the town wished to surrender without a fight and that he knew that British simply wanted the tower destroyed so communications with the capital, Windhoek, were cut off.

Joe also told me that the new British commanders got each of the German farmers to come to town for an interview. If they promised to keep the peace and not fight the British, they would be allowed to go home and keep any arms they needed to run the farm. On the other hand, if they wanted to be antagonistic to the British, they were escorted out the other door and pretty much imprisoned for the duration. A rather nice and genteel story.

Then Joe asked whether I would like to join him when he dropped in to see a lady, Mrs Schiller, who was an old friend of his from when he worked at the deposit, to which I agreed. Joe knocked on her front door, was welcomed, then he explained who I was. She looked at me, raised both arms as though she were pointing a rifle at me and then pulled the trigger with her finger, so as to speak. Probably not a Namibian g'day! By sheer coincidence, it turned out that it was her 90th birthday, and a collection of friends from the town had come for a morning tea celebration. Nine or ten widows and one widower as I recall. And I entered a time warp. They only spoke German to each other, some of the women wore embroidered dirndl blouses, and they celebrated by drinking sekt. And this was 83 years after the German colony became a British protectorate! Since then there had been two world wars, then joint British and German membership of NATO, Russian communism had come and gone as had the Angolan Marxist freedom fighters, Germany had been divided and then reunited, South African control and apartheid had come and gone, and so on. But little Andrew from Australia was still worth being shot apparently. I joined in the food and drink and smiled rather bemusedly at the guests, without being able to engage in much conversation by them. If they spoke any English, it was not done in my presence.

I seem to recall thinking that if I were having communication and outlook difficulties with people who had the same broader European heritage that I had, it may become even more difficult in the future dealing with people of a different skin colour who had a much different cultural background, and much more reason to be carrying historical baggage.

My next experience in Namibia was over a couple of years from 2010. After some corporate manoeuvres, Minemakers Ltd, of which I was Managing Director at the time, was in a joint venture over a very large and promising marine phosphate deposit. Its partners were another ASX-listed company, UCL Resources, and a local black empowerment entity group comprised of a rather well regarded and influential women's group. The last had a 26% equity which was free-carried by the two Aussies, who had an equal share of the rest.

The joint management worked its way to successful completion of a bankable feasibility study on the development of the deposit. It is worth digressing a little to describe that rather interesting mineralisation.

Nutrient-rich currents off the coast of Namibian supported a very large fish population which, in turn, was the foundation for a very lucrative commercial fishing industry based at Walvis Bay. When the fish died, or passed waste, phosphate-rich scales and bones and discarded matter subsided to the sea floor. That material decomposed, then re-composed as a phosphatic carbonate mineral which, under wave action, formed black spherical concretions up to sand grain size. With no rivers of any consequence bringing other sediments into the local ocean to dilute those sands, they concentrated, together with the shells of bivalves which lived in the sand, into very extensive and rich phosphate deposits. The better concentrations were around 100 km south-west of Walvis Bay and water depths in excess of 130 m.

The JV aim was to use a modified Belgian deep-sea dredge to suck up the sands and shells, which were then to be passed across screens so that much of the shelly material could be rejected straight back into the ocean. The predominantly sandy material would then be brought back to port and pumped onshore as a slurry in a pipeline to a processing plant to be built a few kilometres from Swapokmund. The phosphate price was good, banks were keen, and all that was needed were government approvals for mining and onshore processing. Naturally, the usual suspects of greenies, NIMBYs and the ignorant or selfish opposed the plan – as they did with almost all mining ventures. But they came up with a powerful new angle which was to frighten the government that any dredging we did would irreparably damage that very valuable commercial fishing industry.

This was patent nonsense for several reasons. The first was that the phosphate deposits extended into the areas which were the main fishing zones and so obviously the fish weren't affected by it. The extent of the shelly debris in the sands also indicated that these beasties were very happy to live within it. And, as the trawling activities were always tracked and monitored, we knew where they were doing their harvesting and we ensured that our intended mining areas were quite some distance from there. We had gathered supporting expert advice that there would not be any damage caused by our activities, but the problems still remained. There may have been other underlying reasons: tribal, historical, general anti-mining, a desire for greater Namibian ownership – those deeper disturbances below the surface which I have mentioned before – but we were advised that the possible damage to the fisheries argument was holding a lot of sway in government.

A very high-level meeting was eventually organised: we three joint venturers on the one side of the room and, on the other, the President of the Republic, relevant Ministers and senior officials. We did get a good and fair hearing, but I was not sure we had made traction. Then I had a thought: if a picture is worth a thousand words, and actions speak louder than words... .

We had brought to the meeting a jar containing the concentrated black phosphate sand and had passed it around for everybody to look at. I addressed the meeting, saying that we were aware that some people were of the opinion that the material we aimed to produce was harmful, but I assured them that this was not the case, despite all like they might be hearing from those who were either genuinely worried about it, or who had other motives to say so. The jar had returned to me, so I unscrewed the lid, licked my forefinger and plunged it into the sand thus getting a good coating of the material sticking to it. Then I stuck my finger in my mouth and licked the sand off and, with some effort, managed to swallow it all. It was worth it just to see the look on some of the faces as they variably thought, or hoped, that I may have been laid prostrate at any moment. The meeting finished soon after, and we all left, with yours truly not frothing at the mouth or having to be carried out. I have not heard of anyone else ever having to eat their intended ore to gain a mining approval, but, even if others have, it must still be a pretty rare event. The things we directors do for our shareholders!

Anyway, it was still not sufficient to gain those precious rubber stamps.

For quite different corporate reasons, my time at Minemakers finished shortly afterwards and, as I write in 2022, those mining approvals have yet to be granted. I presume that, as usual, those in government and the public servants still draw their salaries anyway, but Namibia could be considerably richer now if they had approved the project for production. And it would be providing a lot of jobs at all levels. But that is politics. I'm sure the deposit will be mined one day, and will make its then necessary contribution to help feed an ever more populated world.

27

Real Men Don't Need Visas
West Africa, late 1990s

Mort Cowan

In the latter part of the 1990s I did a few trips to West Africa assessing properties for gold.

The first trip was to the Ivory Coast (Cote d'Ivoire) and, being a bit green concerning travel to those regions, I booked the flights through a local travel agent. I recall the agent's words very well: "you don't need a visa to go to Africa".

That worked for the first part of the journey to Johannesburg via Harare. I got a shock when the 747 took off from the latter, climbed to about 400 metres, dropped its nose and increased the power. What I did not appreciate was the plateau location of Harare and the lack of topography. It must have been very noisy on the ground below the flight path. From Johannesburg the next flight was via a Russian airline direct to Abidjan, the Ivorian capital. However, the Russian Bear was a day late and was not going to Abidjan after all, it was instead going more easterly to Togo. That country had just emerged from a civil war, and the airport was full of angry people who just wanted to get their airfares refunded.

I was there without a visa and the prospect of going to jail was not appealing at all. There was a flight going to Abidjan, but it was full, so I offered to take the Jump Seat. The service manager looked at me and said: "you have done this before". As luck would have it, I scored that seat and was then in Abidjan without a visa.

The immigration official asked for my non-existent visa, took my passport and gave me a slip of paper to present at the Immigration Office after the weekend. Finally, I collected my luggage and went outside and a large African said: "Monsieur Morton". My life was saved!

We looked at a few exploration properties, but that region is quite strongly lateritized and there was little to see on the surface.

We drove around in an older Chevrolet that had been converted to gas. Refilling involved buying kitchen gas bottles and hand pumping the contents into the car LPG tank. This was something I had not seen before. Driving at night was something again. The colonial French left a lot of bitumen roads but little in the way of street lighting. All the pedestrians walked along the smooth bitumen roads, and there was little contrast between the pedestrians and darkness. How we avoided hitting them makes me shudder.

The next trip was to Guinea. Prior to the trip, I spoke to the local geologist about the project. During the conversation I could hear a lot of background explosive sounds and, upon asking, was informed that the local militia were shooting up the presidential palace with tank artillery. This was not reassuring at all. I flew to London then Brussels and then south to Conakry, the Guinea capital. The mesas in the middle of the Sahara Desert reminded me of similar features out near Telfer in WA.

Being a bit wary about visa requirements, I was given instructions by the client. After landing in Conakry, I was to go directly to the VIP lounge and avoid the normal immigration stream. My contact would arrange my visa and we would proceed. I sat in the VIP lounge like a stale bottle of beer for quite some time. Finally, someone came to the back door and told me I was to collect my luggage. Having done this, I was about to go back into the VIP lounge when the attendant passed me my briefcase and said to get out of there. He summoned a taxi, and off I went as an illegal immigrant.

We drove towards the city and came across a hotel with a high security fence. I thought it would make a good place to hide until things got sorted. All I had was a contact name and a phone number, but the phone was on fax; there was no one answering. This went on for a few days and I was getting very worried. The hotel seemed more like a bordello, there were several attendant young ladies around the pool during the day who seemed keen on increasing their vitamin D levels by exposing most of their bodies to the sun.

Finally, the contact phone number answered, and I was saved.

Conakry, as was Abidjan on the earlier trip, was constantly shrouded in a thick pall of black diesel smoke from all the poorly tuned diesel vehicles. It was very unhealthy.

We took a flight to a town in the middle of Guinea and visited the project areas. At the time there was a lot of artisanal gold mining, mostly focussed on alluvial gold in ancient stream courses. The working areas were like instant villages: all the goods and services and essential supplies were available, including the blacksmith, general vendors and gold buyers. The men worked underground, and the women did the hauling with windlass equipment. Hard ore was crushed in wooden dolly pots before the gold recovery by mercury amalgamation. The amalgam was reduced to gold by open air heating and evaporation of the mercury; there was no proper retorting to recover any of it.

From a technical perspective, the trip was a disaster. Samples had been collected using GPS coordinates but only one value had been recorded. Without a location plan the data was useless.

With the level of artisanal workings and the companies trying to secure exploration or mining title to the lands it was easy to see that land rights issues would come to the fore.

The lateritic profile was quite well developed, with a lot of surface pisolitic materials present. While this could provide a good sampling medium for assaying for gold anomalies, it really only works if the lateritic soil profiles have not been disturbed by erosion, mixed and redeposited. Often, I found that this was unfortunately the case, and the geochemistry was compromised. Often transported pisolites there also contained quartz grit, a further sign of a mixed provenance.

The trip back to Conakry was in an old Russian turbo-prop machine, a bit like a DC-3. All the passengers boarded, including a few animals, and then there was a tremendous roar from the back of the right-hand motor, I thought it was going to blow up. It was actually the Auxiliary Power Unit (APU) used to start all modern turbine motors and mounted at the rear of the motor. The APU was kept running until we were well in the air: maybe it was insurance in case of a flameout.

Back in Conakry we motored past the presidential palace and, sure enough, the walls were pock-marked from the attack mentioned earlier.

Things may have improved over the intervening quarter century but, at the time, my two trips did not fill me with confidence.

The trip back to Perth was uneventful. It is always a good feeling to know you are on your way home.

28

Humility in Botswana

Botswana, 2001

Mark Hughes

Mark Hughes graduated from the University of Wyoming in 1971. He spent many summers on exploration work in Northern Canada and the Arctic in the late 1960s and early 1970s. He spent most of the 1970s in Southern Africa and relocated to Australia in 1981 where he was principally involved in gold exploration. Between 1999 and 2004 he worked for Gallery Gold Ltd on gold exploration in Botswana.

Travel to less developed countries often compels the visitor to appreciate the advantages of life in First World ones. This is especially so for geologists who might find themselves in the back blocks of these foreign lands and in contact with the local hoi polloi. Sometimes said geologist gets a new and sobering view of life ... if not a bit of humility.

2001 found me engaged in regional fieldwork for Gallery Gold in Botswana, working on their Mupane gold project. Local Motswana (which is the name for the people of the Tswana ethnic group of people in this part of Southern Africa) made up the field crew. They were mostly quiet, industrious people who were keen to listen, learn and contribute. Although being not well off, they were generally happy with life.

In Francistown, where our exploration efforts were based, we had an office compound on the edge of town. Our workers lived at their homes scattered around the city. These workers walked to and from their homes to the office each day. This might be an hour's trek – twice a day. For some reason they did not ride bicycles.

I had worked throughout southern Africa in the 1970's, and there were bicycles everywhere ... in their tens of thousands. By the early 2000's there were virtually no bicycles to be seen on the streets of Francistown. I did not and, in fact, could never understand, why.

We had a brief weekly meeting with the field staff on a Friday morning to discuss the work program and to generally ventilate any issues. One week I commented that I wondered why our guys wasted so much time walking to work when they could spend just a few minutes cycling. Somehow in the conversation I mentioned that I had a bicycle back in Australia which I used quite often.

By way of an aside, I had previously mentioned that the fleet of Toyota HiLuxes we had in Francistown was similar to the vehicles we used in Oz, and that I personally owned a LandCruiser.

A week or so later I was going to the work area with Philip, one of our local field assistants. "Mister Hughes," he said. "Mister Hughes. Did I understand you correctly? In Australia, you have both a car AND a bicycle?", with some wonderment in his voice. I didn't have the nerve to tell him that, back in Australia, my wife also had a car and there were at least four bicycles in the family garage!

Around the same period, I was with Phillip returning to Francistown after a day's work. It was hot, dry and dusty. At the edge of town there was a kiosk where I knew we could get something to drink. I asked if he'd like to stop for a cool drink. He didn't jump at the suggestion and looked a bit dubious. To avoid any misunderstanding, I said that I was offering to pay for his drink, and he then welcomed the offer. We stopped and selected drinks, I paid, and we got back in the traffic. I ripped the top off my drink and had quickly gulped down half of it when I noticed he was holding his in his lap, still unopened.

"Aren't you going to drink yours?" I asked.

"Oh, yes," he said, "I'll take it home and share it with my wife and son."

I felt about an inch tall.

29

Africa – East and West
Africa, 1993-2010

Cyril Geach

Cyril Geach is gratefully 69 years old and graduated with Honours as a geologist from Nottingham University UK in 1980, after spending eight former years as an industrial chemist up until 1977. He started his geological career as a mud logger in oil and gas before migrating to Australia in 1981 as a diamond exploration geologist with Anglo American and De Beers.

Based in Perth, he has extensive experience in Australia and in Africa, Europe, Russia and Asia in mineral and oil and gas projects.

I am a lad who hails from South Shields, County Durham, a Geordie sand dancer and, in the immortal words of Bill Bryson, "(someone had to) – be a geologist from there".

Why a geologist? Well, the lure of the coal pit or shipyards was never a fancy as a young man, but the spectacular 100 foot high dolomitic limestone cliffs forming along the North Sea created a passion to understand these rock types (of Permian Age). An unknown urge and captive interest in rocks would lead to a world tour from my twenties onwards, otherwise unrealised in my teens.

The following stories are derived from African experiences gained in trips to four countries, two in West Africa and two in East Africa. They span a time range from the 1990s to within the last decade.

I will start with some general comments about my African experiences. Firstly, overseas expeditions require advance planning. Keeping costs to a minimum was essential for many junior mining companies, resulting in the geologists often flying on budget airlines and having to swag it out in the bush.

Visits to Africa obviously commence with a business reason to make the journey. From the outset, they generally need an African local go-between who sets up a meeting with the national government executive as soon as possible after arrival. The go-between was usually a resident in the country of interest and often served as a representative of the company and was engaged on the company payroll.

Africans are undeniably smart people and often are highly educated, particularly at government level. Quite a number of government officials I met had been educated abroad and held higher degrees. For an international mining or exploration company, the massive mineral wealth in-ground, or potentially able to be found during exploration of geologically attractive locations, could be combined with the general financial incapacity of the host country to set up a win-win situation for both sides.

The visitor represents a cashed-up western world of mining entities looking for "a project". He, or occasionally she, held a key to open a financial door to exploit the untapped mineral wealth that invited gratitude, and so was often welcomed with open arms, to a point anyway. Essentially, put your money where your mouth is and expect to be prepared to be a joint venture partner with the government in some form or another.

First up is the suit and tie first presentation with the government. My observation was that such meetings were always eyed with a degree of local circumspection as overseas businesses plying their trade mostly come and go, often to no avail: they are often on a reconnaissance, or a tyre-kicking, mission at first. A willingness to employ locals was an important lever to break down barriers, and impoverished government officials often would hasten up approvals to carry out exploration and grant tenements if some US dollars enabled the wheels of bureaucracy to be oiled.

I found that sub-Saharan Africa is a continent worth seeing: it is rich in variety, but mostly poor in consumer goods, and is populated with happy souls who are willing to embrace foreigners to their land which is potentially abounding in mineral treasures.

On arrival, one is generally met by a smartly clad person and is instantly greeted with perfect movie star teeth and ivory smile and a firm and friendly handshake. I wondered why Africans have such good teeth: either it's a heredity thing or just good hygiene, more about that later.

Having passed the government test, official business done and cleared to venture out into the African bush, here we go.

Cote d'Ivoire/Ivory Coast, West Africa, 1993

It was a journey that took over 25 hours to arrive at the destination of Abidjan, capital of the Ivory Coast. Perth to London, then London to Paris, and Paris to the Ivory Coast, also known as the Skeleton Coast owing to so much death from malaria. Obligatory pre-visit inoculations are advisable before venturing into malaria-infested countries, which account for about half of the planet's landmass. The risks geologists take are considered part of their psyche – daring, swashbuckling and out for adventure. Sort of! Personally, apart from the quest for adventure, daring and, to some people daftness, there might be an astrological link as I am an Aquarian – born in the Year of the Snake perhaps?

Flying from Paris Charles de Gaulle airport was my one and only time sitting in first class and a feeling of being of importance emerged for once in my life; mixed with a James Bond on a spy trip persona, exciting times lay ahead indeed, methinks? Due mainly to my five years of French lessons at secondary school being of some use, but too long ago to be effective to grasp instructions, I did not pick up that there was a first-class lounge at the airport, therefore I stood amongst the cattle class to board. To my other utter disappointment, it was a smoking flight. Air France at the time had a smoking and non-smoking flight arrangement, at least to the Ivory Coast and, yes, right first time, I got the short straw! Having sensitivity to perfumes and cigarette smoke, blamed on enduring eight years of industrial chemical exposure, set me up for a bunged-up nose and watery-eyed flight. Not just cigarettes but heavily scented French cigarettes, pipe and cigar smoke. The seating was closer than today's executive seating which allowed engagement in popular discussion amongst Ivory Coast passengers who were more than interested in an Aussie geologist visiting Abidjan. Armed with my smattering of schoolboy French, I got through the pleasantries. Thankfully, in Abidjan I was met by a fellow Perth geologist, Max De Vietri, who is gifted in French and Italian. Max was my go-between. Max later went on to excel in Mauritania along with the late great Ted Ellyard.

The objective was to travel north to a diamond area, the infamous Toubabouka and Bobi Dyke (TBD) area in Seguela and Haut Nzi, to see if there was any scope for a business venture that could be funded out of Australia. A bit of geology here: the TBD area comprises sheared diamond-hosting kimberlite dykes which host diamonds, and which alter to soapstone, which is soft and soapy to feel. This soapy feel also compared well to the metamorphosed kimberlite in the West Kimberley and north Yilgarn of Western Australia, which I have also experienced at first hand.

In Africa there is always a personal driver who organises everything for you and intervenes on the occasion of any untoward events. Internationals are regarded as imposters for no reason given to me, but I just had to accept it. Off we scooted along dusty roads, semi-sealed and in disrepair for most of the journey; people everywhere along the roadside, walking machete in hand. It was a worrisome sight to see so many machete-wielding locals, and left me concerned that perhaps this was for defence? It seems that it may have been to use for a more frequent cause such as obtaining lunch –a slice off a banana bunch or a cane to suck on and chew, probably accounting for the excellent white teeth of most Africans. Who wants to argue? A couple of nights spent in an earthen hut and fitted with a bedside mosquito net became a less captivating experience, with interrupted sleep due to buzzing insects and dreaming of perhaps a machete intrusion? Mosquitoes in Aussie are tolerable but the thought of being infected with nasty malaria is always worrisome to say the least.

At Bobi a demonstration of diamond recovery using crushed local kimberlite using a hand-held gold pan was remarkable in that diamonds were recovered from the panning of a few kilograms of material. The sample had obviously been salted for demonstration purposes to ease the embarrassment of not finding any diamonds at all, which would be far more statistically probable in any relatively small amount of kimberlite. It should have taken a few tonnes of it to find a diamond based on the usual grades at Bobi. It was a worthwhile experience for the geologist but, alas, the financial world would not support a standalone prospect such as at Bobi at that time.

The visit was set up for junior ASX junior mining company promotion purposes and for a potential fundraising opportunity, particularly out of the United Kingdom, where investors fancied West Africa in untroubled times as a magnet for those wanting "blue sky" appeal, albeit at high risk. Timing is essential in the business world and the timing of the Cote d'Ivoire visit was not favourable from an investor's viewpoint: interesting but not viable at the time. C'est la vie.

Ghana, West Africa, 2009

Sixteen years later found me east of the Cote d'Ivoire border in Ghana, an English-speaking country where dress sense contrasted to the French safari suit style of the former: city suits, a tie and fashionable brogues being a hallmark of the upper class in the capital, Accra.

Buildings of modest architecture were more prominent and a general wealth was evident, although this was 16 years after visiting adjacent Ivory Coast which may also have changed accordingly. The air was vibrant and busy, with pedestrians filling the sidewalks and obstructing the way forward. The street markets were haphazard.

Ghana was not previously called the Gold Coast for no reason! The job was to spend six weeks north of Accra drill testing unmined extensions and mine spoils over an old mine at Konongo in mid-central Ghana, and which had produced a million ounces of gold. It is located east of Kumasi. A sealed road for most of the journey was potted with many holes, and it was inadvisable to drive it by night. People galore mobbed the roadside along the route, carrying baskets and machetes. Bleached white shirts presented like a flag procession. Road sense was unheard of and many near misses had me jumping for the brake pedal in the rear seat. A landscape of green cane grass and fruit trees with oranges was repetitious. A flat landscape for most of the way, with distant views of elevated hills, made the journey visually interesting and distracted me from the mobs being passed on the roadsides. Due to our slow speed, I caught more than a glimpse of many villages of mud brick, concrete and earth. The bystanders curiously observed who we were as we slowly passed by. Outside the villages there were fruit sellers and individuals also selling roasted field rats on skewers, which were charred still with fur on, and tasty so the driver emphasised, but with no takers from me. Bananas and oranges were a safer bet.

The land that encapsulates the gold mine belongs to the village chiefs of the region, who have given permission, along with the government, to issue a Mining Lease. It was a pleasure to sit at meetings and to address the village chiefs and senior administrators, who were normal suit and tie persons and more than savvy when it came to business matters and protocol. There was a strong sense of British procedure and attention to detail.

The Konongo area is part of the north-east Ashanti Belt and gold generally occurs in schists and quartz stockwork veins. During my stay, diamond core drilling was to be carried out. My supervisory role was to induct two Ghana-born geologists and to establish them as project leaders going forward.

Heap leach stockpiles were a remnant of a 1990's carbon in pulp (CIP) process and were stacked up to six metres high. They were many hundreds of metres away from the well preserved and relatively unrusted CIP gold milling and recovery plant. While I was there, we tested old Carbon in Leach (CIL) heap leach pads for the grade of the remnant gold.

Around the heaps, the smell of almonds from residual decaying cyanide was barely detectable and the plastic liner at their base was occasionally exposed. Rehabilitation of the CIL heaps had resulted in profuse vegetation upon them, with mature trees sprouting from the sides and adjacent areas, and with orange and coffee plantations along the top ridges of the heaps for commercial production. The resourcefulness of the Ghana village people was shown to be remarkable.

The Perth-listed ASX company for which I was working had been astute in that it had become involved in an old gold mine which had a secured and fenced mine and camp site which were still intact, and were now heavily guarded by armed personnel. Many locals, who previously were employed in the 1990's, were housed inside the mine camp in brick and tile houses. These people offered considerable field operational experience with knowledge of where everything was housed. Outside the campsite, the "other" locals had taken no time to become artisanal goldminers and to carry out their own exploration by digging open-hole burrows, to several metres' depth, and as close as possible to the external camp walls, and by recovering weathered bedrock for gold extraction. It was literally a rabbit warren and dangerous to walk about at night for fear of falling down an open hole. A scenic tour one day to a nearby outskirts village of mud brick homes came across locals carrying out gold extraction in rudimentary and unhealthy ways using cyanide and mercury amalgam processes to extract the gold from material recovered from the artisanal workings and carried to the village. The need for the mighty dollar was ever-present and easy gold pickings brought survival to many impoverished locals. Unemployment was unlikely, however impoverished the look, and did not adversely affect the bubbly personality of the local people. I found them enchanting, despite their apparent adversity, and they remained upbeat whenever talked to.

The mine site was home for my stay. I had a separate bedroom with a kitchen, and a personal servant to cook and clean my clothes, and a valet, who was a nice chap with a pleasing personality and who was chatty, which suited me. Each night I had to make sure the insects were cleared from the bedroom, usually as well the lizards that followed the insects. A ceiling fan was my only escape from the humidity. Luckily, at nightfall I was able to get internet and so communicate with the wider world and report progress daily. Breakfast was a cooked meal starting with cereal and finishing with a Vitamin B spread, a favourite starter. Worldly advice is now given – do not take Vitamin B-enriched foods as it could end up with a visit to the GP with internal top to bottom problems – thrush or candida as it is otherwise known. Due to taking antimalarian prophylactic and antibiotic tablets, my internal gut flora had been stripped, and this allowed the yeast in the Vitamin B breakfast spread to blossom, and it played havoc with my insides. Ouch indeed, and quite unpleasant. The local GP suggested that I was likely to be at more risk from taking the prophylactic antibiotic, and thus lowering my natural defences, rather than from the prophylactic preventing me being seriously affected by malaria. "Get malaria and be treated", the GP said. When you hear of locals dying within days of getting cerebral malaria, it is frightening not to take anything, as he advised, particularly when your body is a target for mosquitoes.

Every day I would take a journey outside the mine camp to examine another part of the interests of the gold mine, which was via a pleasant, chauffeured journey through a local village that had more kids than adults. Word had got out that a white fella – "a bruny" –was about and I was greeted like a royal dignitary. Forearmed with bags of sweets I became a welcome visitor for a few days, as I gifted the sweets out to so many grasping, smiling, youngsters jumping and yelling in English that they loved me (as long as I had the sweets). This had also worked well for both sides in Cote d'Ivoire.

Each day a stop at the mine security gate was met with formality and an inspection. Greeted with local dialect ("hom-mut-esay" or wording to that effect) a security guard jumped to attention with military salute each time I went in and out – once again I was treated like a royal dignitary. The importance of getting the salute right was obviously vital, and it seemed as though the guard's life depended upon it, and it was done with aplomb as if I were a Major General. "At ease, Private!"

The humidity at the time of the visit was always high during the day. The region was densely vegetated with broad-leaved tall and lunging tree trunks, with a cane appearance. Behind these tropical trees were hidden many plantations of coffee, banana and orange trees. Ubiquitous machetes again. From a bystander point of view, and unable to defend a machete attack, it was imposing – do not argue with a machete wielding person! Baskets carried on the head held a full day's bounty of fruit to sell on the roadside, and or food for the family, and were commonly seen along the muddy tracks. Wellington boots worn by the field workers were a common observation – things tended to get a bit muddy when it decided to rain at any time of the day. Nighttime rain was best as it provided cooling relief and cleared the air. I found this was especially beneficial when a night shift for the drilling was introduced to speed things up a bit.

In the evening, and occasionally at weekends, a social gathering with the locals was encouraged, and it provided a laugh. At weekends, it was generally arranged to meet the team at the old headquarters, now a club house. Sitting at the whitewashed clubhouse, which was externally derelict, but still functional inside – leather upholstery and carpeted floors. It evoked a memory of the past times when it would have hosted many colonial inhabitants, playing billiards perhaps on the still perfectly preserved table with its baize underlain by Welsh slate. In its heyday, a nine-hole golf course was designed and built. When I was looking out to the final ninth hole from the clubhouse verandah, I could imagine the players with long socks finishing off a round and about to go in for drinks and a meal. Now it's a fairway of weeds, grasses and trees. Picture a regimented, organised lifestyle and many local servants, that had all changed when the mine ceased production. However, the geologist, as the sole remaining representative, is revered as a saviour by that part of the local work force which is being employed. The Ghana people, at least on the mine site anyway, still have a mental record of the past in their minds and maintain upkeep of traditions and administration. This order, stemming from the colonial administration, enabled the resumption of exploration work to be promptly carried out based upon, once freed from the cobwebs, hard copy data scrolls and geological cross sections which had been stored in the dusty rooms, but dutifully preserved – under guard and safely away from the hands of the local diggers outside the premises. All historical diamond drill core was stored and catalogued, a testament to the pragmatism and foresight of the locals who inherited the gold mine.

Central Ghana offers a green land of tropical heat and saturation, combined with snakes and pestilences that from time to time play momentarily on the mind. It's an adrenalin rush that often presents, such as when going through long grass and dense wilderness. Fair to say the only snake encounter by me was one large poisonous brute up a tree next to my digs. It was shot down and disposed of amongst a lot of excitement, poor snake. Malaria seemed to have escaped me – owing largely to the heavy duty full length garments worn and sweated through each day. Little exposed flesh to invite a mosquito and insect repellent was thankfully close to hand and used on a regular basis.

There were two moments in Ghana that will forever stick in my memory apart from the friendliness of, and interest in me by, the local people.

Firstly, funerals are heavily promoted on the TV, and I assumed that the more important the deceased person was, the longer the duration they were promoted in death on TV. It is not unusual to see many promotions about funerals during the week (TV, like the internet, was clearly available only after sunset – sunspot activity seems to be the issue?). Funerals were carried out on a Saturday. In the local town, processions of people would follow with cameras and videos and lots of jolly making – it was a big event.

Secondly, to say that people dressed up in finery when they could was an understatement in Ghana. One chap was seen walking in the main street of town on a Saturday with top hat and tails, white gloves and a cane. It was like seeing a rendition of an old 40's song and dance routine, Frankie Vaughan comes to mind. The local was off to a Rotary International meeting.

It was an enjoyable six weeks of hard work, but the ability to have a few weekend days off provided the opportunity to play a round at the Royal Kumasi Golf Club, with a personal caddy who played off the stick, and on a course that was more ash- than grass-covered. Needless to say, my golf was not up to scratch compared to his!

Mozambique – East Africa, 2008

Travelling from Perth on the east side of the Indian Ocean to the African west side, arriving via Johannesburg in the Mozambique capital, Maputo, was uneventful apart from the long waits at airport immigration counters on arrival and departure, a normal routine for the traveller to Africa from Australia. My objective on this trip from Perth was to assist the company geologist to engage with the high ranks in Maputo and then go into the field mapping to seek coal deposits. High grade bituminous coal was in great demand for energy and steel making at the time of the visit. I thought that nothing should be easier for a lad born and bred in the coalfields of Newcastle on Tyne: like taking coals to Newcastle, is the saying.

A personal driver was on hand to pick us up at the airport and to take us immediately to a hotel on the ocean front. This driver was on loan from a wealthy businessman who also was an ex-minister in the government at the time that the Russians were leaving Mozambique – after a long and bitter civil war between 1977-1992. It is of note that Robert Mugabe from Rhodesia, now Zimbabwe, was also one of the leaders of the Mozambique communist party at that time. My 2008 visit was before another flare up of tensions in 2013 which lasted until 2019. Portuguese is the generally spoken tongue and is mixed with local dialects. I was told to be aware of land mine areas when out walking the field: thanks very much, I thought; should I really be here?

Mozambique is one of the poorest and most underdeveloped countries in the world, as attested by the crumbling or unpainted and unkempt look of older buildings around the capital. Obviously, a Westerner was seen as a rich person and was noticeable: one had to try and mingle but stood out anyway. In the east coastal areas of the country, tourism is popular amongst South Africans and among visitors from neighbouring countries who are attracted by the seaside resorts in those parts, in contrast to the impoverished central areas of the country.

Mozambique extends to the east side of the African Rift Valley, and Malawi lies to its west. The hotel in Maputo was remarkably smart with a four-star elegance and attention to detail. Meeting with government geoscientists to obtain data information was not difficult if a payment for the information was made in US dollars: fair trade. The objective had been predetermined in Perth to visit a number of exploration permits to be assessed for their coal potential in the north-west region bordering west with Malawi and east with Zimbabwe, whilst aided by the purchased data (satellite and geophysical) to assist in detailed mapping of the regional geology using a GPS mapper. Easy enough it appeared. Where is it located – north to Tete on the famous River Zambezi? Hold on, that is where land mines clearing has yet to be carried out.

Another plane journey broken by two stops before Tete; a hot and exasperating journey, made more so for the waiting on the ground at the stops. A driver at the Tete airport met us and delivered us via a bridge over the River Zambezi – a brown and murky river about 500m wide with crocodiles and abundant fish – into Tete itself, a scrappy and dishevelled looking town.

Tete was a stronghold for revolutionary forces during the civil war and a lot of areas were land mined and, to that time, many areas had yet to be cleared. I realised why I was asked to do the job – no one else would! My perception was that the people living up at Tete for the most part had a glary look about them – unsmiling and an immediate sense of being alert – quite unlike those I had met in Ghana.

The hotel was located on one corner of a decrepit pot-holed tarmac crossroad boasting the only traffic lights in the town. I was met graciously by the proprietor, and he explained that I was considered to be of royalty and would always be referred to as Doctor, and would be given the premier room at the top of the hotel. Turns out it was the only room with a brass knobbed, king size bed and an en suite bathroom, with drawing room, plus a poky lookout. Third world charm at its best, with two-star hotel rating. Suited me fine, as long as the air conditioner worked, which it did. Turned out that the rest of the rooms were occupied by a Brazilian coal company and its expatriate staff working on developing a major coal deposit close by.

I was later invited to visit the offices of that coal company and was provided with a technical briefing. Simply put, geologically the coal deposits are contact metamorphosed through volcanic activity heating the margins of the coal-basalt contacts, creating a high-grade bituminous coal which could be used for supply to electricity stations, and for steel making purposes in Brazil and China. The problem was how were they to transport the coal from an interior area to a coastal port? There had been a railroad, but it was abandoned and used for scrap metal during the civil war. Before committing to development of a mine, the factors involved with logistics and risk need always to be considered.

Drinking anything other than sealed bottled water was a no-no, even in the hotel restaurant. In the latter, my colleague and I were looked upon suspiciously, for no reason other than we did not speak Portuguese. A week here was going to be an adventure, and it was.

Out every day to look at five permit areas spread around Tete meant a fair amount of driving on potholed tarmac and unsealed tracks – nothing unusual about that. The landscape was not too rugged, and the exploring kept to farmed land – so we knew any land mines would have been previously unearthed by farming. Peanuts or maize was the main crop at the time.

One day, following on foot a farm track and crossing a river, we could see coal fragments washed up in the river gravels, which indicated coal was somewhere upstream. I walked with GPS and mapping software in hand when, suddenly, my head was under watery mud as I had stepped into a pothole in the riverbed. Fortunately, I re-emerged chest high out of the pothole still with GPS in hand above the head and mouth shut. No change of clothes. Boots off and socks and trousers rung dry as best possible – onwards we go undaunted. Geologists are made of tough stuff. In fact, the cooling effect of water around the river in a humid climate was quite refreshing, the rotting vegetation smells less.

At the end of our evaluation period, it was decided that any coal which existed in the company's tenements would be found buried beneath the land surface, and that drilling would be the best method to determine if the appropriate coal measures and volcanic mix existed.

Google Earth is a fascinating tool and, at time of writing, a satellite image shows that Tete and surrounds have developed enormously with extensive buildings and a new airport. Two coal mines have been developed southeast of it and a new railroad has been built to the coast. Obviously, mining has established a new wealth for the region. "Well done", I say. Coal does have its place in modern society, irrespective of climate change concerns. I look forward to the establishment of more carbon capture and storage facilities, such as are being done at Boundary Dam in Canada and Petra Nova in the USA, so as to sequester the carbon dioxide emissions from coal burning.

Malawi – East Africa, 2010

Malawi is a westerly neighbour to Mozambique and occupies part of the Southern African Rift Valley. Each has an open border with the other and I could really only distinguish Malawi from its neighbour by the dress of the people. It is an ex-English colonial outpost and the people dress differently: a relaxed style in Mozambique, and a smart casual and suited style in Malawi.

Another long trip from Perth to the other side of the Indian Ocean resulted in my arrival in Lilongwe, the new capital of Malawi, via Johannesburg: the trip was incident-free. The older capital, Blantyre, is further to the south, and was also to be visited on the trip.

The mission was two-fold: to visit a granite-hosted Rare Earth Element (REE) joint venture project in the south and a sedimentary-hosted uranium project in the north.

A meeting in Lilongwe with a Presidential senior advisor, a bishop, was enchanting. He was also a Minister of the Government and cousin of the "contact", a legal chap from Lilongwe. An invitation was arranged for the very next day to the opening of the new Presidential Palace, and was eagerly accepted. The palace, a grand affair, was mostly built by the Chinese government as a mark of solidarity, one initially assumes. It was impolite to ask at the time, but were the Chinese, probably first and foremost, interested in accessing REE? I was invited as a dignitary and into the VIP section, mixing with the ministry and the well-heeled. At some stage in the proceedings, jet lag was taking control of the eye lids and my head was starting to nod off as it does under sleep deprivation. Acknowledgements were being handed out with much applause and kind clapping of hands. I woke up out of a minute's slumber and jumped to my feet as I saw a group in front of me do the same. I was upright and alone with a handful of standing Chinese VIPs who were being applauded and I realised that I had inadvertently insinuated myself with the group who were being praised. The look of the Chinese delegation was profound to say the least. Sit down, was a quick mental order. Geologists are never short of testing the boundaries if they want to succeed in new frontiers!

REE rocks are somewhat radioactive. For the reading of radiation intensity in the field one uses a radiometer, always a handy tool to have in the bag. A few REE samples were despatched to Australia for check analysis. When I returned home, I presented a few small specimen samples to a teacher friend, who also plays at the same tennis club, and who frequently asked me for exotic rocks. I had forgotten that these rock specimens were slightly radioactive, and it was not until after a brainwave a few weeks later that I requested the specimens back, just to be to be careful. Wearing a radioactive monitor was essential if working for periods of time alongside the REE rocks. The same applies to uranium-bearing rocks, even when the grade of it is low, as the cumulative absorption of the radiation over time is the issue. Short, localised visits are fine. I often wonder about the local villages that are resting along the sides of the REE and uranium host rocks and the time lag effect of radiation on their bodies? Nothing unusual reported? Fortunately, uranium minerals tend to easily oxidise and become soluble and so can be leached by rainfall from the near-surface ground. Generally speaking, the radiation intensity at ground level is relatively low.

Malawi is blessed with varying topography. It ranges from the rift valley floor along Lake Malawi, at about 470m elevation, to over 2000 metres up on the African Plateau. As one ascends, temperature and flora changes are noticeable. I noticed it particularly in the change of tree types in the south near to the previous capital of Blantyre. Large timber pine trees, like European forests on the higher landscapes, are encountered above 1000 metres elevation.

The National Park northwest of Blantyre conserves a remarkable mountain-top environment which is far different from that of the lower slopes. At about 1200m elevation it is totally forested and there is a tree-tops hotel of four-star rating set on the top of a range. The ears popped going up and downhill. The air was pure with conifer smells and the views are absolutely picturesque looking out towards several pinnacles of similar tree-covered mountain ranges as far as the eye could behold: at sunset and sunrise, they provided a photographer's dream.

A journey from south to north was undertaken at break-neck speed by an initial driver who had no fear of an impending accident, and who did not slow down going through the villages. A replacement driver was introduced which avoided the white-knuckle ride of the beginning. Still harrowing, however, as the roadsides are covered with people and motor bikes. Brilliant white shirts, colourful garb and fewer machetes, compared to neighbouring Mozambique, relaxed me.

Another stopover residence in the central north was on the west side of Lake Malawi. A small holding with four elevated cabins is situated on the edge of the lake with its own sandy bay which was about 30 metres wide. It had a cosy open-air wooden bar and lounge with ice cold beers and, best of all, a bottled dark sweet stout, which went down a treat or two or three. It was owned by an English couple who were not there at the time. It was declared that my room was previously slept in by a famous Scottish actor. I explained after my visit they can say it was slept in by two famous people, but the humour went down like a lead balloon with the local managers. Very pleasant overnight stay; the food was delicious, cooked up in a 4m x 4m pantry open grill to the side of the bar, and a few steps down. The water's edge was a matter of five metres away and a two-metre drop from the lounge bar. Soft calcareous sands underfoot with a gentle ripple from the edge of Lake Malawi, quite romantic for the so minded. Would not cope with a high or low tide or a seismic disturbance – situated along a rift zone, the latter may occur one day. There was company, a cycling man and wife team from, you guessed it, Perth Western Australia. They were on an across-Africa trip as a sponsored venture for a charitable cause in Africa. Good on them, I thought, and paid for their board, as we Aussies are always good at that sort of thing.

An interesting incident at the end of the Malawi trip occurred in the far north, when returning to Lilongwe for the trip back to Australia. A village was being passed which had a steep bank ascent on its far side. A sign on the road displayed "Slow Down" but no speed limit was advised: my dutiful driver did so. Concealed, until the crest was reached, was a tall senior police officer in smart attire and with scrambled egg peak hat and lanyard, suggestive of high rank. He also had a speed radar. He stopped the vehicle and asked us to get out as we were speeding. Apparently, the speed limit was 15kmh but, as previously, no one was to know that. He asked for a fine to be paid on the spot. The driver, if unable to pay would be immediately prevented from driving for three months and his car impounded. Unfortunate man: he had no money and so was unable to pay the US$100 fine. The only way out of this situation, other than lose my belt to the senior officer who was admiring it, was to pay the fine, which was done on the spot by yours truly and off we went. I wondered that, if I were a local, rather than a "visitor", the penalty would have been somewhat less? And why would a local driver be carrying US dollars anyway? I was told that Malawi had only two speed cameras then and we were caught by one – or should one say conned by one.

Wrapping up, Malawi is set up for tourism along Lake Malawi itself and the most famous of fish to be caught and eaten is tilapia, which is also found widely in the southern half of Africa and tastes delicious: it is best barbequed. It is a type of perch.

Africa is a place to visit to experience variety and enjoy the hospitality of its people. Malawi is certainly a place to revisit as it has many places of interest to see with good infrastructure.

THE AMERICAS AND
BEYOND

30

Ambush in Peru
Peru, 1992

Brian Levet

Brian Levet was born in Kenya in 1953, educated at Allhallows School in Lyme Regis in the UK and completed a degree in geology through the University of London in 1974.

His interest in geology began by collecting specimens while on safari throughout Kenya in the 1950s and 60s and was cemented in England, where the school he attended not only offered geology as a high school subject, but was located on a remarkable coastal sequence of Jurassic rocks, made famous by the fossil discoveries of Mary Anning in the early 1800s.

Throughout his 45-year career, Brian has maintained this passion for geology, exploration, discovery, and wildlife. His exploration philosophy is summed up by a quote from the Hungarian Nobel Prize winner, Albert Szent-Gyorgoi, (the discoverer of Vitamin C), who famously said that: "Discovery is seeing what everyone else has seen but thinking what no one else has thought."

He emigrated to Australia in 1983 during a mining and exploration downturn, he was unemployed and seemingly unemployable. A friend at the University of Western Australia commented that he was perhaps only employable in war time. However, a three-week contract with Newmont later that year turned into a 27-year career with the company, working all over the world with direct involvement in a number of team-driven discoveries.

Thick cloud blanketed the Andes in northern Peru, as the vehicle began the tortuous climb from the town of Olmos up the Andean pass to the watershed. The road followed steep-sided ravines, with multiple hairpin bends, up to the divide that separates the rivers that flow west to the Pacific from those that flowed east into the Amazon basin before continuing the long journey to the Atlantic Ocean. Five geologists made up the team travelling in the vehicle, four from the American company Newmont and the other from the Peruvian company Buenaventura. The team was on a reconnaissance trip to explore the mountainous area between the town of Cajamarca in Northern Peru and the Ecuadorian border.

Such a trip in normal times wouldn't be considered unusual, except this was early 1992 and the communist insurgency group, the Sendero Illuminosa or Shining Path, was at the height of its ascendancy and exerted considerable influence in the region. Competing with the Sendero was another terrorist organisation, also with communist leanings, known as the Movimiento Revolucionario Túpac Amaru (MRTA). The MRTA's aim was to provide an alternative to the more radical Sendero, which placed them in direct competition with the latter for support in many areas of the country. The Shining Path's strength was largely based on rural support, which forced the MRTA to focus their effort on their urban and more middle-class support base.

Sendero carried out 37 car bombings in Lima in the first seven months of 1992, which did help build support for autocratic measures taken by the then president, Alberto Fujimori. The Tarata bombing in the business area of Miraflores, an up-market district of Lima, where Newmont had their office, was by far the deadliest attack carried out by Shining Path during their period of insurgency.

Peru had become a dangerous and difficult place to visit, let alone to consider carrying out exploration for gold. There were few mining companies working in Peru, a country which, up until then, had been considered more favourable for silver rather than gold discoveries. The bomb blasts, stories of a geologist being executed by Sendero in front of his work team, and threats against Newmont's project manager in Peru added to the difficulty in trying to justify continuing mineral exploration in that country at that time.

When Newmont employees arrived in Lima, they were usually met at the airport by personnel from a security company known as Forza. It provided an armed escort, usually in the form of a lead vehicle and a pursuit car, with the visitors travelling in the middle of the convoy in case of ambush. In the field, Forza security guards were generally dressed in black, wearing balaclavas and carrying guns. They often appeared like 'Ninjas' and were stationed on the higher ground above the areas where the geological teams were working.

Newmont had made a very promising silver/gold discovery in Peru at a place called Yanacocha, through a joint venture with the BRGM and Buenaventura, via an agreement which was originally signed in 1984. The company had constructed a pilot heap leach plant to test the metallurgical recoveries on the million-ounce gold resource the company had identified. The project was, understandably, in mothballs due to the security threat created by the terrorist insurgency.

As Newmont's newly promoted Exploration Manager in South America in 1991, I was based in Santiago in Chile. Exploration had been focussed in the Andes of Chile and Ecuador, but did not include Peru due to the political instability. Our brief in South America was to structure an exploration program modelled on the company's Indonesian experience, largely based on Newmont's proprietary stream sediment sampling techniques and, consequently, the ability to explore large tracts of land efficiently and quickly.

On visiting the Yanacocha project in 1991, the extraordinary potential of the project was evident, and this was immediately conveyed to Newmont's senior management. Our recommendations were largely based on a visit to Japan in 1988, to look at the Nansatu deposits, which were being mined by the Japanese for silica flux, and from which gold was extracted as a by-product.

Keen to supplement the falling gold production from their Carlin operations in Nevada, where oxide reserves were being depleted, the Newmont Directors wanted to know if the Yanacocha deposit was a discovery that could justify the expense of development in such difficult circumstances. Exploration in Peru was quickly managed from Santiago. We sent in a due diligence team to assess the opportunity. The team reported directly back to Newmont's Chairman and President, committing in writing that Yanacocha was likely to become a world-class gold mining camp and was likely to become the largest and most economic gold mine in South America. The deposit was also expected to become the premier example of an economic high-sulphidation gold system anywhere in the world. In hindsight, the team's estimate of at least a 10 million ounces gold potential proved to be shy by nearly an order of magnitude.

Having recognised the extraordinary gold potential of Northern Peru, Newmont's exploration in Chile was largely suspended. The entire exploration team and effort in South America was focused on the safer areas of Northern Peru and in Southern Ecuador, with the aim of being able to identify and secure other areas of potential during a period when there was little competition from other companies.

The geologists, travelling up and over the Andes on that day in 1992, were scouting out an area to establish a secure base from which to conduct a helicopter-supported stream sampling program. A number of targets had been identified by remote sensing techniques and the group set out to establish the feasibility of conducting such a program.

On deciding not to use the services of Forza on this occasion, the exploration team took a much lower- key approach, so as to not to attract too much attention. In retrospect this turned out to have been a very smart move, or the team might have been involved in a major gun battle.

Arriving at the top of the Andes, we broke through the thick cloud and into perfect sunshine. We noticed a broken-down vehicle, or at least one with its bonnet up, parked outside a small local home, or *casita*. This did not cause us undue concern and we continued on our way to one of the river crossings where we planned to take orientation samples from the stream that was sourced from one of our target areas. While four of us took the sediment samples in the creek, our veteran geologist, Bill Gemuts, traversed up a hill to try to find the source of the interesting rocks we had observed at the sample site.

On completing the sampling, the group was waiting on the bridge for Bill to return when a couple of young men, dressed in jeans and runners, in contrast to the more traditional Peruvian clothing, came up to ask us what we were doing. We explained we were looking at rocks but before we knew it, we were surrounded by at least half a dozen people with semi-automatic rifles and sub-machine guns. We realised that we were in serious trouble. Having called Bill back to the car, we were stripped of our coats, jackets and geological gear, bundled into a vehicle and taken back to the small casita that we had passed previously. We were herded into it, where we found a Peruvian woman and her children who had been taken hostage much earlier in the day.

Our team was treated with respect, even when they found the pistol that one of our Peruvian geologists was carrying. Most of the group were high as kites, no doubt through a combination of adrenaline and whatever else they had been taking. They informed us that they were members of the MRTA, that they were about to ambush a convoy that was transporting one of their captured leaders over the Andes that day, and what we were about to see would be like Hollywood. They told us, if we remained calm, nothing would happen to us, constantly saying "tranquilo, tranquilo". We kept the conversation focussed on our families as we had been instructed to do in such circumstances.

We were held in the casita for some time before being suddenly bundled into the back of a car with an armed guard, and taken back down the road into the thick cloud. We were taken to a small quarry, which was little more than a cutting used to source road stone. At this point I thought that we were to be executed. Strangely, I had no fear of dying. I was much more concerned with how Newmont would break the news to my wife, and how the company would get the family back to Perth.

We were kept in the car with the bonnet up, obviously a well-used tactic so as not to attract attention. In the meantime, our own car was driven up to where we were being held and was parked across the road to be used as a road block. Time seemed to drag. Suddenly we were yelled at to get out of the car and lie down in a nearby ditch: we immediately complied. Our hostage car was then driven off and almost immediately the shooting started. The bullets going over our heads sounded like angry wasps, hitting the cliff face above us. I thought they might be trying to shoot the Gringos, as they didn't need witnesses to this kind of activity.

Noise of the shooting, the collision and the wrenching of metal as our vehicle was hit by a security van; and the screeching of tyres, gave way to an eerie silence. We were alive, unhurt and suddenly free, so all five of us ran down the side of the ravine, going about 20 metres before stopping. We had no coats, no vehicle, were in the middle of nowhere, and descending into Sendero territory. Wisdom prevailed and we made our way carefully back up to the road.

There was absolute silence and, in the gloom of the swirling cloud across the road, we could see the bumper bar of our vehicle and empty cartridge cases on the side of the road. Determined not to be captured again, we walked cautiously down the road. After some time, we heard a truck approaching slowly towards us. We hid in the bushes and, as it approached, we could see a terrified single driver, who became even more terrified when we stepped out into the road and asked for a lift. On getting into the back of his truck, we found our bumper bar, obviously a prized find. We stopped after a few hundred metres as there was something lying in the road. It turned out to be one of our mapping jackets, covered in blood.

Further down the road and around the corner, the scene that greeted us was indeed just like Hollywood. There, lying on the side of the ravine and balanced on the precipice, was an armoured security vehicle, the sort that banks use for transporting money. The back door had been blown off. The bandits had gone. The gunman who initiated the ambush had been remarkably accurate, there were five bullet holes in the bullet-proof windscreen of the vehicle, but none of them had penetrated. The whole thing was exactly like the scene from the 'Italian Job', where the bus is precariously balanced on the side of a cliff. In this case, everything was shrouded in cloud. One of the security guards from the armoured vehicle was badly injured having been shot through the shoulder. There was also a very frightened old Peruvian couple in an equally old car that had been travelling up the mountain before being stopped. The bandits had removed their keys and thrown them over the side of the ravine. Remarkably our car was there also, worse for wear after having been used as a roadblock to stop the armoured vehicle. All four tyres had been shot out, with an entry and an exit hole in each. It meant that ultimately eight punctures needed to be repaired before it could move.

We travelled back down to Olmos in the back of the truck and were held at the Police Station for the next nine hours, released later that night after questioning. Our vehicle had been retrieved and the tyres repaired.

We never determined whether the group that kidnapped us were really the MRTA or a just a gang of bandits intent on a bank heist. A number of the group had scars on their faces, something of a badge of honour in the Peruvian prison system, we discovered. There was a possibility they might have been ex-army or police. Either way, we had been treated well and were unharmed and we realised just how lucky we were.

The incident did change my philosophy on life, to which my wife can testify. Exploration was no longer the most important thing: family now took priority.

A few months later, on 6 July 1992, genuine MRTA fighters staged a raid on the town of Jaen, a jungle town located to the north of Cajamarca; it was very close to the site of our ambush.

The MRTA achieved further notoriety with a prolonged hostage saga at the Japanese embassy in Lima, before their demise, and that of the Sendero, which followed soon afterwards. Peru then went through a remarkable period of growth, which saw intense exploration activity and new discoveries. Although Newmont and Buenaventura made no further discoveries in this immediate area, they did make further discoveries in the Yanacocha district and that gold mine went on to become everything we had predicted.

31

The Cerro Negro Story
Argentina, 2000-2007

Neil Stuart

Neil Stuart originally gained qualifications in agriculture, before turning to geology. He has a BSc from the University of Melbourne and an MSc from James Cook University in Townsville. He began exploration in Nickel Boom times, initially working for Utah Development Company and searching for copper, uranium and coal in Australia, Africa and Indonesia.

He then turned successfully to coal exploration in Queensland, managing the coal section at Marathon Petroleum Australia, and finding the Darling Downs deposit, which supplies coal to a large power station in Southeast Queensland.

In 1980, he established a consultancy and also identified and acquired projects which led to the establishment of a major gold mine and a large lithium brine operation, both in Argentina. He has been a director of several ASX-listed companies over many years.

I was a founding director of a junior exploration company called Rimfire Pacific Mining NL, which listed on the ASX in 1996 with, mostly, Queensland gold-copper projects. Such companies change/morph over time – change of directors, projects, shareholder base and etc. By 2000 Rimfire had undergone such changes, and I found myself 'on the outer'. Consequently, I submitted my resignation as a director in exchange for a cheque covering my outstanding director's and consulting fees over the preceding six months or so. I was more or less full time with Rimfire and I then found myself looking for something to do – in the mining-exploration space.

This narrative concerns the strange way fate, enterprise and corporate intrigue can result in a world-class ore deposit.

As it happened, I bumped into Reg, who was an older geologist/prospector I had met a few times previously. He had been working for many years in Mexico and was back in Australia for a break. He suggested that I should go to Mexico and look at numerous prospects that he had knowledge about and would be available for Farm-In/Joint Venture or purchase. So, being at a loose end, I agreed to come over and have a look. In due course I jumped on a few planes and eventually landed in Tucson, Arizona and caught up with Reg who lived in a van in a trailer park and stayed with him there for a week or so. He did have a good 4x4 vehicle but, somehow, I expected his assets might have been more substantial. Apparently, he was divorced from his Australian wife, but had a couple of 'wives' in Mexico. Anyhow, over the week Reg took me around to all the interesting sights in Arizona – including some of the mines. But after a week I had to ask: "what about Mexico?"

"OK, yeah, we better go". And so we did.

We crossed the border at Nogales and proceeded within Mexico travelling in the Province of Sonora looking at various prospects – Cu-Au porphyrys, Cu shear zone systems and etc. These prospects were held under tenements owned by local Mexicans and sometimes showed evidence of old workings, but very little, modern, sophisticated exploration seemed to have been undertaken. Some of the old workings probably went back a hundred years or more. The locals who held the tenements were often small-scale farmers/graziers in the general area. Some of the tenements looked quite prospective, and I asked if they had tried to farm-out or sell them to any of the mining companies that were active in the country. Some had approached companies like Barrick, but when I asked for details, they would show me what they had been sent – such as a fourth generation of a photocopy of a fax of an assay sheet. They were surprised that none of the companies had got back to them with offers, nor did they get visits from the company geologists. Surprise, surprise.

So I undertook to write up a report detailing the geology, petrology, assays and etc. for each of the prospects. I did 'quicky' geological maps, collected numerous samples that I took back to Australia for analyses and so on. I promised that I would send them the completed reports after returning to Australia and they could copy them and send them to the mining companies – and, if they did a deal, they could flick me a few pesos.

I duly had the samples analysed etc., completed the reports and sent them to Mexico. I never did receive any pesos, but a month or so later, back in Australia, I received a call from MIM Holdings Ltd's (MIM) Brisbane exploration office. They had received from their Mexico City office a copy of my report on the Mexican prospects and asked if I would come in and discuss them. I duly arrived at its office. They informed me that they were quite interested in the projects but, because of budget constraints, they suggested I ask my Mexican friends to drill a hole or two to firm-up the prospectivity of the projects. I replied that my Mexican friends would be pushing to afford a taco, let alone a drill hole – and such work should be the responsibility of the farm-in partner which has access to the necessary funds. "Never mind – and, by the way", they replied, "we have this epithermal gold project in Argentina Patagonia, which we need to sell/farm-out; and you, being in the junior mining game, would you be interested?" So, I spent the rest of the day going through the digital data and reports on it– the Cerro Negro epithermal gold project.

Cerro Negro looked interesting, but I was not an epithermal gold expert, so I invited a real one, Max Baker, to come and have a look. I had met Max years before, at James Cook University of North Queensland when I was doing an MSc. in Exploration Geology and he was doing his PhD. Since then (about 1980) Max had been consulting widely on epithermals and porphyry style deposits – commonly with the well-known 'guru' Dr. Richard Sillitoe. The period 2000/2001 was another one of those periodic slow times for geologists and even Max was short on work and would be free to work for us. In due course we started up an unlisted public company, Oroplata Limited, did a farm-in deal with MIM and raised sufficient funds (mostly among our directors) to pay a deposit to it and to fund our due diligence activities. Shortly after that, Max and I departed for Patagonia.

Being a top consulting geologist, Max always flew Business Class, but Oroplata had limited funds and we convinced him to fly economy (with me) and to take some of his fees in Oroplata shares. This he agreed to, with the proviso that he determined the flight bookings – he wanted to garner extra Frequent Flyer points with his airline, United, which, of course did not have direct flights from Australia to Argentina. Consequently, we travelled – Brisbane-Sydney-Los Angeles-Houston-Panama City-Lima-Santiago de Chile-Mendoza and, about three days, later landed in Mendoza – without our baggage of course (it couldn't keep up with us). We did get our luggage a couple of days later in that city where we had been meeting up with the MIM people.

A few days later, accompanied by a couple of MIM staff, we were on our way flying to Comodoro Rivadavia in the Patagonian Province of Chubut where we hired two 4x4 utes and headed into the Province of Santa Cruz and out to the site of the Cerro Negro Project – a good three-hour drive. The last hour and a half was on dirt roads where 4WD was advisable. These roads/tracks had numerous wire gates – similar to the ones in Australia – just wire, tied up to posts in some ingenious ways. These gates took a bit of time to open and close and commonly marked estancia boundaries. Estancias are our equivalent in Australia to cattle or sheep stations.

Eventually we arrived at Estancia El Retiro, where arrangements had been made for us to stay as it was adjacent to the prospect area – we ended up staying there for about three weeks. I remember getting out of the ute and being rapidly approached by a middle-aged lady who gave me a big hug and kisses on each cheek. I thought to myself that surely I must have met this lady before, but no – it's just the Argentinian welcome (you get the hugs from the guys as well). That lady was the cook, and later we met two other ladies – the widowed owner and another helper. They looked after us well with good food, wine and comfortable beds.

The MIM geologists stayed with us for a couple of days to help us orientate ourselves and then we were on our own with our maps and geological and drill hole data. Over the next few weeks, we proceeded to inspect all the different prospect areas and any core from diamond-drilled holes. Many holes were non-core, Reverse Circulation (RC) ones). The core was held in a garage in the nearest town – Las Heras, about one and a half hours away. The drill hole sites were still well marked in the field and plotted on the maps we had – however, there were a couple of holes that were mis-plotted. We also discovered why there were a few holes that were completely devoid of any assayed gold mineralisation. The vein systems were well mapped at surface, but sometimes there was little structural information about their orientation. On the ground we saw some vein systems, which were clearly dipping in a certain direction, but the drill holes were located on the wrong side of the vein and consequently the holes were drilled sub-parallel to the vein dip and would have had little chance of intersecting the mineralized zone. This explained some of the dud holes we were seeing in the data. We asked the estancia owner about this, and how the geologist was managing the last program. She replied that on that program, or part of it, there was no geologist present, and the field assistants would contact the Mendoza office when they had finished a hole to find out the co-ordinates of the next few holes. This explained the dud holes and, from our point of view, the project started to look better and better.

Much has been said about the winds In Patagonia. Westerly winds generated over the Pacific Ocean get lifted over the high Andes range along the west coast of South America – in this case, southern Chile. With altitude, these moisture-laden winds drop their rain, which mostly falls west of the range in Chile and then continue over the range into Argentina Patagonia as strong, very dry winds exceeding 70Km/hr. This situation produces the semi-desert conditions so well known in Patagonia. Consequently, when out-bush there, when one needs to relieve oneself, one has to orient the stream in such a way as to avoid vortexes – which could easily lift and swirl it into your beard if one is not careful. We found that angling yourself with a wheel/mudguard/whatever of your vehicle seemed to work – although, even here, a few trial runs are recommended. Once after manipulating this manoeuvre, we opened the doors of the ute on a straight section (we were facing East) and let the car go in neutral – the open doors acting like sails. We reached 30km/hr (I must add that there was no traffic here). Another thing not to do with car doors in Patagonia is to open two doors at once while having your maps and things laying around on the seats – all of them get sucked out – some to be never seen again!

Anyhow after completing our field due diligence, we drove back to Comodoro Rivadavia and, from there, flew to Buenos Aires where we visited Government Departments and the Geological Survey and purchased reports, books, data, etc. From there we flew to Toronto to attend the Prospectors and Developers Association of Canada ('PDAC') conference. As we had decided that the project was worth going ahead with, we started to explore possibilities by talking to possibly interested parties re funding et cetera. However, the conference was affected by the general downturn in the industry and, at the time there were only about 1200 delegates (in 2020 there were 25,000!).

On arrival back in Brisbane we (Oroplata) confirmed the go-ahead with MIM; prepared a fund-raising document, at 10c/share, to raise a minimum of A$300,000 and went on a 'Road Show'. Our company Secretary and Director, Paul Crawford, and I were nominated for this job, and we took our show to Brisbane, Sydney, Perth and Melbourne visiting brokers, friends, geologists and any other group who we thought might be interested in investing in a "ball-tearing" project in Argentina. This was early 2002, and things were grim for mining/exploration companies trying to raise money for exploration and development, as the gold price was relatively low. People were polite enough and we were usually offered coffee etc. while we unrolled our maps, diagrams and so on, in those pre-PowerPoint days, and gave our spiel. I recall some beaut comments from people at the time – "If this project is so good, how come you've got it?"; and, "If you don't speak Spanish, how can you possibly work in Argentina?" We did not raise a cent from brokers or fund managers and the like, but we did raise above $400k from friends, associates and particularly from geologists. One of our best events was in Perth where we invited (with the help of my friend Rob Duncan, who lived there at the time) local geologists to attend a "drinkies, nibbles, talk and slide show with physical samples of gold-bearing specimens from Cerro Negro" at the Irish Club in West Perth.

A short time later, having raised in excess of A$400,000 for the company, a modest 10-hole RC drilling program was organized for Cerro Negro. As we needed to limit expenditure, we targeted a couple of the prospect areas which we could test with relatively shallow holes of around 60m and which had easy access. The assay results from those holes were encouraging, but not spectacular. Interestingly, at this time there was a financial crisis in Argentina. Previously the Argentina Peso was pegged to the US$ 1:1, but there was obviously quite a lot of inflation in Argentina: the Government refused to change the exchange rate and insisted that this would not happen. Then, overnight, the Government changed the exchange rate to 3 Pesos to 1 US$, and also limited the amount of money Argentinians could withdraw from the banks. Of course, there were those 'in the know' (many of which were associated with those high up in the government) and who had exchanged their Pesos for US dollars before the edict was made. The drilling company was US-based, so we could pay them direct from Australia, but our local contractors, with their earth moving machinery for drill pad construction, needed to be paid in Argentina. Putting money into their local bank accounts was of little use as they could only access a certain miserable sum per week – the government was afraid there would be a run on the banks.

As a foreigner I was not subject to these regulations, but at the time Oroplata did not have an Argentinian bank account. Consequently, every few days I would drive into town (Las Heras – which only had two ATM's) with my three or four Australian credit cards and withdraw as much cash as possible; and then wander over to the contractor's office/workshop, hand over some money, and hang around for a while to chew the fat while sipping communal 'mate' – a tea-type brew drunk from a straw. We still owed them money, but they trusted us.

After the drilling program had finished and the rig, trucks etc. had left, the Oroplata geologists (including myself) hung around for a few more days tidying things up, surveying and whatever. So, some days later, we were driving on the dirt tracks on our way back to Comodoro Rivadavia and came across the drilling crew and their trucks stopped at a gate. One truck was towing another. What was going on? They should have been in Esquel, Chubut Province, dropping off the samples at the assay laboratory there by now. Apparently, on their way back to the main highway it had started raining and they were worried that the big rig and trucks might get bogged – so they called into an estancia and asked if they could park their trucks there while they drove their 4x4 utes into Las Heras to stay the night. "Sure, no worries" – or words to that effect. Next day they came out to collect the trucks, and there was one missing – the one that had all the drill samples on it! My heart dropped: "we might have to re-drill *everything*", I thought. Luckily the truck they were towing was the one with the samples after all and they said they were all intact. "Phew!". Apparently, a group of young indigenous youths had stolen the truck and gone for a joyride, but didn't know how to change gears, etc, and the vehicle broke down and was abandoned. Enquiries over a day or so resulted in the vehicle being located and the samples eventually finding their way to the assay laboratory.

Although the drilling program was modest, it was sufficient to adhere to our staged expenditure requirements of our farm-in agreement with MIM. In the meantime, there were corporate manoeuvres and the large Canadian Mining company, Xstrata, was taking over MIM and, in this process, it was divesting the MIM projects that weren't sufficiently advanced for it to bother with – Cerro Negro was one of them. With our farm-in agreement, we had a 'First Right of Refusal' to purchase the Cerro Negro project – that is, we could match any offer that came in, and we would be given the right to buying the project at that bid price. We thought we could buy the MIM interest for about A$800k, but a Canadian company (which had a geologist who had worked for MIM – and perhaps knew of the project's potential) put in a bid for USD1.5 M. We had discussions with this group but could not come to a satisfactory deal and we ended up merging with an existing, ASX-listed Australian junior mining company, based in Western Australia. It had gone 'dot com' during that little boom; went bust, and then was resurrected by some entrepreneurs as a junior mining company again. It was Kanowna Consolidated Gold Mines Ltd: they were a shell looking for a good project, and we were a good project looking to list on the ASX. Being already listed on it, they had a much better ability to raise the A$ equivalent of the USD1.5M needed to purchase Cerro Negro. What also made it easier and helped us was that, under the deal proposed by the Canadians, payment was staged over a period of 18 months. A merger/take-over deal was done and all Oroplata shareholders received Kanowna shares which, of course were ASX-listed, and could be traded. I think the share price at the time was 2.5 cents or thereabouts. Being a bit naive at the time, we got a bit of a haircut going through, however we did have shares in a listed company which had the ability to raise funds and a good project to pursue.

Anyhow Kanowna raised the money and started re-exploring Cerro Negro – I worked with them in the early days as a consultant geologist, as I was actually the only person in the company who really knew the project, how to get there, with good relations with the Estancia owners, how to hire Argentinian geologists, assistants, contractors, etc. Original Oroplata shareholders held about 35% of the company and we were promised one or two Director positions – but this didn't happen. Instead, some outside new directors were appointed, and the new board issued themselves new options and whatever. Oroplata shareholders were not happy, and we thought about calling for an Extraordinary General Meeting to force some representation on the Board from the ex-Oroplata group.

However other events intervened – a Private Equity group, Sentient, out of the Bahamas, wanted to invest about AD9.0M into the company (which was desperately needed) and two of their people were elected to the Board and one of the original Kanowna directors retired. One of Sentient's director nominees was Gavin Thomas (of New Guinea Gold fame), who I knew and in whom I had confidence. Later, I joined Gavin in Argentina to guide him around the project when he came there to carry out Due Diligence for Sentient. Over a three-week period, there was plenty of time in the evenings (we were staying at the Estancia) while sitting around enjoying Argentina's famous barbeques, sipping their famous Malbec red wines, for me to have a bitch about Oroplata not having Board representation in Kanowna. And I talked about the possibility of calling for an EGM to sort it out. He replied that, "if there was a shit fight at board level Sentient might quite likely pull the pin on the whole deal". I didn't want that to happen, so I assured him that, as long as he was there, we would have confidence in the direction and quality of the exploration programs. Consequently. we did not call that general meeting. Gavin was happy with the results of his due diligence and, with my contacts in Argentina, we were able to put together a good team to advance the project further.

Under the new geological management and with adequate funding, the Cerro Negro project continued to advance and impress. Also, the company changed its name to Andean Resources Ltd and, over the years, developed gold resources in the millions of ounces. Later, Andean was taken over by the large Canadian miner, Goldcorp, in a deal worth USD3.5B! At that stage, the resource was some 10M ounces of gold and the project employed about 4,000 people.

Postscript

Some of the people who first started Oroplata Limited went on to establish and list Orocobre Limited on the ASX in 2007. Orocobre started off by exploring for gold and copper deposits in Argentina but moved to exploring and then developing the Olaroz lithium- bearing salar in Jujuy Province in northern Argentina. This project has now been producing battery-grade lithium carbonate for many years, is profitable, and presently (under its new name of Allkem Limited) is capitalized at some AD6.0B.

Hat Tips

The original Directors of Oroplata were: Ken Foots (Chairman), Neil Stuart, Gregory King, Paul Crawford (also Company Secretary). The original directors of Orocobre were: Neil Stuart (Executive Chairman), Richard Seville (Managing Director), Paul Crawford (also Company Secretary), Dennis O'Neill and Jack Tan.

I'd also like to acknowledge inputs from Jim Clavarino (who organized the first drilling program at Cerro Negro for Oroplata), Nick Sheard (who was our contact man in MIM), the Argentinian geologists, our Argentinian accountants, lawyers, consultants and many other contributors.

In particular I would like to thank Mirko Brkljacic, a soul mate of mine. I met him within the first week of coming to Argentina. He was my interpreter and translator, both in the early years with Oroplata and later with Orocobre. Mirko helped me soak up the culture and the language (this only to a certain extent!) for over the 20 years I've been going to Argentina. He came to Argentina as a refugee from Croatia post-World War ll. Apparently at the refugee camp they were asking people where they might like to go – they started alphabetically. Argentina was first, so he put his hand up. The next country would have been Australia. He could speak about eight languages. He passed away in 2021, in his nineties. He will be sadly missed.

32

What – No Piranhas?

Venezuela, 1992

Lynda Frewer

Lynda Frewer graduated in 1985 from Curtin University in Western Australia. Before and during her studies she worked for a junior diamond exploration company in South Australia, and in the Western Australian Kimberley and goldfields regions. After graduation, she worked for Freeport of Australia and then Normandy Exploration. In 1988 she opened a diamond sample processing laboratory in Perth, 'Diatech Heavy Mineral Services', to cater for the rapid growth in Australian diamond exploration: it still operates today.

It has been 37 years since I graduated as a geologist from Curtin University in 1985 in Perth, Western Australia. That time was the heyday of diamond exploration in Australia. The Argyle Diamond Mine had recently been discovered, and opportunistic exploration companies bristled with excitement. West Perth, the city hub for exploration and mining offices, was gridlocked with 4WD vehicles, and office floor space was snapped up. Companies large and small raced each other to peg ground in search of a woman's best friend. Australia had been thrust onto the world diamond stage. Geologists suddenly earned more than doctors, every available helicopter was leased, and building shark- and crocodile-proof cages to dive into our northern waters for diamonds made the front page of 'The West Australian' and 'The Wall Street Journal'. Everyone, it seemed, wanted a bit of the action, and I was no different.

I worked with a junior explorer before graduation and learnt the fine art of how to identify the elusive diamond indicator minerals, the key to finding the 'mother lode'. Upon graduation, in what was largely a male-dominated industry, I had also learnt:

- how many slabs of beer could be packed onto a pallet for field trips?
- how to change a tyre on a fully laden 4WD
- how to 'split the rim' to repair the tyre
- how to 'bleed' a diesel engine, and
- how to pee in the bush and not on the boots whilst retaining some dignity.

Male geologists were sometimes concerned about the lack of 'toilet facilities' for females in the field, and some even tried to use it as an antiquated excuse not to employ a woman. Gritty Aussie female geologists were not put off by this chauvinistic nonsense.

Now 60, Lynda is very thankful not to have shrivelled up in a swag somewhere in the great Australian outback. Because this happens – geologists of either gender head off for adventure in the field, and never come back to the 'real life' of friends, family, children. So, in 1988, to move away from extended periods in the field, I opened my own mineral processing facility in Perth specifically for diamond exploration samples. Tonnes and tonnes of dirt in neat calico bags made its way to this facility from all over Australia, and then from every corner of the world.

I still love Australian field trips and meeting the people and characters who make it the 'Outback'. But working overseas can test even the grittiest Aussie geologist, male or female. Here is my South American story.

London, 1992

"You will have to take it back in the morning!", exploded the man in the front seat of the near-new Rolls Royce. "I've taken this damn car back twice now to have the suspension adjusted and it still isn't right!" I sat in the rear of the car with an older woman beside me, his wife, and was attempting to make polite conversation with her between the man's angry outbursts. The neatly uniformed driver who was the target of the older man's diatribe tried not to flinch and took it on the chin, obediently driving us through the dark rainy streets of wintery London. "It just doesn't ride well!" the man continued to complain loudly. "Can't you feel it, rides like a horse!"

He bellowed on and on, oblivious to anyone else in the car.

And so, we continued on like this to the local Lebanese restaurant, *apparently* bouncing around too much in the back of this second-rate brand-new Roller.

I tried to make small talk with this older woman, but nothing seemed appropriate. I was a female geologist from Australia. I was 30 years old, and I had spent my short working life enjoying fieldwork in the great Australian outdoors, in places like the Kimberley and South Australia. Bashing around the Goldfields of Western Australia in 'Tojos'. Sleeping outdoors in swags, cooking in camp ovens, and enjoying the pub food back in the nearest town – where the menus were always the same, and 'Chicken Parmi' or 'Steak Diane' (all served with salad and chips) were standard fare. I had worked for structured exploration companies with dutiful employees. I was one of those dutiful employees. Four years earlier, I had left the bosom of a large company to establish my own mineral processing facility. Mainly to have a break from fieldwork for a couple of years (which is hilarious because I was only 26 at the time, and 34 years later I still own and run the lab), but to more importantly make sure I didn't wake up one day and find myself 40 years old and shrivelled up in a swag somewhere, or as a permanent fixture in an outback mining camp. With this temporary move back to the city, and the opening of the laboratory in 1988, came unexpected bonuses.

I was suddenly a desirable independent and not attached to any *one* company. And, with this independence, followed irresistible offers from the different people or organisations who passed through the laboratory, with work opportunities in far-flung and exotic places. And those places were not restricted to the (usually Australian) exploration areas that a single company employer might have in their portfolio. Not restricted to Kalgoorlie or to Kununurra. But opportunities to overseas locations. I couldn't resist a field trip to China (1989, just two weeks before the Tianmen Square shootings) (as recounted in her story *100 Year Old Eggs and Warm Coke* in *Rocks In Our Heads*, Ed.), and also Zimbabwe for good measure. But those are other stories. Although I needed to be in Perth running the laboratory and building the business, I was just too easily tempted to try an adventure to South America.

I would have to say this woman and I had almost no common ground whatsoever. But there *was just one thing*. When I turned to speak to her, I struggled to keep my line of sight politely on her face and not drawn to her bosom.

For around her neck, there were countless carats of uncut diamonds. The diamonds, each equal in size and bigger than a pea, had a single hole drilled through it for threading. Laser cut? I don't know. Maybe I should have asked. At a glance each diamond was perhaps 4cts. Not of great quality. They were threaded onto a loop long enough to go over her head, and then to be tied into a knot that sat at cleavage level with some left over to swing below the knot. Charleston style.

My laboratory back in Perth was specifically a diamond exploration laboratory. And here I was driving through London with this woman who considered wearing a string of uncut diamonds knotted around her neck quite normal to wear to visit the local neighbourhood Lebanese café, in the family Roller, for a mid-week meal of flat bread and hummus.

The older man and woman were the parents of my client who wished to explore for diamonds in Venezuela. Their son had contracted me for the work. They were a Lebanese family who had lived for decades in Sierra Leone. Their two sons were born there. The father was favoured by whoever was the president at the time, and obviously had influence during his time there. Even down to advising the president as to which London-made suits to wear to look snappy. They had fled Sierra Leone for London just before the brutal civil war that broke out in 1991, which lasted a decade and killed 50,000 people.

After our flat breads and hummus, the parents, their two sons, myself, and my associate Peter Ferrick (mining construction manager turned goat farmer from Gidgegannup) were taken to The Ritz London casino for a couple of hours. The father insisted on giving us all £500 each to fritter away on bets before we left. As you do mid-week.

Although I was tempted, I couldn't bring myself to ask for a photograph of the jaw-dropping loop of diamonds. It is one of my great regrets, but it was clearly not appropriate at the time. This was 15 years before Leonardo Di Caprio starred in the movie 'Blood Diamond' which alerted the world to the seedy underworld of diamonds mined in war zones, which were then were sold to finance conflicts and to profit warlords and companies – at great human cost. Looking back, I have no idea how these particular diamonds were acquired.

I was quite oblivious to any possible connection anyhow. I was just a young Aussie female geologist heading off to South America to conduct a sampling program for diamond deposits on behalf of the son of a rich Lebanese businessman from Sierra Leone. What could possibly go wrong?

Perth, six months earlier

"Peter, how would you like to join me on a trip to Venezuela to do some diamond exploration?", I casually asked.

There was a short pause. "Sure" replied Peter Ferrick. "I guess a man can be driving a tractor on his farm in Gidgegannup one day, and diamond exploring in South America the next. When do we go?"

And so our adventure began. I had previously met Jasar, the Sierra Leone-born and raised son of the Lebanese businessman, when he was sent out into the world from London by his father to seek his own fortune with a generous bag of start-up money. He had two good friends who would be his business partners in this quest, and who I would meet later – the very likeable Diego from Colombia who lived in the United States, and the equally likeable Eriko who was born and raised in Venezuela. Jasar, who was the leader of this trio, had already decided that diamonds were the go-to commodity. I would hazard a guess that this had something to do with his family's connection with diamonds in pre-1991 Sierra Leone. And he had decided that Venezuela would be his target area. Later I had the distinct feeling that Eriko and his family were more than well connected in Venezuela, and Diego's family wasn't short of a few dollars either.

They were all aged between 25-28 years old. Formal qualifications? Well, none that I knew of. Certainly, none directly related to diamond exploration or mining, but they had a drive and enthusiastic determination to make up for it. And connections. And money.

Jasar had left London months before to seek out diamond industry-related people and equipment to help in his quest. He eventually ended up in Perth. And then at my Perth laboratory. He was intrigued by the diamond indicator minerals that the lab recovered from river samples to follow a trail along a river system to locate a primary diamond deposit (and hopefully the mother lode). I showed him some ideas for practical exploration gear that could be used in Venezuela to process these types of indicator samples for himself, on site. At one stage we stood peering into a rusting and mud-splattered diamond sample pan in the Normandy yard in Welshpool with Rob Jordan (who manufactured X-Ray sorting machines for separating diamonds). The pan was about the size of a small caravan. Jasar was dressed to kill in his fine London business suit and very shiny shoes, without a hair out of place, and Rob and I looked like a couple of Aussie bumpkins in our practical working clothes. And then it rained, and all of our clothing was drenched. The distinctive yellow mud called 'yellow ground' (the weathered product of diamond-bearing rocks) had collected over years in and around the base of the now decommissioned diamond pan and now ironically plopped from the ground up onto those immaculate shoes. Jasar wasn't put off at all by the mess. His search was apparently over.

I was hired.

My first thought was "Oh shit. What the hell was I going to do now?" Call Peter of course.

Three years earlier, Peter had called me in a similar fashion "How would you like to join me on a trip to China to do some diamond exploration?" How could a girl say no? Now I was returning the favour. Peter was a down-to-earth diesel mechanic by trade who had worked for many years with Roberts Construction as Operations Supervisor. Diamond projects were no stranger to him having arm-waved tonnes of diamond-bearing dirt over a period of seven years at the Ellendale Diamond Project, the Argyle Diamond Mine, and the Bow River Alluvial Diamond Mine, all in Western Australia. He was a practical and hands-on Aussie man. He could fix or make anything. And he was also a flexible adventurer, a goat farmer and a windmill builder. He had my back in China, and now I needed him to do the same in Venezuela. My partner (now husband) was also Peter's friend and trusted he would watch my back. Or at least bring me back.

Our brief – to locate a promising diamond project. Equipment supplied – none. Maps – none. Budget – the subject was evaded. A few thousand? Hundreds of thousands? Millions even? No one was saying. This was a recce mission without the usual prerequisites on hand.

"OK Peter, we leave in a few weeks. The area of interest is on the border of Guyana and Venezuela, where the Rio Venamo forms the border in the east of the country. Plenty of water. Piranha in the rivers apparently. No power. No vehicle access. No maps. No tenements in place. We need to help select a promising diamond area after this trip. Without a clear budget so we should aim to do this on the cheap. On the bright side, we do have a helicopter. I am thinking we should take stream sediment samples and winnow them down to concentrates somewhere on the riverbank, then send the residues back to Perth to study for indicator minerals". Sounded easy enough.

Silence.

"Don't worry Peter. The client has asked for a pleitz jig (*a portable but heavy machine designed in Africa to winnow away unwanted minerals but leave behind coarser diamond related minerals*) to be built here in Perth and sent over, and they will arrange to get it near site. They will also arrange for a large cement mixer to be near site to mill up the muddy samples before trying to sieve them. They have mesh screens and plenty of manpower over there. Can you get some of that black rubber sluice matting to put in your suitcase? Just roll it up. We can make a sluice over there to concentrate the finer material."

Venezuela, 1992

Caracas airport was busy and smoke-filled as people lit up their Marlboros in the long queues. Anti-smoking campaigns had not reached this part of the world. As we stepped off the plane and approached the bluish smoky queue for customs and immigration, we were immediately intercepted by Jasar and introduced to Diego, and a well-dressed man in a suit. "Welcome, may I take your passports?" said the man. He disappeared in one direction with our only form of identification, and Peter and I were ushered through various side doors away from the long smoky queues, and were suddenly in the arrivals hall, unprocessed. We could sense some flexing of influential muscle here. We were apparently very important.

Peter leaned over and whispered to me. "Not bad for a goat farmer".

Our baggage quickly appeared at our feet, and we were ushered to a large black RV with 'no see in' black tinted windows and a driver. Soon, the well-dressed man was at the driver's window which opened a few centimetres, he slipped in the passports, and the window quickly closed. We drove off, down the Avila Valley, which sandwiched the city of Caracas at high elevation, and between some seriously high mountains.

It took two hours to drive down to sea level in the heavy traffic to the multi-storey 'Melia Caribe Hotel' which competed for space with other equally large luxury hotels and resorts squeezed along the Caribbean beachfront. We were deposited there for three days to assist getting over jet lag. But we weren't jet lagged at all and we definitely didn't want to sunbake. We were itching to get to work in the jungle. So we filled the days trying to formulate an 'Amazonian Jungle Exploration Plan' whilst overlooking the Caribbean beachfront strewn with plastic sun lounges and American tourists, and a lifeguard tower emblazoned with Marlboro cigarette advertising.

(In December 1999 this hotel was largely destroyed when much of the coastline was devastated by days of torrential rains, and the resulting landslides wiped out 100km of coastline. Thousands of people died as they were carried along in the massive debris flows, in what was called the 'Vargas Disaster'" It is listed as one of the top ten natural disasters on Earth).

Jasar and Diego visited us briefly at the hotel for quick meetings. They seemed otherwise preoccupied organising logistics.

"Have you found any maps that we can use to plot out some sample sites?" I asked them.

"No, sorry, they all belong to the military", said Jasar.

"How about aerial photographs?" (this was very much pre-Google Earth days).

"No, *definitely* military property. But the jig has arrived in Tumeremo. And we have a cement mixer!" chipped in Diego with obvious excitement.

Great. My tourist map would have to do for now.

We flew from Caracas eastwards to Puerto Ordaz, and then on to our base at Tumeremo. We flew over the massive Orinoco River, one of the largest rivers in South America, and forever enshrined in peoples' minds by Enya's 1988 song 'Orinoco Flow'. Sail away, sail away, sail away....I would listen to it on my mini-portable CD player as I peered out of the plane window. One of those Leunig moments.

Tumeremo, our base. Population of maybe 10 000 people. Seven degrees north of the equator. This is the last stop before the jungle. To the east, south and north, it's all jungle. Not far to the east is the river border with Guyana, and we were soon to visit that border every day to do our work. A long way to the south, is the geologically intriguing Mt Roraima. And there's just one road that goes anywhere out of Tumeremo, and that rough road beats a path through the jungle southwards to the Brazilian border past that mountain.

Our hotel, 'The Miranda', is about the best in this town of mainly single-storey decaying cement-box buildings painted in varying shades of dirty white, with rusted corrugated iron roofs, no gutters on the rusty overhangs, and rusting iron-bar grilles on windows. Our hotel was two-storey, with an open-roofed concrete deck area on the second storey that led to each single bedroom down a passageway. This was where we would have our daily work meetings and eat most evenings, sometimes buying take away food from 'MacDonalds' across the road – a small food van with a badly painted drawing of Donald Duck on the side. We could order 'Hamburguesas' done two ways – 'especial' or 'normal'. There was no shortage of Marlboro cigarettes and Pepsi. Our view from the deck was of a very old water truck with 'Agua Gratis' ('Free Water') painted on the rusted tank, and beyond that, a sea of rusty iron roofs.

Which is quite ironical because the water supply for showers was intermittent at best, and the first Spanish phrase I leant was *"no hay agua aquí"* ("No water here!"), when the water stopped completely mid-shower and I was still soaped up trying to remove a layer of Amazonian mud from my naked self. More than once I had to wrap my wet and still muddy body up in an almost-white towel and charge across the concrete deck, dash downstairs and yell it out to the owners with my Aussie accent *"No hay agua aquí!!"*

We met Eriko, the third member of the business partnership, in Tumeremo. And also, 'El Capitán', (The Captain) – our chopper pilot. I never did learn his real name, and he was always referred to as 'El Capitán'.

Peter busied himself on arrival with the construction of the sluice, or 'the sloo' as everyone called it. Peter built a deck from plywood, to direct the watery sludge over the ripple-grooved surface designed to catch any fine heavy minerals (such as diamond indicator minerals, diamonds or gold). It works in the same way that fine gold is collected in a gold pan, but on a planar surface. It is tilted to allow a slurry of muddy gravel to flow over it.

The jig arrived in Tumeremo on the back of a pickup. I have no idea how it got there after I had dispatched it by airfreight from Perth airport, destination Tumeremo airport. But arrive it did, and at least now we had some equipment, and we looked semi-professional, ready and able to do some work.

Our broad area of exploration was along the Rio Venamo, about one hour's helicopter ride from Tumeremo directly east on the border with Guyana. We were to do this trip daily.

On our first day, we went to the helipad. Peter and I rolled up in our typical Aussie field gear – navy-blue or khaki trousers and shirts, steel caps and with Akubra hats to shade the sun. There was no such thing as 'high vis' in those days. Jasar, Diego and Eriko were all dressed in military-style camouflage clothing and army boots. I tried not to look alarmed. Four months earlier, there had been an unsuccessful coup d'etat in Caracas by the Hugo Chávez-led Revolutionary Bolivarian Movement as they tried to overthrow the President. Chavez was now in jail. But was this type of clothing wise, I wondered to myself?

"This was all we could find for field clothing" they offered. At least Eriko wore a bright orange T-shirt so we could spot him in the jungle setting. There was some sense of excitement amongst the trio of partners. This was the start of their Boys Own adventure with a financial upside if successful.

The chopper was a decent size. I don't know what type it was, but it carried a passenger up front with El Capitán, and four in the back facing each other. Peter, Diego, Jasar and myself.

"Where's the handgun?" said someone suddenly as we gained height.

I hadn't expected that.

"Don't worry" grinned Diego, "We just need to have it within reach in case we go down. TIGERS. Oh, and always keep your boots on for the same reason. You'll need them in the jungle".

So I guess that was the safety talk.

Tigers? He was referring to jaguars, but the locals called the big cats of the Amazonian jungle 'tigers'. Was the gun to shoot an attacking tiger after the helicopter was downed in the jungle and we had somehow survived this experience (with our boots on)? Or was it to kill for food to survive until help came, I wondered?

We passed quickly over farming fields to over jungle. From these fields there were enormous scars fingering out into the jungle, across high ridge lines and reaching into the valleys below where the trees had been stripped out by loggers as they encroached further and further into the jungle. I must say I felt very sad to see this level of destruction every day as we flew over. This scene gave way to a seriously thick carpet of green jungle. We could only see the treetops which must have been 40-50m or higher and, every so often, there was a towering yellow-flowering kapok tree of about 70m, dominating the forest. There were plumes of disturbed bird life as we flew over.

We finally reached the Rio Venamo, and El Capitán flew very low along the river. Not just low, but with the nose dipped forward for good effect, military style. We could see piles of sandy 'tailings' in the large river where the locals had dredged the river bottoms, collected the heavier minerals on board the dredge, and dumped the waste material behind their dredges as they moved along the river. For gold? For diamonds? I wondered.

But my heart sank as we flew. The jungle was a bastard. It grew right up to the edge of the Rio Venamo, and then tumbled over into it, so it was rarely possible to even see a riverbank. The bits that tumbled over were covered in thick tangled mats of vines. There was no sign of any smaller streams or drainages coming into that large river that we might want to sample for diamond indicators. There wasn't even anywhere to pull up in a boat if we wanted to move along the main river and sample the tributaries as we found them. The jungle obscured everything.

There was the occasional small patch of open ground clinging to the edge of the jungle and bounded by the river, cleared by the locals to live on. Or that they had excavated into. Into metres of thick gloopy yellow-brown mud. The holes they dug were everywhere, and filled with soupy yellow-brown water to match. It was a quagmire down there.

The 'helipad' came in to view. A very small square of the yellow-brown mud on the side of the Rio Venamo, held together with some freshly cut timber, constructed just for us. There were a handful of corrugated iron shelters and buildings clinging to the riverbank, and some raggedy children and men came out to meet us. It was a settlement of sorts, more like a campsite than a village.

A line of yellow muddy men led us along squelchy foot tracks into the jungle. They wanted to show us their 'mine'. We had Miguel up front with a rifle. For tigers. He was the leader of the raggedy bunch. Peter was handed another rifle. He was a farmer, he would know what to do with it. Diego had the pistol in a hip holster. With this arsenal we set off. We could hear the screeches and squawks from unseen animals in the upper storey. A lot of monkeys and birds were up there, but in all the time we spent in the jungle, we never saw them. After 20 minutes, we reached 'the mine'. It consisted of a high-pressure hose which blasted the yellow mud into a slurry that was then fed over their own primitive sloo to collect the gold from the slurry. The place was a right bloody mess. Rusted 44-gallon drums, sludge pumps, hoses, and yellow muddy pools everywhere.

Gold. Gold? Where were the diamonds? I started to quietly panic.

The partnership trio were *very* excited by all of this activity. They ALL wanted to explore for GOLD now. Just like that. Day one. WHAT? Gold? I didn't want to be a killjoy at that point and tell them I knew very little about gold exploration (well OK, nothing). So that piece of news would have to wait.

But we worked it out and, in the excitement, it was planned to take large mud samples at depth heading inland from the workings, hoping that the gold had travelled along a hidden tributary now covered in mud. At least it was a plan for the next few weeks. There was no way we could do any widespread sampling for diamonds using gravel taken from tributaries flowing into the Rio Venamo – that we couldn't see or even get to.

We had the use of a largish wooden flat-bottomed boat with an outboard and various 'tinnies' to zoom up and down the 'piranha' river in. We collected the jig and the cement mixer from a road access point a long way upstream, strapping them onto the boats. They were hauled into the jungle with wooden poles strapped to them for carrying. A shelter was built using a framework of freshly cut wooden logs, with black plastic sheeting secured onto the roof. Next to this we built the milling and 'screening area' – a shallow hole maybe 3m in diameter with knee-deep muddy water where our 'screening team' stood most of the day covered in yellow mud and wearing gumboots filled with muddy water, sizing the slurry to pass over the jig or the sloo. This was our worksite for the next few weeks.

We gathered a large enough team of enthusiastic local field-hands quite quickly – 'path clearers' whose job it was to go ahead with both chainsaws and machetes to clear foot paths deep into the jungle, and the job description included being stung by swarms of angry wasps; 'diggers' who would use pickaxes and shovels to cut into the mud to gather samples; and 'mud movers' who moved the mud in large plastic tubs from A to B around the site. It amazed me how quickly the jungle undergrowth grew. It started to grow back the moment it was cut by machete, and Miguel was skilful at judging, by observing the amount of growth, how many days it was before us that someone else might have used the same path.

Over the weeks, if we were sighting lines using a compass straight through the virgin jungle, a few more words came in handy "un poco a la derecha" and "un poco a la izquierda" ("a little to the left", or "a little to the right"). Or "DETENER!" ("STOP!") With a lot of arm waving from the rear of the line, and me furiously taking notes and making sketches so we knew where the hell we were. We created our own localised maps in this way.

The main danger on the paths wasn't jaguars – but old pits filled with muddy water, flush with ground level. They were hard to see, yellow abutting yellow, and usually right on the edge of a track. Stepping into one meant a neck-deep plunge, and anyone who went in had to be helped out over the slippery, muddy rim – and then had to walk around the rest of the day completely caked in yellow mud.

Newer pits had their own danger. They hadn't filled with water yet and, although they might be quite shallow, they were a great trap for all sorts of jungle life that fell into them each night. These pits would collect countless frogs and insects, not just plain brown ones, but ones with brilliantly coloured and patterned bodies in red, blue or green, including the infamous yellow and black poison dart frog. Not to be outdone, caterpillars were bright red too, and there were beetles in shimmering rainbow colours. It spelled out one thing – danger, poisonous.

It wasn't long before other miners along the riverbank heard about our activity and wanted an 'assessment' of their own gold prospects. Apart from happily processing our samples, our days were spent doing several reconnaissance trips along the river in our 'tinnies' visiting other prospects. We pushed a wheelbarrow full of gifts (or were they bribes?) to the helipad at Tumeremo most days – including live chickens for the locals at these prospects.

The people who lived at these prospects were poor. One day we were offered a beautiful baby macaw ('guacamaya') to buy. It was already quite large, and had superb red and blue plumage. It couldn't stretch its wings in the small cage. It was brought out of the cage but it wasn't very active, and I was devastated it might be dying. The bird had been captured for us – purely to secure an opportunistic sale. Of course, we didn't buy the bird as it would only encourage more nest pilfering. It was so sad. But we still gave them chickens to eat.

As we made our way up and down the Rio Venamo we came across them – the diamond diggers. Local people standing in the large river, poking about under the water.

Not in large, organised groups, just one or two opportunists digging here and there. They would float a large plastic gold pan/dish like a boat on the water and shovel up what gravel they could from the bottom of the river into it. Before the dish sank, they would skilfully pan it down right there in the river, and if they found a diamond, it was placed into the end of bird quill with paper used as a stopper. And this was where I learnt my next Spanish phrase *"Daimonds en venta?"* (Diamonds for sale?) I would yell from the tinny, and if the answer was *"Si!"*, we would go over and negotiate a sale. One man was kitted out with a scuba face mask and fully clothed in the river. He didn't appear to have been attacked by piranha so maybe someone had been pulling my leg about this. We bought a few diamonds in this way to study them. All had the tell-tale greenish tinge of diamonds that had had a long history once they had left the parent rock. Maybe they really had been reworked from the ancient Roraima sediments and flushed into the present rivers, as geologists widely believed. Maybe our now-abandoned search for diamond indicator minerals to track the parent rock may well have been fruitless, even if we had been able to get it started.

I still have those few diamonds, and they came in handy later in the Perth lab to test equipment and the like. It is now illegal to take diamonds from one country to the next without a 'Kimberley Process Certificate', an international process of certification that was introduced in 2003 to stop the trade of conflict (blood) diamonds that Leonardo so skilfully plied in the movie.

At one of the more sophisticated gold prospects, there were shallow underground workings and a hammer mill, as opposed to the usual mud quagmire and water hoses. Bingo. Hard rock containing gold had been found by the local miners. There were people pounding up quartz rocks in crude metal mortars ('dolly pots') with metal pestles. There was a large sloo which we attached to our own, to check how much gold was missed in the crude treatment. There was plenty of obvious gold being lost over that sluice. There was a man who splashed mercury about like water as he poured it onto a shovel held over a fire to attract the tiny gold grains into a button. Thin plumes of menacing white smoke rose up. The man continued puffing on his Marlboro. No masks or safety gear here. He was only wearing shorts. I intimated something about mercury poisoning and they laughed.

This 'advanced' project was negotiated for and quickly became the second prospect in the exploration portfolio.

I actually had no clue about how these things were negotiated. There didn't seem to be any mining law to refer to, or tenements to be applied for. Just chickens exchanged.

We were invited to stay for lunch one day by the owners. Next to the whole wild pig hanging from a tree with a bloodied bullet hole in its neck, and the jaguar skin pinned out to dry, there was a freezer, and out of that came a large, skinned animal with its head and head fur still intact. 'Capybara' – the largest rodent on earth, and native to South America. "Oh yum", I thought. I had eaten dog liver before in China so what could be worse? It was actually a local delicacy and we were to be honoured. This one was the size of a wombat. The capybara is a lazy vegetarian and it kicks an armadillo out of its hole after the armadillo has finished all of the hard work digging a nest. And the taste – well like everything to us Westerners, it tasted like chicken.

When I arrived home from this first trip, I immediately rang a geologist friend who specialises in gold. "Help!! What do I do?" I asked, trying not to scratch the numerous tiny red bites still all over my torso. ("Have you tried digging into one of those?" suggested my Perth doctor). Later I realised they were tiny tick bites – but not before Peter had to pull a larger bright red tick from my butt cheek when I discovered it in the shower in Tumeremo.

"Try cutting a mag survey" suggested my friend … So, four months later, Peter and I arrived in London en route to Caracas for the follow up exploration program, this time with a hand-held magnetometer. The night before we were to fly from London, we received a message – don't come right now, wait a day or two. A second coup attempt had been made on 27 November 1992 while Chávez was in prison, but was directed by a group of young military officers who were loyal to the Revolutionary Bolivarian Movement. When we finally flew in to Caracas airport, a helicopter with a bullet hole through the windshield was the only evidence of this new attempt to overthrow the president.

The magnetometer went into the daily 'bribe barrow' bound for Tumeremo airport, along with the chickens which sat on top of it. We were greeted like long lost friends at the gold prospect, and completed a tough but thorough magnetic survey over the following weeks to try to 'see' with the instrument what was happening below the mud that overlaid the hard rocks. Our trio of clients were very happy. We had yet another gadget to march through the jungle with, and another excuse to fire up the arsenal for tigers and have another adventure.

The new black two-door Mercedes sports car was darting in and out of the London traffic in front of the Rolls Royce on the way to the Lebanese restaurant. Driven like a race car by the 21-year-old younger brother of Jasar, Peter Ferrick was upfront and smiling like a Cheshire Cat.

The Gidgegannup goat farmer was having a ball. "Bit faster than my tractor", he laughed.

So how did it all end up? Well, as so often happens with consultants, we were basically hired to do a specific job and when it was completed, we had nothing more to do with the project. Peter and I never went back to Venezuela. We had two field trips to the jungle and that was probably enough. The trio had their maps to work from, and their two gold projects in the jungle mud. They were very happy.

The country was relatively affluent when we visited but has since declined after years of economic mismanagement and corruption.

Our parting gifts to the trio were three Australian Akubra hats.

This story is dedicated to the memory of Peter Brian Ferrick (1947-2019).

33

The Treasure of the Sierra Madre – For Real

Mexico, 2004-2006 and 2015-2017

Jason Beckton

Jason Beckton graduated in Geology from the University of Melbourne in 1993 working in the Olary Block under Ian Plimer and then gained a Masters from the University of Hobart in 1999. He worked in Broken Hill, Kalgoorlie, Queenstown and Gympie from 1992, Mexico in 2004, Chile and Argentina from 2006, China in 2007, Laos in 2009, Slovakia from 2014, Finland in 2017 and in other locations. He is currently Managing Director for Prospech Ltd (PRS:ASX) in Slovakia, and is a director of Lode Resources (LDR:ASX) as well as assisting DGWA Frankfurt and Baker Young Adelaide on corporate matters.

2003 – Bolnisi Gold NL – Sydney Office

Before the Bolnisi team released the first one million-ounce estimate for its Mexican project to the ASX in 2004, and another three million-ounce resource in 2005, there were a few earlier steps.

An introduction from my Melbourne Uni alumnus, Sharif Oussa, to the Thomas Mining recruiters of London led to an interview for Bolnisi Gold NL in Sydney. It was a privilege to meet the late, great Ken Phillips (of Bougainville fame) and directors Norm Seckold and Peter Nightingale on a trip there. The interview started with Ken showing me a picture of a rolled over drill support truck for a reverse circulation (RC) drill rig in Mexico. It was a converted Canadian log skidder, thought to be perfect for the winding mountain tracks.

The introductory photo handed to me by the grinning Ken was one of chaos with 6m long, 300kg rods sprawled over 100m of hillside; and an unfortunate offsider looking sheepishly at his boss, the hyper-masculine Nico, who was striding down the hill towards him. Apparently the offsider driving this 30 tonne monster had left the handbrake off while unnecessarily consulting with the driller and site geologists, Jonathan Cordery and Richard Hatcher (both of whom I would later work with again in other companies). When the experienced RC driller looked at the offsider and told him to get back on the support truck to reapply his foot on the brake, it was too late, with the big monster rolling slowly backwards down the track which had been carved into a jungle hill, before flipping over the edge of it and onto its back. The skidder was later able to be flipped back over with the use of an excavator and ran fine apparently – Canadians know how to make robust bush gear.

Bolnisi had been in Mexico for some years, diversifying from the Georgian production asset after which the company was named. The role offered me was to assist the company on resource drilling on the Ocampo JV. Shortly afterwards, Bolnisi withdrew from that JV. I called Ken, hoping a position in Mexico was still available, and luckily he mentioned a new project the company had optioned, thanks to the efforts of the Bolnisi generative team, and it was called Palmajero.

Company chair, Norm Seckold, mentioned a drill out of the old mine was in order, as it was a classic example of extensive high-grade workings which had not been subsequently drilled. The team on site had already completed underground trenching, and surface work had been done so, in my view, the way to go was straight into 80m-spaced pierce points drilling.

Palmarejo

The guys had already drilled at least six RC holes with some success, based on encouraging surface and underground trench sampling undertaken by Hall Stewart, Richard, Ian Laurent, Andy Dacey, Jonathan Cordery and the supporting team. One takeaway point I learnt for the future was Bolnisi Sydney hiring experienced expats early so as to really focus on gaining quick economic results before the team grew significantly in the drill out stage.

Ian and Andy had been Bolnisi veterans for some years, including stints in Georgia and Sweden, and introduced me to the job of running the drilling campaigns and peripheral projects. Aldo Aguayo, the Chihuahua-based Administration Manager, drove me to site and introduced me to the team, including the site Geological Manager, Jonathan.

Conditions were comfortable but rustic, with basic communications by satellite phone. Food was excellent of course, thanks to the local ladies and I can't remember in all my time in Chihuahua, or later Sinaloa, ever having had a bad meal.

There was a baptism of operational fire. On review of the drill plan, I thought we could save some metres by drilling multiple holes from the same pad. The RC driller Nico and his boss in Hermosillo, Leo Wurtz, cautioned against drilling steeper than minus 60 degrees but I thought I knew better and suggested more steep holes from the same pad. Loss of a barrel on day one after my planning adjustment – lesson One, it's a team effort and listen to drillers who have been onsite more than the 24 hours I had.

The local field staff were led by Israel, who was the manager and care-taker for the local Mining Lease vendor in Mexico City. The labour demands of Bolnisi grew to over 80 people, including cooks, and the local staff were over 90% of the team in the end. There was a delicate task in determining fair wages, making sure we did not pay the staff an excessive income compared to the locals 'informal farming' income, and this was a key point of balance with the local town leadership.

Israel was introduced to me by the Jonathan, who was doing an excellent job in starting what would become a much larger drill campaign. Jonathan and Hall Stewart completed the key and difficult task of unravelling the layered stratigraphy which was very subtle. The rocks were featureless and difficult to map, apart from a key unit, the amygdaloidal andesite, which seemed to be more coherent and either on the hanging wall or footwall of the best mineralised zones (shoots). We used this stratigraphic work to determine the relative up or downthrown blocks of rock, and to see which parts of the faults had slid or moved the most, thus allowing mineralised fluids to enter and travel more easily. In these types of systems there were veins all over the place but only a few 'parent' veins containing most of the juice. Rather like a braided river stream, there is always one channel that is deepest. In this case it was the 076 Shoot. Originally siting the hole location was a mild point of contention as even a tough field supervisor was wary that it was a *'lugar de Brujas'* – a place of witches, with dolls hung upside down in trees – Blair Witch style. We had permission to enter, so I instructed our D8 to keep to the surveyed line and plough through the whole thing, much to the amusement of the Mexicans and, so far, so good. Underneath the witches' area was Zone 076, which alone was host to at least 1 million ounces of gold equivalent, which we discovered with drillhole 76 a few weeks later in March 2004. Sydney office had been quite patient and supportive as prior to this we had encouraging, albeit not bonanza style, hits.

The early operational issue of drilling into a myriad of underground workings told me we needed accurate survey support. The old plans were excellent, but just not sufficiently well registered in three dimensions to allow accurate void models. Luckily for us Leo Wurtz of Layne Drilling in Hermosillo knew Ray Roripaugh, a surveyor and geologist who continues to spend time in Northern Mexico and Southern US. Ray came from Hermosillo, where he lives to this day. Jonathon and I picked him up for the long drive back down to the expat town of Los Alamos in southern Sonora (an old German-English town), then the winding further eight-hour drive, after crossing the border west into Chihuahua and Palmarejo. It is a very dangerous drive and sadly dotted with cantinas serving alcohol; some years later we were to lose some people to accidents who were driving home to Hermosillo.

Ray and the local team he hired quickly made a difference to the accuracy of the maps of the workings and to my ability to model them in three dimensions to avoid drilling into them unnecessarily and losing more precious drill barrels, of which were in short supply. He affectionately named his team 'victimas' as he would sometimes lower them into stopes with ropes around them – safe but scary work. This kind of work resulted in an accurate void model we could then use with confidence in later resource estimates. I recommend people, even in remote circumstances, recruit not just geologists but surveyors early on.

Building the team once the rigs started to arrive was possible with Chihuahuan and Sonoran geologists, excellent ones generally. We were joined on site after six months of drilling by a Uni/Kalgoorlie mate in Glen Masterman to add experience of seeking out hidden systems; he used cutting edge technology which later led to discoveries in the southern projects, including at Guadalupe. Pretty soon we became a well-oiled machine with up to 11 drill rigs on the hill organised to drill the two main structures. The rigs had to have roughly separate areas of operations, while not bumping into each other on the tracks that were being created by a parallel team of earthmovers led by the very capable and precise Robert Mark Anthony ('Three First Names') for a time. In a side anecdote, Three First Names was prospecting in nearby Sinaloa some years earlier using a Cessna to reach a remote airstrip when some *narco traficante* 'borrowed' his plane at gunpoint for a mid-air drug drop over the border in Arizona, promising to leave the plane intact in a northern airfield, this they did apparently.

We soon recognised we could increase the quality for the resource models by a visit to veteran geologist Tawn Albinson, based in Mexico City. We proposed to purchase some old plans from him to help in the upcoming resource drillout. Tawn said: "Well I think Palmarejo has finally met its Waterloo with you Aussies, so happy to sell the maps". That night Glen and I stayed in Mexico City and went to a restaurant recommended by Tawn. The taxi from the hotel to the leafy suburb of Polanco dropped us at the front door, which was guarded unusually with men making no effort to hide the Uzi sub machine guns under their oversized jackets. We were assured by Tawn this was a sign of a good restaurant as there must be some dignitaries inside.

Back at site, the resource soon grew bigger and more complex than I could handle in-house after the initial two resource ASX releases which took us to three million ounces gold equivalent. For the next stage, I was asked to travel to Reno, Nevada to work with the consulting resource geologists at Mine Development Associates. MDA are a hard-working crew who, along with independent consultant Keith Blair, enabled me on weekends to visit amazing places such as Squaw Valley on Lake Tahoe and the old mining town of Virginia City just outside of Reno.

A few weeks later the reality of trying to keep a ten-rig program going adjacent to a village sometimes ran into snags, including the time on a Sunday afternoon when one of the geologists informed me that some notorious brothers had gotten drunk and left their truck unintentionally in the way of the drill water truck access, which was critical for our diamond drilling. No one wanted to visit them in the town square as they enjoy their traditional Sunday afternoon drinks with occasional young and old stumbling to the middle of the square to fire weapons into the air. I hasten to add it was only a few of the town folks. On going there, accompanied by the tough Ray Roripaugh, I must admit to sweating down the back thinking I was about to use up my ninth life. I knocked on the door of the 'brothers' home but was met but a lady of the house who said they were not in, but would move the truck, which they did later in the day. I needed a siesta after that tense anti-climax.

A day later a D6 bulldozer rolled off a steep track onto its side threatening to roll down a 200m slope onto the road below. I am still struck by the relaxed attitude of all there as they waited for the other dozer to slowly make its way from the other side of the project hill. I was starting to wear out from the constant stream of '*novedades*', or logistical fires, needing to be put out, on top of the technical work of planning a menu of planned holes for the rigs to work on. To that end we started training the Mexican geological crew in 3D software in order for them to plan and propose their own drill holes. They took to it like a duck to water and, in the end, most of the team had input into drill planning, rather than just waiting for another high priest expat geologist to direct things.

The investors death march

Macquarie Bankers Warwick Morris and Richard Crookes had supported the company from its early days and asked for a site visit in 2004. I was quite new to the management for shareholders and site visits, so what followed was a complete disaster as described by all involved. The first mistake I made was trying to show them everything in one or two days. In summary Rolando, the Tucson-based company advisor, escorted the visitors along with Bolnisi Director Peter Nightingale and Bolnisi Chief Geo Glen. I and Jonathan hosted some light lunch at the site offices, and we then headed to the top of the main project hill, above the village of Palmarejo, walking the drill tracks which basically followed contours. I suggested we walk around from one point off the road, going cross country to another point of mineralisation known as Halls Clavo. I had not so cleverly planned a route I had never traversed before in nearly two years of operations. I laid out the plan and Rolando declined saying he would ferry the vehicles to the planned pick-up point.

In hindsight he was lucky he didn't come. It soon became evident with spikey bushes, cactus and knee and throat level vegetation that it would be a long, sweaty haul. It was in the order of four hours with instances of rock falls on my boss director's ankle and, being covered in blood and tears, it was very unpleasant and unproductive hike.

After finally making back to the cars, Nightingale famously said, "Beckton, that was fucking hopeless". I agreed with him, and investor trips were somewhat simplified post this self-inflicted debacle.

The Sierra off the beaten paths was an unforgiving and hostile bush, with much of the foliage designed to repel and injure, such as cactus of all varieties. The generative geologist who brought us into the project in the first place, Hall Stewart, famously sat on a cactus and asked his field assistant to remove the tens of spines from his behind, which he stoically but reluctantly did with tweezers (a key piece of field gear for any future Mexicans out there). For such an experienced operator to sit on a cactus still baffles me, he must have fallen.

Histoplasmosis in bat caves and a near miss

We were made aware that our practice of mapping and generally working in dry underground conditions was dangerous when we heard of the sad death of an Australian geologist from histoplasmosis on a property in Mexico some years' prior. That disease is a lung infection caused by exposure to a fungus which lives in soil that contains large amounts of bird and bat droppings. It became news to us when one of our team members contracted it, and this required significant hospital treatment using antibiotics. We made the decision that there would be no more underground sampling and mapping to enable drill targeting, as Palmajero was a long way from aid.

In one instance we did manage to assist a village elder who was also a *velador* (night watchman) for the now defunct processing plant, and who had suffered a stroke. We mobilised a Medivac light plane from the closest airport, Los Mochis in Sinaloa, to take the gentleman to hospital in hours rather than days. To some extent the local folks, who previously had not been convinced of the benefits of the 'Australian' company, were then anecdotally fully supportive.

A new Mexican

In terms of the family and personal life, after a starting in early 2004 on a short period of a six-on two-off FIFO roster, it was agreed I could move my young family, which was based in Southeast Queensland, to Chihuahua city. On the flight over, settling into the Sydney to Los Angeles leg, I noticed a movie that was called 'Man on Fire', and was all about a blonde kid being kidnapped in Mexico (something that never happens) and nervously looked across at my wife, who was aghast as she nervously pondered our two blonde boys who were aged under six. On arrival at Tucson, Arizona, my host and legal expert Rolando Corona gravely asked me if I was fluent in Spanish to which I replied negatively. He paused carefully then said: "I think we have a problem". He must have seen my crestfallen look and burst in laughter, a great way to break the ice.

Settling into Chihuahua country club housing estates took a few months, albeit I was prone to driving on the wrong (right?) side of the road, only for concerned but patient Mexicans to wave. We were also assisted in settling in by all-round gentleman, David Groves, a long-time local resident and who was high up in Mexican mine producer Penoles.

Preparation for encounters in Spanish mainly involved practising with my Aussie nasal monotone but it took at least a year before I could participate entirely in conversations which were fully in Spanish, as we were in an environment where there is almost no spoken English. The trick I found was to prepare for the next encounter in terms of the sentences required, with the inevitable errors; the main point was to keep trying and gradually build up Spanish fluency and not worry too much about grammar. Still now 16 years later, I probably sound like a 'new' Mexican.

On arrival in Mexico, we did discover my wife was pregnant with our third child and we pondered planning for the future birth. We were lucky enough, thanks to friends at the Canadian School, to be introduced to an excellent Doctor Chavez who was able to guide my wife through the next months. There were some difficult times in the middle of the pregnancy, and we were relieved not to worry about having to travel to the US (no point in an Aussie trying for US dual citizenship as not possible back then). Our daughter was born in Chihuahua and afterwards we attempted to get her passport photo done. There, the steps were Mexican passport first, then Australian passport from the Australian Embassy that made them in Washington. The point of levity was she had a fair complexion like her parents, but most Mexicans do not, hence the white background in their passports and the blue background in Australians, for some contrast. The very patient staff at the Chihuahua passport office took ten photographs (nine more than normally taken) before a photograph of her with baby eyes only just visible, was permitted for use. Her Mexican birth certificate had had to have two surnames, which is the normal tradition to preserve the name of the mother's family, so initially it was Beckton Beckton.

We toasted with Aldo and our Mexican friends later and I noticed there is no sculling of tequila as in films and popular culture, just sipping, especially if the high quality *anejo* is available.

B LNISI
GOLD NL

SERVICIOS AUXILIARES DE MINERIA S.A DE C.V.

12 de Marzo del 2005

Ponciano Parangaricutirimucuaro Arriaga,

Esta es un pequeño aviso para agradecerte por tus esfuerzos durante el año 2004. Un ajuste de salario ha sido autorizado en reconocimiento a todos los esfuerzos que pusiste al estar al servicio de la compañía durante todo el año pasado. Por lo mismo se te informa que tu salario neto mensual, ha cambiado a $ 811,210 Pesos MXP.

Se te recuerda que tu salario diario, es información confidencial entre la empresa y tú, por lo tanto no esta permitido discutir o platicar acerca de eso con tus compañeros de trabajo.

Con los mejores deseos,

Jason Beckton
Gerente de Exploración – Palmarejo

Calle Diego de Vilchis # 3303   Col. San Felipe   C.P. 31240
Tel: 52-614-413-85-85 y 413-77-22   Fax. 52-614-414-44-21
Chihuahua, Chihuahua, México

Figure 1. Giving a payrise of $US 40,000 to a volcano, an early lesson in project management.

We would have some family trips north to Juarez and then El Paso about three hours north. On one occasion I did manage to make a wrong turn resulting in us driving through some of the dodgy suburbs of Juarez (recently featured in a film 'Sicario') complete with burning tyres on a side street. Before the days of GPS, it was a relief to get back onto the city bypass and into bright and shiny El Paso. Thank God: as the Mexicans commonly say 'no pasa nada' – nothing happened.

When ten does not equal ten or $811,210 nothing

Every year the able Rolando Corona would escort myself and other family members or colleagues for the normal process for work visa renewal south to Nogales, which straddles the US/Mexican border. Once we did this and were ahead of schedule with Ray and Rolando suggesting from the front seats of the car an early lunch at a Tucson 'establishment' before our flight back to Chihuahua later in the day. Ray and Rolando were sniggering like schoolboys, as they pulled the car into Tens – a Tucson strip joint.

Tens served steak and coke, which is all we wanted on a Monday morning being healthy mining type lads, before my host warned us the Monday girls were not very pretty. All I wanted was a meal but my hosts insisted that killing an hour before the flight was best spent at Tens and not at the Tucson airport departure lounge. I remarked to one of my hosts that this establishment should be renamed Fives but felt the girls' efforts should at least be tipped as we ate our steaks. I was about to hand over a US$1 single when my host told me "not to feed the bears" as they come back then and should be encouraged to quit to make way for more attractive girls. I suggested that in a busy near-full employment city like Tucson, there was little chance of the establishment living up to its name of Tens.

Back in Chihuahua, invoices and approvals were coming across my desk at such a rate that our Administration Manager Aldo was worried I was not checking things correctly. He slipped a page into the pile for the annual pay rises in 2005 from me addressed to 'employee' – Ponciano Parangaricutirimucuaro Arriaga (a volcano named after a politician, and which is an Indigenous name) for a pay rise to a net salary a month of 811,210 pesos (USD40,000). We both agreed I should have checked things a little more and I still have the signed original on my desk wall.

Tragedy in the Sierra – Africanised Killer Bees

Once we established that Palmarejo would be the main resource area in the short term, we still had a long list of targets in the broader Trogan tenement. So we started to put together a regional program that required up to date aerial photographs. These had to be flown by aeroplane in 2004, which also required accurately surveyed large white crosses to be placed on the ground so as to be visible in the photographs which were then used to register them digitally for field use. A very capable consulting group in Hermosillo brought a team of surveyors led by Hector Soto to site, and he in turn was working with our surveyor Ray. Glen, myself and the geological team headed to another project we had in the north, Yecora, with a visiting consultant, Greg Corbett. In the morning, Ray called me with the chilling news that Hector was dead – having been bitten over 150 times by Africanised Killer Bees.

It seemed Hector's horse had kicked a ground nest of bees, which emit a pheromone that provide its attack location when it bites. Like all bees, they die after biting and the trouble is the next one tries to bite the same location. The two young local men tried manfully to swat the bees off him with hessian bags only to be attacked themselves. In the end they were driven off, but led the team to recover Hector as soon as possible but, sadly, first aid was no use. The crew and authorities were very professional in handling the matter, including Hector's employers in Hermosillo.

After that the only good bee was a dead bee and we hired extermination suits and used poison, I and a valiant Chunel, our leading hand, having to don them first. A nest was located close to the main Palmarejo camp but there was no way to confirm it was an Africanised Killer Bee nest and not one of normal bees. Wearing a white hazmat-style suit we gingerly sprayed the hole in a tree and bees came swarming out African-style but sadly for them all fell to what must be a toxic concoction. Soon afterward a conservation officer in Temoris correctly expressed his concern with our program of bee slaughter and we set this practice to one side, only to react when staff members were attacked in the immediate project area.

Jail bird in Mexico

In order to increase my Spanish fluency, we were attending one on one language classes the company provided. One day, after a getting up a bit too early and so doing some reports I wearily drove to the class at the instructor's house, completed what was an exhausting language class and climbed back into my Dodge Durango. In order to get back to the office, I had to go across a major thoroughfare and managed to cut off a car that had right of way. Its driver saw me at the at the last minute through the cellophane plastic windows on his passenger side. I managed to pull the Durango out of the way, in a flat-out wheel spin into a nearby driveway, only to witness the car I cut off smashing through a 1950s-style white picket fence into a tree. In the middle of the nice part of Chihuahua, a crowd soon gathered. Luckily the driver got out shaken but ok but then, to my horror, I noticed his passenger was his heavily pregnant wife, who was also apparently uninjured but clearly distressed. Within minutes we were surrounded by a mob including very professional ambulance and transit police who sketched out the scene and agreed I was at fault despite there having been no actual collision between the cars (an important point for a later quirk). Our Aldo informed me: "Ese ('mate' in Sierra slang) I think you are screwed". "What do you mean?" I asked.

Well it turned out that, in any accident where someone is injured, in this case possibly an unborn baby, then the other party is *automatically detained*, until the health of the non-faulty party is determined. The transit policeman asked me to drive and follow him to the police station for questioning and detention while the health report of the entire family was generated over probably eight hours, maybe more. After some initial processing and then being introduced to a lawyer, aka the opening scenes of 'The Castle', I was then asked to follow the policeman into a lounge-type room complete with TV. There were a few other worried-looking 'driver at fault' type people. Aldo was there with me the whole time and duly informed Tucson and Sydney offices of my predicament. I was allowed to keep my phone and I texted my wife that I had been held up and would be home late... In any event soon Aldo passed me a hot dog through the bars and was laughing that the policemen in charge of the 'traffic cell' had related a funny story to him. Some weeks earlier a wealthy lady had refused to be detained. The door to the cell is in fact unlocked for practical purposes, but you are not allowed to leave. This lady had taken it upon herself to make a break for it – Butch Cassidy style, but was caught and sent to a real cell with concrete lounge suite and no TV. I was duly chastened and I think the look on my face caused Aldo even greater mirth, splitting his sides. Soon enough unnamed executives from Sydney and Tucson were calling advising me through tears of laughter to pack condoms for the long haul and try to stand up like a tough guy when I went back into the general population, all very funny lads. On a serious note, the young lady and her baby were fine and I was let out about 11 hours after the event with the long suffering Aldo helping me process out. One postscript to this story was the insurance company was not required to pay out for the clearly impoverished young man and we found some cash to buy a replacement car for him and his family.

Contractors' fun

The drill targets soon grew, after our instant success from hole one onwards, of thick silver-gold intercepts. Along with the reliable Layne Drilling, we needed additional contractors. In 2004 there was a shortage of qualified drillers, and we started to notice turnover of the US-based drillers as they quit to go back to the equally paid, but higher quality of life, operations in the US.

One particular group from New Mexico even had an owner wearing a hard hat shaped like a cowboy hat when he came for a site inspection. We had primitive accommodation for the teams, including lack of reliable running water, unreliable electricity and other food-based problems. It was suggested by one contractor that we build a brothel to reduce driller attrition: I said I would ask Sydney office about which budget cost centre this non-starter idea would be under – drilling?

One of the strict rules we had for the all-male crews, was for zero fraternization with local ladies. We had no appetite for the difficulties we knew this kind of issue could result in, particularly given the informal ways the farming community had for protecting their family structures and their young women in particular. There was one example when this rule was broken with the US-based drilling supervisor calling me on a Sunday while I was back home in Chihuahua with my family at a BBQ. His opening line was "Jason, sorry I quit but I value my life and some crazy guy I haven't seen before pointed a pistol at me and I don't want to die in Mexico". I replied this seemed totally out of character, "where was he, and could he come by to my house in Chihuahua?" He replied he was already at home in Oklahoma, having passed through Chihuahua and all of Texas. That must have been a twenty-hour drive at a minimum, someone really must have put the wind up him. On my next trip back to site I asked about this unusual gunplay as we had seen none in the two years to then, and was told informally that for sure this supervisor had a girlfriend who sadly was also seeing a local narco boss. They had hired a *sicario* (heavy) from a neighbouring state to scare him with an empty gun and thought it was hilarious. I asked that it be passed around our team of at least eighty members that we intended to handle such matters internally and not Hollywood-style.

On another occasion another driller did get in some serious legal trouble and was arrested and taken to the nearest jailhouse in Chinipas de Almada, a beautiful historic town a short drive west. The magistrate dealt with the case in a fair and reasonable way, for me an example of a robust overall legal system in Mexico, despite the propaganda to the contrary. One of the local police told me he was lucky he was not subject to Sierra Madre justice of '*Llena – Bote – Plumo*'. I asked what this meant, and he related that sometime in the past, particularly in those days of less formal legal proceedings, offenders may have been offered a choice upon their arrest. The first choice, '*Llena*', involved getting beaten up by wooden planks or 4x2s as they are known in Australia – the preferred choice for most. The second was '*Bote*' which meant a term in a cell in a usually decrepit, damp and diseased cell, crowded with others. The final selection, '*Plumo*,' which I think for some was not a choice, was being given a head start running away zig-zag style from whatever detachment had arrested the felon, but then being followed by a hail of lead from automatic weapons.

We had an excellent earth moving contractor, Luis Terrazas, who flew me out once to site in his Cessna: it proved to be an example that even hardened Sierra Madre operators can get scared. As we approached the strip at Chinipas, it was covered in wet season cloud with the pilot taking some time to finally locate a open sliver in it. He dived through it Stuka-style and I will never forget looking around from the front seat to see a gritted-teeth Luis exclaim "chingaso" a term of non-affection for the pilot who was as cool as a cucumber.

Chinipas airstrip itself was overseen by a bored Mexican Army platoon, including bilingual soldiers coming home from temporary stays in the US (*otra lado* – the other side) to do their mandatory service. The airstrips were used by the informal farming community so at times the army, not wanting to destroy these municipal facilities, would guard them from non-legal use.

2006 end times in Palmarejo

In the intervening period the corporate team led by Norm and Jim Crombie in Montreal had managed to dual-list the company in Toronto as Palmarejo Gold Corp which was a master stroke for all concerned. I learned a lot from Jim on how to conduct investor tours for genteel folks of Vancouver and Toronto more efficiently, that is "It's not a foocking commando course Beckton!' The merits of an organised trip but sometime events overtake us, as happened during one investor lunch onsite which was interrupted by the familiar and awful rattle of an AK47 being fired into the air a few hundred metres away. Justifiable worried looks resulted from the investors and Jim in my direction. Jim asked if they needed to be worried. I replied: "No, just kids playing with guns", which it was. He kindly hosted my family in Montreal toward the end of our time in the company.

I had passed my used by date as project manager after the third resource estimate and was helped into a new role in Chile with Exeter Resource Corp. At my send-off party the local people arrived, including a band and the village leadership, which was very kind and required spicy hangover recovery food the next morning. Not always smart cards to have a stomach full of yummy Mexican food and be driving eight hours in the Sierra the next morning.

The real winners, and I sincerely hope they view it this way on balance, are the local people with jobs and community development going on since 2004.

2015 – playing for Santana in Sinaloa

Staying in Sierra Madre, I later worked in 2015 to 2017 for Santana Minerals, part of the same group as Bolnisi, drilling silver projects in Cuitaboca Valley up the hill from Los Mochis. Using a technique I had learned about in Argentina from the Exeter guys at La Cabeza, we cut channels, with a power saw, into the sides of the road cuttings as the best exposure to mineralisation. We took samples and assayed them, resulting in a resource estimate at the Mojardina project: it remains open, which means we have not found the ends of the mineralisation yet. Santana management kindly let me name in our ASX releases one mineralised zone we later drilled as the 'Evangelina' shoot. It was named after my daughter who had been born in Mexico some 15 years earlier. My sons protested but I reminded them deposits named after men bring bad luck, as per maritime superstition with ships.

All of the employees, bar the geologists, were related to people in the valley and were familiar with its informal farming practices, which had recently switched from marijuana to opium growing due to the legalisation at the time in the US. I did hear an anecdote that an injured soldier had been assisted in his helicopter medivac by the locals, which

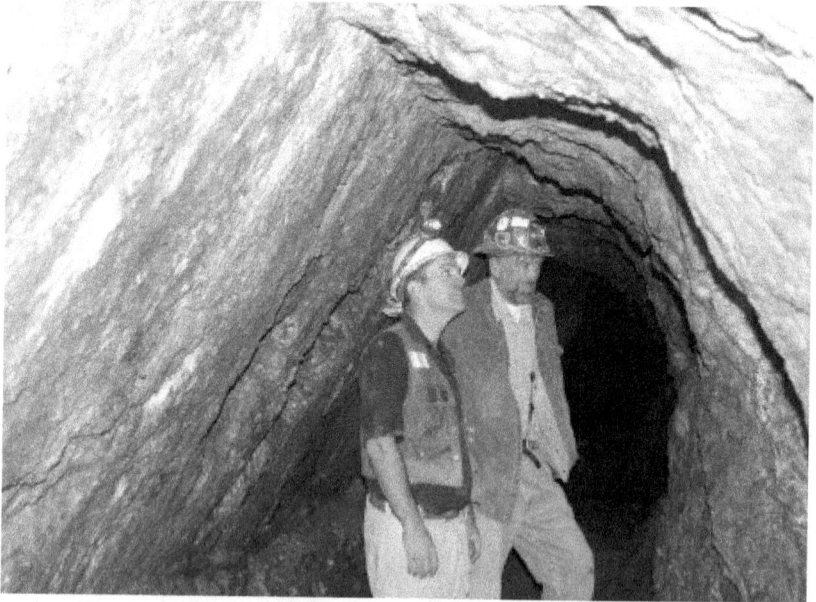

Jason Beckton and Ray Roripaugh examine old silver mining at Sierra Madre.

included clearing a critical helicopter pad, so tensions seemed quite low at the time. Young soldiers in particular would ask to use our wifi to communicate with family: we just asked that they not live stream soccer to wipe out the bandwidth. We did get buzzed once by a gunship which escorted a herbicide helicopter and we must have been a curious sight as we were saw channel sampling a road cutting at the time. We were wearing orange, so not trying to hide from anybody. It was still a little unnerving to have a Huey chopper complete with M60 machine gun, with crew peering at us through binoculars. I thought I could diffuse the situation by being the only gringo in the storm of dust and noise as they continued to hover about 50m off our beam in the valley as we were sampling the road cut at Mojardina. I raised my hat, exposing non-Mexican hair and gave them the peace sign – what's the worst that can happen! The chopper broke of and went about its herbicide business much to the chagrin of our field team.

The local people who agreed to work for us were excellent, super fit to handle the steep terrain. They quickly learnt sampling protocols on trenching and RC drilling, being well trained by the guys at Layne Drilling. The rig was mounted on Canadian log skidders, as is best in the Sierra, and it turned out it was the same chassis as was used in 2005 at Palmarejo. It was like déjà vu in Spanish.

A highlight was an invitation to a 100th birthday party of a lady named Epitacia. We managed to provide electricity for the night but the highlight was a cultural dance at the end of the party.

I had the opportunity of working with great local geologists, including the leaders Efrain and Javier, who did an excellent job, but the project slowed down due to a low silver price. Despite all the negative media, Sinaloa is a fabulous place to work with happy and welcoming people, as was the case in Chihuahua State; and I would recommend both jurisdictions to investors, workers and tourists alike.

Dedicated to Martin Kralovic a superb family man and Field Supervisor who passed suddenly, 6 January 2022, at the age of 43. Martin worked for us in Sinaloa on secondment from Prospech Slovakia.

34

"The World Is Your Oyster, Son!"
Many Places, Many Times

Ken Russell

Ken Russell has had a long and varied career involved in many facets of business that have seen him involved in various types of projects around the world. He was trained initially in electrical and electronic engineering, specialising in IT and associated products manufacturing. He then became involved, due to a growing expertise in work, study and the application of statistics in time management, in product manufacturing, including consumer products. A desire to see the world however saw him take a career change with him moving into mining exploration and construction; and oil and gas exploration, construction and production around the world. A growing expertise in these fields through his consultancy work with a range of major and junior listed stock exchange companies saw him become involved in the corporate world, leading to various directorships of listed companies.

I grew up with a father whose job in the aviation industry took him to many places around the world and I always looked forward to hearing about these places when he returned from his business trips. This sowed the seeds of inquisitiveness in me, and probably rubbed off as I, too, have enjoyed travelling around the world, meeting and making lifelong friends with many people from different cultures and walks of life, as my career developed through involvement in the many facets of predominantly the mining and oil and gas exploration and production industries. It has taken me to most continents, from freezing cold zones to hot and steamy places in the tropics.

As I sit here and think about this journey and reflect on some of the situations I have been involved in, I must be honest and admit that, in reality, I am very lucky to be able to be here at my computer and to reminisce! To give you an idea of what I'm thinking of; I've been taken prisoner by heavily armed rebels in Angola during its civil war, I've been on an offshore oil rig that lost buoyancy and tilted at extreme angles while the crews worked tirelessly to stop the vessel turning over in the middle of a storm, and on others where fires have broken out.

I've been on a supply workboat in The Philippines that was tied to an oil storage tanker and couldn't be released due to a trapped oil transfer hose lodged in its propellers, while a storm suddenly came out of nowhere and turned the seas from a mirrored surface to a raging surge of waves in 15 minutes. Who would have thought that the hose's outside diameter was the exact same size as the recess in the propeller? That storm caused the vigorous sideways-rocking movement of the supply boat's funnels, which are located mid-ships on the side of the vessel, to slice into the wing tanks of the storage vessel. Have to say the captain of the tanker was not a happy man: his voice could be heard, even though there was a howling gale, as he looked over the side of his vessel in the wind and rain gesturing and shouting abuse at us as water sprayed out of the ruptured ballast tank. It didn't stop there however, because once we managed to get free from the tanker, we manoeuvred away only to be hit with a complete engine failure on the workboat. As this was at night, with the storm continuing, it was another of those moments when you're unsure of how this is going to end as we rode out very large and greenish waves while the ship's crew struggled to get the 'black start' generator going (when all else fails it's the last lifeline!) or we would have been back into the side of the tanker again.

I've been in a small Bell 206 helicopter where the windscreen has blown out during a rainstorm, again in the Philippines, and we all donned plastic garbage bags to keep dry as the pilot who, it has to be said was himself a lucky man in this situation, struggled to get us back to shore and able to land in a safe area. In those days there was no separation of freight and personnel and we could be sitting inside with various pipe sections or valves or bags of foodstuffs held on our laps, any of which could have impacted us had we ditched.

That reminds me of another helicopter episode when in Russia, on the island of Sakhalin just after the separation of the country from what had been the Soviet Union: I'll describe that in a while.

I've been in mining locations such as Sishen, a mighty iron ore mine that's still running today in South Africa, where a team of us were contracted to provide support to the local company personnel. We had a separate accommodation camp, made up of questionably suitable caravans, located outside of the main plant. It was run for us by a couple who were both inflicted with an over-indulgence in alcohol problem, so whether we got our meals or washing back on time was a case of pot luck.

We always seemed to get called out in the evening when a major problem developed and, on one memorable night, I was summoned over the radio with: "the jaw crusher has stopped working", and with no other information given to me. When I arrived on site I found a very large Euclid truck had backed right into the crusher, with fully loaded tray raised, and it had ripped the main support structures and wiring for lighting, etc, to pieces.

Then, in what was once the largest open cut copper mine in the world when it came on stream in the early 70's, Bougainville Copper, on the beautiful island of that name which is to the east of Papua New Guinea, also had, as a component of its processing plant, the then largest ball mills in the world. As part of the construction procedure, liner plates needed to be installed inside each mill and a new generation of hydraulically-controlled liner handling machines was employed to do the job. The only problem was that at one time it suddenly went berserk due to blockage of the hydraulic system. It caused violent waving of the handling arm which swung around inside the ball mill: luckily, no one was injured but, boy, could the team of men inside that ball mill move quickly. It was subsequently found out that the hydraulic valves that had caused all the problems were supplied by a French aircraft manufacturer and were usually used in fighter jets!

I also remember an earthquake that took place there and, from memory, the only person who got injured was one who jumped from an upper floor of the accommodation premises thinking it would be safer on the outside, but he broke an arm on landing. I had been at an electrical substation away from the mine when it occurred and can remember all the white pebbles that surrounded the building bouncing like ping-pong balls as the vibrations from the earthquake took place.

There are many other things that come flooding back to mind but, as my Dad would say: "Well Son, that's what's called 'character building'". Yes Dad!

So, where to start with that 'character building'? My capture by rebels in Angola is probably a good place. I was working in the oil producing enclave of Cabinda in the mid '70s: it is that part of Angola which is located on the west coast of Africa and on the northern side of the mighty Congo River. I say mighty because by the time the river has reached the coast, it has a massive force of energy, and I witnessed a commercial diver being retrieved from working in it and the support harness he was wearing had been broken and only held him with one strap!

At the time I was there, I was young, carefree and making lots of money, and was ignorant of the potential impacts of the civil war, which was in full action, with various groups vying for control of the country. There were the MPLA, the FNLA and the FLEC. Each thought they would take over and run Angola. 'Hiding' in the background were the usual players in turmoils around the world at the time, the Chinese, the Americans and the Russians, who each supported one of the combatants; and, at this time based in the enclave, were also the Cuban military who were of course working together with the Russians.

We would sit outside the mess room at night, perched on a hill in a cool breeze having a cold beer, or three, listening to the BBC World Service and Voice of America radio stations describing the ebb and flow of the civil war in the country, but hardly being affected by it ourselves. That was soon to change, however.

I had arrived in Cabinda following a very quick and simple interview by a Texan based in Johannesburg. On the phone he had said, "I'll ask you three questions, 'boy'", as Texans do. "How old are you"? "What passport do you have?" and "How soon can you be available?" Obviously, I must have given him the answers he was wanting to hear as, that afternoon, I was in his office and next morning was on a small, very sleek and expensive seven-seater Learjet flying from Jo'Burg airport to Victoria Falls, where we were refuelled, and then directly into Cabinda. As the jet departed Jo'Burg and took to the skies at a very fast pace, I can remember thinking to myself that I could get used to this life very quickly! As I mentioned before … young and carefree … and not knowing as much as I thought I did!

It was not long after my initial arrival in Cabinda, that the use of the Learjet was cancelled for crew changes due to management's fear of it being 'requisitioned' by the rebel groups and so we had to fly by commercial airline. It was usually Pan Am into Brazzaville in the Congo, a country famous at that time for experiencing its own series of troubles and for being a place swarming with troops. They were markedly visible all along the roads from the airport to our company house where we would have our overnight accommodation.

The next day we would go to the airport and have a small twin-engine plane fly us west across country to an airport at the end of the Congo River at a place called Banana. That is, if its passenger door would close correctly as it had a habit of springing open as the plane hurtled down the runway. At Banana we would get picked up by a Petroleum Helicopters' JetRanger chopper that would fly us out to sea until we were over the horizon so we could not be seen from the shore anymore. Then we could enter Cabinda directly from offshore to avoid the immigration requirements as we had no visas. This was due to the civil war, and nobody knew who would or could issue them. Those helicopters were piloted by American ex-Vietnam pilots who were very skilled people.

As an example of their abilities, and as an aside to this story, at one point a new oil storage production vessel was positioned at the mouth of the Congo River and a grand opening ceremony was planned with the President of Congo, Mobutu Sese Seko, scheduled to fly out to the new tanker and declare the field open. He flew into Banana with his military helicopters and then flew out to the tanker. He was not a happy man when his pilots couldn't land on the moving tanker, and they had to return to Banana.

Petroleum Helicopters' pilots were summoned to come to Banana quickly to pick up the President. Sure enough, three bright yellow helicopters appeared in the sky in perfect formation and landed in line at the airport all touching down at the very same time. The President was flown out to the tanker and all three helicopters landed on the small helipad. His pilots had to watch all this, and I've wondered over the years what happened to them as Africa, in those days and especially in the Congo, was not a forgiving place!

There were a number of single-well platforms located in the offshore field and we would fly out to these in a G47 helicopter – think of the helicopters in the TV series MASH – when we had to do any work on them. Local fisherman would be out in that area as well and would wave to us. Then they would duck quickly because they knew what was coming, as we would sometimes carry a few party balloons blown up and containing a mixture of flour and water that we would drop on them, occasionally hitting our target.

One of the G47s developed a mechanical problem when it flew out to a single-well platform and was forced to do an emergency landing on the small helipad on top of the well. It was decided that the only way to get that helicopter back to its base onshore for repairs was for it to be towed. The plan was for a pilot to start it up and then land it straight onto the water in front of the platform, without waiting for the engine to warm up or to go through the normal checks, as they were uncertain about the mechanical problem with it. It could then be towed back to shore.

So, on the following day this took place: it was a fast start-up and then the helicopter was plonked down onto the water on its floats and the slow, laborious journey back to shore commenced; the chopper being pulled by a small crew change-type workboat (like a small PT boat seen in war movies). At first it was a relaxing tow but, after a while, the pilot who had just landed the G47 onto the water decided he wanted some lunch and, as there was nothing on board the boat, he called up his fellow pilots on the radio and requested it. Not long after, one of the G47's could be seen approaching us and it positioned itself on one float across the stern of the crew change boat which continued along at its steady pace, while the lunch was handed across to us. Skilled flying at its best, and a very good lunch. The other good thing was that we weren't flour bombed!

There were approximately ten expats working at the oil installation in Cabinda during this period, and three of us were in a Toyota-type ute coming back from the local village when a group of heavily armed rebels with massive machine guns and bullet belts jumped out from the bushes into the road in front of us yelling for us to stop. Well you would, wouldn't you, when faced with that scenario! As the rebels got close, we could see from their bloodshot eyes, the fags in their mouths and the pungent cloud of smoke around them, that they were very high on drugs. There was lots of yelling and shouting and gesturing and two of us from the front seats were made to get into the back of the ute with some of the rebels, while others jumped into the front seats. Our driver was then forced to follow a side track.

We went for approximately fifteen minutes along this track, in a great cloud of dust for those of us in the back of the ute and came upon a well maintained house with a veranda that looked quite nice apart from the fact that on either side of the front door were machine gun posts. We could also see walking around the grounds other heavily armed soldiers and a few mortar posts. The ute stopped in front of the steps up to the front door and one of the rebels went inside while we all waited nervously outside, unsure what was happening.

After approximately ten minutes, the rebel came back with an officer, who was immaculately dressed in full military uniform. He was a tall African man and he walked up to the side of the ute and, in perfect high class educated English, asked where we were going and what we had been doing in the village. After our explanation he said, again in perfectly spoken English, that a terrible mistake had been made by his solders and asked us to come inside and join him for some tea. You can imagine the double take that we three made following this, but we complied and went inside. Sure enough, he got one of his men to make tea and he then started to chat to us in a very friendly manner. During this conversation, he told us he was in charge of the MPLA in this area and that he had attended military college in the UK and was trained at Sandhurst!

We spent something like an hour there with him enjoying tea and biscuits. He then provided us with special passes so that, if we were stopped again, we could show these and we would have trouble-free travel. He then got his solders to escort us back to the main road and we were sent on our way, still trying to take in what had just happed. The thing was that whoever won the civil war wanted to keep the oil flowing, so it was important to them that we could move around. A funny sequel to this story, however, is that as the weeks went on, I finished up getting special passes from the FNLA and FLEC as well. So, the main thing to keep oneself out of trouble was to be sure to show the right pass to the right group of rebels!

The second pass came from the Chief of a local village where FLEC solders were being trained by the Chinese. Part of our workforce were locals who generally were uneducated. A lad was assigned to work with me so I decided I would try and teach him a few things and let him do more than he probably would have been allowed elsewhere. This, it came to light, was really appreciated by him because one day he came to me and told me that his village Chief wanted to thank me for looking after him, and I was invited to the village to meet him.

After some discussion, it was agreed that I and a couple of Oceaneering divers, both ex-UK No 1 Paratroop regiment people, who were also based in Cabinda, would go to the village on a Saturday night. We got some cartons of beer and drove out to the village, not sure what we were going to see or be involved in.

When we arrived, we were introduced to the village Chief and handed over the cartons of beer to him and he was very thankful. Then we were informed that we would have a feast of guinea fowl and he would introduce us to the villagers. So, at the front of his hut, we stood as all the villagers filed past. Nodding and shaking our hands. Every so often the Chief would spot a pretty girl and, through my local lad, tell us you can take her home, don't have to bring back!!!

The feast of guinea fowl was prepared and was very, very spicy. For days afterwards I thought my lips would never ever return to normal. The next day back at camp my local lad presented us with special passes for the FLEC. His brother had obtained them from his Chinese military advisors as a thank you for us.

Helicopter travel in the resources industries is the main way of getting to the various locations, especially so in the oil and gas world. At one point in my career I was asked to go to Sakhalin Island by Ted Ellyard, famous in his own right for some very good resources deals and adventures himself. He was looking at several projects on the island and between us we were putting a deal together for a venture there. Getting to the island was also interesting as we flew to Japan, then across to Russia, and then took a Russian Aeroflot plane from the Siberian mainland to the island. That plane sticks in the memory because the in-flight entertainment on it consisted of a typical household-type TV set positioned in the walkway of the plane on a wheeled trolley and videos were played on it.

On Sakhalin Island we needed to get a helicopter to fly us around to look at the various installations we were trying to acquire: what was required to hire one in Russia at that time was a cash deal. Once the pilots could see the cash all laid out in a pile on the table, the helicopter was organised. It was probably the only one on the island that actually worked, and there were a lot sitting at the airport in various states of collapse.

Having a strong liking for aircraft, I got to talk a lot with the pilots and, at one stage during our flight around the island, the pilot brought the helicopter to a very low hover and the co-pilot jumped out and walked around to the front then started to jump up and down. It seems that they needed to be sure the helicopter wouldn't sink into the soft ground. It looked quite comical but, if they had got stuck, there was no way it could be retrieved so they were being extra cautious.

I do remember the discussion I had with the chief pilot at the time too, where I was looking at the instrumentation of his helicopter and compared it to the various American helicopters I had flown in which were far more sophisticated. As most helicopters in Russia at that time were military configuration, our conversation moved into that. The Russian pilot had a smile on his face when he told me that, although the Russian helicopter was basic compared to the sophisticated American ones, the Russians could build ten helicopters in Russia for the price of one American one and there was no way an American could knock out the full ten before one of them got him.

We also landed the helicopter for refuelling at the restricted military airfield where the Soviet jets that attacked and downed the Korean passenger 747 jet KAL 007 in the Sea of Japan in 1983 took off from. Places that 6 to 12 months earlier would have been off-limits to any Westerner but where we were made welcome.

Getting to and from various installations has thrown up its various challenges but the actual installations themselves have also done the same. Fires can be hypnotic to watch, and most readers will have seen video of flaming flares coming from the sides of oil rigs as oil and gas wells are tested. A good healthy flare is a sign of a good producing well, which is what investors want to see, but getting to that point can be a troublesome exercise.

As an example, I was running an offshore oil field in Indonesia. I lived on a floating production, storage and offloading (FPSO) facility which, in this case, was a storage tanker located in the middle of the field. I had a very nice room on the vessel with an adjoining office and the canteen people would ensure I wanted for nothing – freshly made cakes, continuous hot coffee, and local fruits, et cetera.

In the early hours of one morning the fire alarms went off. Even though we practise for such things on a regular basis, it does get the heart pumping when you're on a storage tanker holding many barrels of oil. I was out of my bed quickly, ran through my smoke-filled office to the outside as the various fire crews and personnel sprang into action. As I asked for reports of the fire, it slowly came to light (no pun meant) that the fire had actually started in MY office. It seems the canteen staff had not filled my coffee pot as usual, and the thing had melted due to the heat and ignited. Now what do you write in a morning report about that!

Another problem with offshore oil rigs is that they are quite often positioned in regions where the weather is not the best. When a well is being tested to determine its flow rate, there is a metal pipe that runs from the seabed through to the rig floor and above it. When people see this at first, they think that, because the rig is so big and feels stationary, the pipe is actually moving but, in reality, it's the other way around. So, if you are in the middle of testing a well and the rig loses buoyancy on one side, a critical state develops quickly. Although you have various safety systems that isolate the well, if this problem cannot be fixed quickly, the situation can develop into a full-blown major incident as you have the safety of approximately 80 to 100 personnel to think of on board the rig.

This happened during a well test in the Bass Strait. That region is notorious for some of the storms that can develop there, and that's what happened to us. The buoyancy was being lost on one side and the various rig personnel had to work very quickly to stop the rig tilting as, if it couldn't be levelled, it would have ripped the pipe coming from the seabed to the rig floor apart and could then have resulted in the rig turning turtle in the storm. Luckily, the rig crew managed to fix the issue just in time, but it was a touch and go situation for a while.

There is one thing that will stick in my career memories forever though, and it was my first trip to work in Italy. I had been sent to work on a rig off the east coast and was told that I would be picked up early in the morning from my hotel in Bologna by a group comprising the senior man on the rig, the Italian company man and a couple of crew members. It was in winter and bitterly cold. I got a call when they arrived and went down to meet these new people. All very welcoming, and then we stepped outside to get into his car. There we were, four extra men, all tall and not the slimmest, and we were about to squeeze into the company man's Fiat 500.

We all managed to get in, as we pulled in our stomachs, and off we set. Being Italians there was lots of talking, little of which I understood, and lots of hand gesturing. We hadn't gone far when we pulled into a petrol station and all got out, me thinking that he was going to put petrol in the car: so wrong, we were stopping for a coffee and a Grappa! This was the first of a number of stops as we drove down the coast to the port at Ancona where the objective at each stop was to have more coffee and much more Grappa. I've always had a liking for Grappa ever since.

So the moral of the stories mentioned, and those not mentioned, is that: travel is a wonderful thing; be nice to the people you meet; and look fondly on the things you've achieved, however large or small, as you go through life.

35

Biggles Goes Exploring
Various Places and Times

Phil Fillis

To get to many of the places where geologists work, they can't just jump on a No. 17 bus. It quite often entails a convoluted journey which at some stage involves air travel on pretty dodgy aircraft and airlines. Ever wondered where the Ilyushins and Tupolevs of 1960's Aeroflot (Aeroflop) go to when they die? Travel to some of the murkiest areas of darkest Africa and you'll find out and, at the same time, have some of the more exciting experiences that life has to offer.

Most geologists working overseas, or indeed within Australia, will have undergone the last part of their journey to project site on local airlines or in a light aircraft. Here are a few of my experiences in getting to site or exploring by air.

Early days

My first experience of a bush aircraft was in Canada in 1974. Meeting up with a field crew in Red Lake, we had a boisterous induction at the Red Lake Inn in the evening. Next day, I came to, somewhat disinterested in life for some reason, and prepared to be taken out to some remote lake and to set up my own camp. Still barely conscious, I boarded the float plane for an hour's trip to an unknown lake in the Great Unknown of Northern Ontario. Upon setting down on the lake we taxied towards shore, then I and my field assistant, Jack, were pushed off the plane and swam the remaining 30m or so through leech-infested waters with the instruction that our first job was to build a dock for the plane to tie up to. The pilot said he would return in a couple of hours with another crew to help us build the camp and he would also bring us our tent and sleeping gear. He lied.

We spent one night in virgin forest with an axe as the only piece of equipment. No food, no tent. But we were not alone. There were others.

Mosquitoes I seem to recall. So big they'd land on the lake and swim up to the camp. And blackflies. More than one. Help did arrive next day, but too late – we were already close to death. After the useless help crew (still hung over) departed, Jack and I were due to spend two weeks at this particular location. On the appointed day of our departure, we pulled the camp down and were ready to go. We heard our aircraft approaching – excitement was in the air – and there it stayed. The plane buzzed around in our vicinity for half an hour or so – then cleared off. Another two nights sleeping under a mosquito blanket. On the third day a more experienced pilot came and picked us up. Turns out that our pilot from three days before was on his first solo bush trip and thought he knew vaguely where we were but couldn't find us, so he returned to base and civilization (well Red Lake anyway). Leaving us to die.

Learning on the job

Moving on to 1975 and South Africa. I was exploring for uranium in the Cape fold belt based in George, Cape Province, where they had a flying school. At a party I met the flying instructor, who was somewhat down at heel as he didn't have any pupils. Feeling sorry for him, although it may have been the beer, I signed up there and then to train for my pilot's licence. I did have previous form – I had been gliding in my mid to late teens. How hard could it be to pilot a powered plane? I found out. Nevertheless, four months later, armed with my pilot's licence and a whole 40 hours' worth of flying experience, I was based at Niekerkshoop, which is located in one of the more miserable areas of the Northern Cape, looking for iron ore, or peace and quiet. One of the two, I can't remember which. From my days in iron ore at Newman in Western Australia in 1970 I recalled tales of Lang Hancock discovering the Pilbara deposits by flying around in his light aircraft. Yes! That's what I'll do – up my flying hours and get U.S. Steel to pay for it. A 'Cunning Plan' with one minor flaw – I didn't have a plane. But surely, the local Boers of the Northern Cape had planes on their properties. I asked around but with little success. That came later, and when it did, it was somewhat unwelcome.

After a few weeks I was returning to camp, somewhat grubby having spent ten days traversing out in the field, when I was stopped on the road from Prieska to Niekerkshoop by the SA Police. "Are you the geologist with the pilot's licence?" they asked. I admitted that I was.

"Get in. Now!"

"But, but what about my vehicle?" I stammered.

"Never mind that", they said, "we'll take care of it". On the way, they explained that there had been a car accident. A little girl was severely injured and needed to get to the hospital in Kimberley as soon as possible. A plane was available on a nearby farm, but they didn't have a pilot. Well, as far as they were concerned, they did now. But not one of the cool, calm, collected variety. More of the 'Blue Funk' ilk. In full panic mode. They dumped me on the airstrip in front of the plane – fortunately a Cessna 172 on which I had done my training. Amazingly, through the red fog of panic, I managed to note the frequency the radio was tuned to and told the police to relay it to Kimberly Air Traffic Control. That was to be the last of the cool calm collected actions.

As the police car raced off, I realized the enormity of the situation before me. Even my panic was panicking. Alone, and confronted by a strange aircraft on a strange airstrip – hitherto I'd spend days planning my evening runs from George to Knysna and back again, making sure I had all information, route maps, wind speeds, frequencies. Here I had nothing: apart from serious problems, which now appeared over the horizon in a cloud of dust. The ambulance screeches to a stop by the side of the plane. Out spills hysterical mother carrying child, and doctor carrying oxygen cylinders and goodness knows what other heavy equipment. All they see is a plane and someone standing by it which, in their state of urgency, they mistake for a seasoned qualified pilot. They take no notice of the cloud of terror floating over the airfield. The speed with which they bundled themselves into the aircraft forced me to do the same – without doing a 'walk around' or safety checks. Somehow the engine started, but as we taxied it spluttered – hadn't turned the fuel on. I quickly remedied that mumbling "safety check". Ever generous, clearly, I was spreading my panic around. Full throttle as we bounced down the runway. Except it wasn't. A taxiway. And a short one at that. We took off with the stall warning going and only by the grace of God, and surely not wing lift, that we cleared the boundary fence. The only extra weight

I'd carried before was the instructor. Here I had a fully laden plane with three other bodies plus heavy equipment. It was a miracle we got off the ground at all. Still, the experience does come in handy on hot nights – almost 50 years on, I still wake up in a cold sweat with the stall warning ringing in my ears.

So, we're up. And on to Kimberley. Wait a minute, where the hell is Kimberley? And where are we? Calming down, I knew the main road went in a straight line from Douglas to Kimberley. All I had to do was follow the Orange River to Douglas then follow the main road into Kimberley. Simple. But wait, why is the sun so low on the horizon? Sunset! Great! 75 minutes to Kimberley and the sun is setting. Flying and landing in the dark – another new and exciting experience. That created another problem – the plan was to follow the road which, because it was dark, I couldn't see. No problem, I'd just follow the lights of cars. This is the mid 1970's. One car every two days. We reached Douglas as the last light glinted off the Orange River and I pointed the plane towards what I thought was Kimberley. Never having flown at night and with darkness all around, I had to ask the Doctor who was sitting up front with me to fumble around and see if he could find the lighting for the instruments. I mumbled something about the switches all being different on different aircraft, but I couldn't help notice that the mother started crying and crossing herself again. At this point the voice of a SAA captain on a jet on its way to Cape Town boomed in to my frequency, gave me the frequency of Kimberley Air Traffic Control and wished me good luck in the sort of tone that suggested "abandon all hope". Now under Kimberley control, they guided me in. "You should see the main runway now".

"Can you see the runway?" I asked the Doctor, "Cos, I can't see one". Then asked him to try to find some landing lights, use of which I believe is the done thing. While this is going on I could hear many requests over the radio for approach instructions from other aircraft, which were all met with the same response: "Clear the circuit, we have a medical emergency". That cheered me up. A bit. I still had to find, first the runway, then the ground. Hopefully, first in a timely then orderly manner.

We got down. With much crying and crossing herself from the back of the plane. No reaction from the Doctor. Either he'd had a few stiff whiskies before the whole episode began or he'd never flown before and thought this was normal procedure. I taxied to the apron where an ambulance was waiting. In two minutes, they'd offloaded and were gone. It was dark, quiet and I was alone by the aircraft. The sound of heavy footsteps approaching. Maybe it's the press I thought, come to interview me – the hero of the hour. Maybe a television appearance and international fame and fortune! "Is this your aircraft?", the keeper of airport law and order asked me in his clipped Seth Efrican accent. I acknowledge this to be so, adding as modestly as I could that it was a medical emergency flight. "Is that so?" he said. "That'll be R12.50 landing fee and another R10 if you're staying overnight". Yes, I came crashing down to earth after all.

I flew back to the farm near Prieska the next day. The Doctor was very game – he came with me. If I were in his shoes, I'd have opted to have made the return trip by bicycle. Maybe he'd recharged his whisky levels. And yes, the little girl did survive. The whole episode dampened my enthusiasm for iron exploration from the air and the 'Cunning Plan' never was realised. I took my last flight as a pilot one week before my wedding. Probably a good thing – marriage saved me from an early death by pilot error.

Under attack

Its 1988 and I'm on my way to review BP Mineral's Masupa Ria prospect in the geographical centre of the island of Borneo. There's no runway at site – the only access is by chopper which I board in Muara Teweh. There's just the pilot and me on board. This is in the day when there is still pristine rainforest covering much of Borneo – it will all disappear in just ten years. Halfway through this trip there's a loud bang and the chopper lurches. The pilot, who has had a bad experience in the Vietnam War, knows full well that we are under attack. Panicking, he sends out a "Mayday" distress call while attempting to ditch in the thick forest. Fortunately, at the last minute he spots a small creek with a break in tree cover, and he manages to land in the middle of the river. Now safely down we examine the damage. The missile that hit us turns out to be an unsecured 44-gallon drum that fell over in the cargo hold. Note to self: must remember to fly only with peacetime pilots.

Rush hour

Also, in 1988 I went to review a gold and nickel project with John Martyn in Xinkiang Province in China. This was in the day when the Sleeping Tiger was just having his morning cuppa. Having completed our mission and being held prisoner at site far longer than the project required, we managed to escape and were waiting at Urumchi airport for our much-delayed flight back to Beijing. Having explored all the delights that Urumchi airport had to offer (two minutes) we went walking around the perimeter of the airfield discussing the merits, or otherwise, of CAAC (China Airways Always Crashes) which did not have one of the highest safety records at the time and on which we were due to fly back to Beijing. As we watched, a CAAC Boeing 707 came thundering down the runway on take-off, rotated, cleared the deck, then aborted take-off, crashed back down and skidded to a halt at the end of the runway (we hoped). Just the thing to steady our nerves as we waited for our similar plane, delayed and probably stuck on the end of another runway elsewhere in China. But, desperate to escape, we were prepared to endure a little more excitement. When boarding time finally came, it paid to be young and fit. On the word "Go" or whatever the Chinese equivalent is, all passengers race across the tarmac pushing each other aside to be first up the steps and into the aircraft. The poor young flight attendant stationed at the foot of the steps was an early casualty, flattened in the crush. The concept of pre-booked seats was still a few years off – and didn't make that much difference even after its introduction.

Spatial awareness

Seat selection on Aeroflot was also a hit and miss affair in the late 1980's early 1990's. I was flying from Moscow to Chita in Siberia to review the Udokan copper deposit. I'd heard that it was a long uncomfortable trip on Aeroflot. Having been assigned my seat and now on board, I could see why. There was no space for my legs and my knees were under my chin. So, everything I'd heard about Aeroflot was true. Then I looked over at the other side of the aisle where people in my row had about two metres of leg room! The trick apparently is to get on board early with a spanner and move the rows back and forth. Must remember to travel with a spanner in future.

Deserted

Between 2006 and 2011 I worked on various gold and base metal projects in Mauritania for Afcan Minerals. The first gold project was Conchetta Florence in the far northeast of the country near the Algerian border – very remote. This is bandit country, but we became increasingly concerned during that period as there had been a number of kidnappings of Westerners. For safety and security reasons we decided to find a place to land an aircraft. Bill Brodie-Good was the exploration manager and one of his first tasks was to find a suitable site. From the air, it looks very easy – all flat and no vegetation, but in practice very difficult to find a stretch of at least 1000m of firm substrate. Eventually Bill found a spot around 25kms from the project site. Having had the site checked and certified for use as an airstrip, Dave Netherway, and I, accompanied by office manager Bechir and newly hired exploration geologist Moctar, arranged to pay Bill a site visit at the end of the field season to check on progress.

You're not exactly spoilt for choice when it comes to private aircraft hire from Nouakchott particularly when you want to land on an ungazetted stretch of sand in the desert. The only plane available with a willing pilot was owned by a particularly shady character to match his appropriately dodgy plane. The first attempt to get in was almost a complete disaster. Our refuelling stop for the trip from Nouakchott to Conchetta was the iron ore mining town of Zouerat. We landed ok but, on taxiing, a tire burst. It would have been quite a different story if it had burst on touchdown. This was a pointer to the overall condition of the plane, which we blithely ignored. Following repairs, we made our second attempt to get to site – this time with a young French girl, Ghislaine, as the pilot. Clearly the dodgy owner sensed impending doom and bailed out, willing to sacrifice one of his younger pilots. Eventually we landed safely at site, Bill picked us up and we reviewed his project over a couple of days. On the day of leaving Bill drove us (Ghislaine had camped by her plane) back to the airstrip. We said our goodbyes and Bill roared off back to pack up the camp to leave the next day at the end of our field season.

We took off, but ten minutes out, one engine failed. Ghislaine didn't panic – which would have been my default mode – but turned the plane around and headed back to the strip. Coming in on final approach, the other engine cut out. Despite that, Ghislaine did a fantastic job and got us back safely on the ground. She also managed to get a message out requesting help while we still had some altitude. With no communication on the ground, we could only sit and wait and hope that help was on its way.

No food, little water and no shade other than under the wing of the aircraft. As a precaution, since Bill and his crew were not that far away, we considered ways to alert him to our plight. The piste over which Bill was likely to come the following day was a route around one kilometre wide and about two kilometres from the end of the strip – not that there was any guarantee that he'd travel over that bit. There is no set track in this part of the desert close to the land-mined border with Western Sahara and the piste is a compromise route between natural drainage channels and landmines. Over the targeted piste, Moctar and I dug shallow trenches with a geopick and dropped brightly coloured clothing and instrument boxes over the 1km width. These really stood out in the desert – how could anyone possibly miss them! This whole exercise took around four hours and, as Moctar and I were on our way back, we saw a dust cloud in the distance. Bill! He'd ripped the camp down in record time and was escaping. We ran back towards our marked piste and even set off a flare, to no avail. We reached the marked piste just in time to see the last vehicle of the convoy disappearing in a cloud of dust. At his trial and on interrogating Bill and his team back in town a few days later, one of the young geologists said, "I did notice the trench and clothing and asked Bill if we should investigate".

"Nah." Bill said, "It's probably nothing". That was over 12 years ago, and I still haven't forgiven him.

Meanwhile, back at the plane, everybody was huddled under the wing trying to escape the heat and rationing the water. We simply didn't know whether Ghislaine's message had got through and whether help was on its way or not. With Bill and his crew now enjoying the nightlife in Bir Moghrein, we were alone in the desert. Poor old Bechir, our office manager, is not cut out for the bush and has rarely been outside Nouakchott – he gets very nervous out of the city. He was particularly upset by his impending demise, and I tried to get his mind off the situation by suggesting he use the time profitably by writing his will. Despite my best efforts, this didn't seem to help.

As the setting sun touched the horizon, and we had given up all hope of being saved, our rescue plane arrived in the nick of time and brought an aircraft mechanic. The mechanic and Ghislaine spent one more night in the desert before the repaired aircraft made it back to Nouakchott and thence hopefully to the scrap heap. We returned on the rescue plane, only just managing take-off before complete darkness fell, and arrived back in Nouakchott late at night. I agreed with Bechir that it was indeed a most exciting day.

Getting to site has certainly been an interesting and sometimes exciting part of my career as a geologist. That was then, and now a walk around the park is all the excitement I can take these days.

www.ingramcontent.com/pod-product-compliance
Lightning Source LLC
Chambersburg PA
CBHW061614220326
41598CB00026BA/3762